Computer-aided Maintenance

Manufacturing Systems Engineering Series

Series editor: Hamid R. Parsaei, Department of Industrial Engineering, University of Louisville, USA

The globalization of business and industry and the worldwide competitive economy are forcing business leaders in the manufacturing and service sectors to utilize fully the best equipment and techniques available. The objective is to have efficient control of the organizational structure in order to produce high quality products at lower prices within a shorter period of time.

Since the introduction of computers in the 1950s, Manufacturing Systems Engineering has experienced tremendous growth. The development of the discipline has helped industry to become more productive and to make more efficient use of resources. Manufacturing information systems, total quality management, facility layout, material handling, value engineering and cost analysis, safety, computer-integrated manufacturing, and production planning and shop floor control are just some of the areas in which manufacturing systems engineers have been traditionally involved in order to help improve understanding and awareness in the manufacturing and service sectors. The recent emphasis and concern about the environment and product recyclability and re-usability have brought new perspectives and more challenges to this ever-growing engineering discipline.

The aim of the *Manufacturing Systems Engineering Series* is to provide an outlet for state-of-the-art topics in manufacturing systems engineering. This series is also intended to provide a scientific and practical basis for researchers, practitioners and students involved in manufacturing systems areas. Issues which are addressed in this series include, but are not limited to, the following:

- Production system design and control
- Life cycle analysis
- Simulation in manufacturing
- Manufacturing cost estimating
- Industrial safety
- Fuzzy logic and neural networks in manufacturing
- CAD/CAM/CIM

We would welcome proposals to write material for this series from colleagues and industry leaders around the world. We hope that researchers both in academia and government, as well as private organizations and individual practitioners, will find this series informative and worthwhile.

1 **Manufacturing Decision Support Systems**
Edited by Hamid R. Parsaei, Sai Kolli and Thomas R. Hanley
2 **Integrated Product, Process and Enterprise Design**
Edited by Ben Wang
3 **Occupational Ergonomics**
Fariborz Tayyari and James L. Smith
4 **Rapid Response Manufacturing**
Edited by Jian Dong

Computer-aided Maintenance

Methodologies and Practices

Edited by

Jay Lee

National Science Foundation
Virginia
USA

and

Ben Wang

FAMU-FSU College of Engineering
Florida
USA

KLUWER ACADEMIC PUBLISHERS
DORDRECHT / BOSTON / LONDON

A C.I.P. Catalogue record for this book is available from the Library of Congress
A catalogue record for this book is available from the British Library

ISBN 0 412 62970 4

Published by Kluwer Academic Publishers,
P.O. Box 17, 3300 AA Dordrecht, The Netherlands.

Sold and distributed in North, Central and South America
by Kluwer Academic Publishers,
101 Philip Drive, Norwell, MA 02061, U.S.A.

In all other countries, sold and distributed
by Kluwer Academic Publishers,
P.O. Box 322, 3300 AH Dordrecht, The Netherlands.

Printed on acid-free paper

Typeset in 10/12 Palatino by Cambrian Typesetters, Frimley, Surrey, UK

Printed in Great Britain by St. Edmundsbury Press, Suffolk

Contents

Contributors

Bruce Bryant Ford Research Laboratory, Ford Motor Company, Dearborn, MI, USA

Mo-Yuen Chow Department of Electrical and Computer Science, North Carolina State University, Box 7911, Raleigh, NC 27695, USA

Hsin-Hao (Tom) Huang Industrial Engineering, National Yunlin Polytechnic Institute, 64 Wun-Hua Road, Hu-Wei, Taiwan.

Lawrence Holloway Center for Manufacturing Systems, 414F CRMS Building, University of Kentucky, Lexington, KY 40506–0108 USA

Gerald M. Knapp Maintenance and Reliability Research Unit, Industrial and Manufacturing Systems Engineering, 3128 CEBA Building, Louisiana State Univesity, Baton Rouge, LA 79893, USA

Christopher Kok-Hwee Koh School of Mechanical and Production Engineering, Nantang Technological University, Nanyang Avenue, Singapore 639798

Jay Lee National Science Foundation, 4201 Wilson Boulevard, Room 585, Arlington, VA 22230, USA

Yuan-Shin Lee Department of Industrial Engineering, North Carolina State University,
Raleigh, NC 27695–7911, USA

Chang-Ching Lin Department of Industrial Engineering and Management, St John's and St Mary's Institute of Technology, 499, Sec. 4, Tam-King Road, Tamshui, Taipei County, Taiwan

Kenneth Marko Ford Research Laboratory, Ford Motor Company, Dearborn, MI, USA

Kazuo Mori Department of Manufacturing Systems, Mechanical Engineering Laboratory, Agency of Industrial Science and Technology, MITI, Namiki 1–2, Tsukuba City, Ibaraki 305, Japan

Anthony Okafor Department of Mechanical Engineering, University of Missouri-Rolla, Rolla, MO 65409, USA

Shashi Phoha Information Systems, Applied Research Laboratory, PO Box 30, Penn State University, State College, PS 16804, USA

Asok Ray Mechanical Engineering Department, The Pennsylvania State University, University Park, PA 16802, USA

Tomohiko Sakao Department of Precision Machinery Engineering, The Graduate School of Engineering, University of Tokyo, Hongo 7–3–1, Bunkyo-ku, Tokyo 113, Japan

Yoshiki Shimomura Mita Industrial Co. Ltd, Tamatsukuri 1–2–28, Chuo-ku, Osaka, Japan

Julie K. Spoerre College of Engineering, Southern Illinois University, Mailcode 6603, Carbondale, IL 62901, USA

Farhad Tadayon Technology Assistance Center, College of Technology, 2409 Scanlan Avenue, Salina, Kansas 67401–8196, USA

Shozo Takata Department of Industrial Engineering, Waseda University, Okubo 3–4–1, Shinjuku-ku, Tokyo 169, Japan

Tetsuo Tomiyama Department of Precision Machinery Engineering, The Graduate School of Engineering, University of Tokyo, Hongo 7–3–1, Bunkyo-ku, Tokyo 113, Japan

H. Joel Trussel Department of Mechanical and Computer Engineering, North Carolina State University, Raleigh, NC 27695–7911, USA

Alice Tsai 7400 S. State St, Apt 13303, Midvale, UT 84047 USA

John Tsai 7400 S. State St, Apt 13303, Midvale, UT 84947 USA

Yasushi Umeda Department of Precision Machinery Engineering, The Graduate School of Engineering, University of Tokyo, Hongo 7–3–1, Bunkyo-ku, Tokyo 113, Japan

Ben Wang Department of Industrial Engineering, FSU/FAMU, 2525 Pottsdamer Street, Tallahassee, FL 32310–6046, USA

William J. Williams Department of Electrical Engineering and Computer Science, University of Michigan, 1301 Beal Avenue, Ann Arbor, MI 48109–2122, USA

Preface

Equipment reliability and maintenance drastically affect the three key elements of competitiveness – quality, cost, and product lead time. Well-maintained equipment holds tolerances better, helps reduce scrap and rework, and improves consistency and quality of the part. The recent rush to embrace advanced sensors and control systems in modern equipment has further increased the use of relatively unknown and untested technology. Typically, when a production system goes down, only a small fraction of the downtime is spent repairing the equipment that causes the failure. A substantial part of the downtime is spent on locating the source of the problems. Difficulty in identifying the causes of system failures has been attributed to several factors, including system complexity and a lack of knowledge and adequate troubleshooting tools.

This book introduces modern theories, techniques and tools such as artificial intelligence, expert systems and neural networks for effective maintenance engineering. In addition, it covers recent advances in reliability, design for self-maintenance machines, digital condition monitoring, signature analysis, and life cycle engineering. Two issues regarding computer-aided maintenance are addressed: the first issue is the fundamental knowledge, including both theories and methodologies required for practitioners to perform maintenance activities effectively. With the revolutionary advances in information, telecommunication, and computing technologies, we forsee needs for new approaches in design, process, and the use of materials, technologies, and human resources in the modern factory; the second and equally pressing issue is to understand how to implement the right maintenance tools and techniques, based on the introduced theories and methodologies, to solve problems in a very short time frame in order to guarantee success.

The book is divided into two parts with a total of 18 chapters. Part One (Chapters 1 to 11) of the book introduces new methodologies which lay a foundation for readers to understand the recent advances in maintenance techniques and methods; Part Two (Chapters 12 to 18) provides readers with 'how to know' knowledge and practices by illustrating the introduced methodologies using seven case studies. Chapter 1 introduces the fundamentals of maintenance, including the definitions of preventative maintenance, predictive maintenance, deterioration process and machine diagnostics. Chapter 2 introduces fundamental sensory systems related to maintenance engineering, including sensors, signal processing, sensor fusion, and operation algorithms. Chapter 3 builds upon Chapter 2's foundation and addresses applied sensory systems for machine monitoring and diagnostics, including the use of expert systems and an artificial networks system for maintenance practices. Chapter 4 focuses on the parameter modeling which deals with the exponentially weighted moving average (EWMA) model for monitoring a process standard deviation as well as the autoregressive (AR) model for the estimation of process parameters. Chapter 5 emphasizes the reliability modeling for machine performance including machine performance degradation modeling and fault detection. Chapter 6 introduces the design methodology for self-maintenance machines, including the systematic understanding of fault monitoring, fault recognition, repairing planning, and self-recovery strategy. A particular example of a self-maintenance copier is examined to illustrate the design practices. Chapter 7 addresses an integrated prognostics, maintenance, and life-extending control of continuous-time production processes. A methodology dealing with condition-based maintenance for a complex continuous system is introduced. Chapter 8 introduces a methodology for root cause identification using a fuzzy neural network approach. A Boeing door assembly process system is used to explain the applied quality control practices. Chapter 9 illustrates maintenance using activity-based costing (ABC) for the system justification of deployment of a modern maintenance system. Chapter 10 introduces a life cycle maintenance management system. Chapter 11 examines life extension of operating machinery using the national information infrastructure (NII). Chapters 12 to 18 demonstrate the implementation of the introduced methodologies in the previous chapters for a number of applications, including motor fault detection, automotive engines diagnostics, machine degradation monitoring, tool failure monitoring in milling, fault detection and isolation in the stamping process, fault monitoring in a manufacturing system, and monitoring of surface roughness and bore tolerance in end-milling.

In summary, this book takes a user-oriented, hands-on approach for practising engineers or managers to develop initial expertise in computer-aided maintenance methodologies. It is hoped that *Computer-aided Maintenance: Methodologies and Practices* will bridge the gap for industrial users, eventually making world-class maintenance a requirement for modern factories to remain competitive and productive, as more and more companies embrace the modern computer-aided maintenance methodologies.

The editors wish to thank the authors and reviewers who made this book a reality. The editors also express their appreciation to Dr Hamid R. Parsaei, the Series Editor, for the opportunity to edit the book.

Jay Lee
Ben Wang

PART ONE

Methodologies

CHAPTER 1

Fundamentals of maintenance

Gerald M. Knapp and Ben Wang

1.1 INTRODUCTION

Managing production and power-generating assets will be the overriding concern of the manufacturing and utility industries as we move into the 21st century. Manufacturing and utilities will continuously be challenged to show insight in optimizing current plant assets and anticipating future asset requirements. The advanced intelligence provided by diagnostics can be the deciding factor in effective asset management through maintenance strategies and improved operations aimed at heading off availability losses and forced outages. Intelligence is especially important when large capital investments in equipment replacements are ruled out.

Maintenance costs are a major part of the total operating costs of all manufacturing and production plants. Depending on the specific industry, maintenance costs can represent between 15% and 40% of the costs of goods produced. For example, in food-related industries, the average maintenance costs represent about 15% of the cost of goods produced; while in iron and steel, pulp and paper, and other heavy industries, maintenance represents up to 40% of the total production costs. When we consider that United States industry spends more than 200 billion dollars each year on maintenance of plant equipment and facilities (Mobley 1990), the impact on the US economic competitiveness, productivity and profit that is represented by the maintenance operation becomes clear. Some of the benefits derivable from a comprehensive, planned, and active approach to machine monitoring include the following:

1. Improved quality. Machine deterioration leads to production of reduced quality goods and an inability to meet high quality standards.

2. Improved productivity. By significantly reducing unscheduled down-time, and providing adequate lead-time for proper scheduling of maintenance and repair, unproductive down-time is reduced.
3. Improved customer relationships, by improving consistency of machine performance, and thus the ability to stay on schedule.
4. Improved operator safety, by reducing and predicting machine failures in advance.
5. Reduced costs. Unnecessary maintenance and repairs can be eliminated. Costly catastrophic machine failures are prevented. Safety stock can be reduced or eliminated.

In this chapter, we will review the general requirements for and capabilities of modern maintenance planning.

1.2 MACHINE FAILURE

Machine failure occurs when changes in one or more of its components render its intended function unsatisfactory or unsafe. A machine failure can be decomposed into its causes and effects. Effects are the changes which occur on the machine. Failure modes are those effects which are primarily responsible for the machine failure; possible failure modes include component deformation, fracture, surface changes (such as cracks), material changes, displacement, leakage, and contamination (Alban 1985; Lyon 1987; Nisbet and Mullett 1978). Secondary effects, or symptoms, often occur prior to total machine failure, providing indicators for predicting failure onset. Important and frequently used indicators are vibration signals, noise, heat generation, and particle wear levels as measured in machine lubricants.

Failure causes are conditions of the machine and its operation that contributed to a particular failure mode (Alban 1985). Possible failure causes include:

1. Design deficiencies, such as inadequate lubrication, or poor bearing design.
2. Material defects, such as rapid corrosion or wear.
3. Processing and manufacturing deficiencies.
4. Assembly errors, such as misalignment or improper tightening of bolts.
5. Off-design or unintended service conditions.
6. Maintenance and installation deficiencies.
7. Improper operation.
8. Normal wear.

More specifically, failure mechanisms are causes directly responsible for the onset of a failure mode. This term is used to describe the metallurgical, chemical, or tribological process leading to a failure mode.

Characteristics of a failure mechanism include the forces applied (and whether steady, transient, or cyclic), time frame, temperature levels, and other environmental conditions (Alban 1985; Smith 1983). Typically, a failure will not be the result of a single cause, but instead might represent a chain of cause and effect relationships leading to the failure (i.e. an effect in turn becomes the mechanism for another effect). As an example, gear tooth defects can create periodic lateral forces which misalign the shaft, causing accelerated wear of the shaft bearings. Often, the first symptom of such a failure is accentuated bearing noise as the bearings begin to deteriorate. The gear teeth may in turn have deteriorated due to improper lubrication.

1.3 THE MAINTENANCE FUNCTION

The primary role of maintenance is to control the condition of plant equipment. Wireman identifies the following objectives of a maintenance organization (Wireman 1990):

1. Maximum production at the lowest cost, highest quality, and optimum safety levels.
2. Identify and implement cost reductions.
3. Provide accurate maintenance and repair records.
4. Optimize maintenance resources.
5. Optimize capital equipment life.
6. Minimize energy usage.
7. Minimize inventory on-hand.

The maintenance function consists of a number of important subsystems, primarily planning, scheduling, inventory, purchasing, and reporting. Inventory planning and purchasing are concerned with insuring that replacement parts and materials are available when needed. Reporting provides maintenance information to the related manufacturing functions, such as operations, engineering, inventory planning, accounting, and upper management. In this research we are most closely concerned with the first two of these functions, planning and scheduling. Planning is concerned with the development and management of maintenance. Problems which are typically addressed by the maintenance planner include determination of inspection frequencies, overhaul intervals, replacement policies, reliability analysis, and maintenance crew sizing (Jardine 1973).

Scheduling is concerned with the matching of labor and material resources to requests for maintenance resources, such as for repairs or planned preventive maintenance. In performing scheduling, many factors must be taken into consideration, such as (Wireman 1990):

1. Work priority.
2. Work in progress.
3. Resources available.
4. Work backlog.

In many cases, low priority work, such as routine maintenance actions, will have anywhere from a two to eight week backlog. In a predictive maintenance environment, an important issue is the ability of scheduling to implement repairs before machine breakdown; such information must be considered when developing predictive maintenance models.

Successful maintenance control is as much a philosophy as it is a function. Total productive maintenance (TPM), a concept introduced by the Japanese, shares many concepts with total quality control (TQC) (Dale 1990; Hacker 1991; Maggard *et al.* 1989; Wireman 1990). The goal is to develop a culture in which continuous improvement in the maintenance process is sought. Thus, maintenance seeks to accomplish more than simply keeping the machines running; it should try to derive insights which will improve performance into the future. To achieve this requires operator and staff involvement and training, and incorporation of automated technologies.

The maintenance function must be involved in the move towards automation and integration of manufacturing systems. In an age of minimal inventories and highly automated complex machinery, automated maintenance systems are required to truly accomplish the goals of computer-aided manufacturing (CAM) and computer-integrated manufacturing (CIM). An automated maintenance planning and diagnostics system can provide a rapid planning and scheduling response to changing system conditions even when faced with enormous quantities of sensory input. Furthermore, automated maintenance systems can be directly integrated to MRP and other management information systems. In this research, we take a step towards achieving this goal.

1.4 MAINTENANCE POLICY

Plant maintenance is a critical function of manufacturing and is in a period of rapid change due both to increased customer demands for better quality and lower cost, and to increasing automation of machining. As plants are increasingly automated, and sophistication and complexity continues to proliferate, it is important that plant maintenance be seriously considered, carefully planned, and thoroughly implemented. Three approaches have been used, often in combination, throughout industry (Bloch and Geitner 1983):

1. Breakdown maintenance.
2. Preventive maintenance.
3. Predictive maintenance.

Breakdown maintenance is the maintenance and repair of machines only after a failure has occurred. It is not by itself a productive maintenance policy, although it has often been used as such by default. Breakdown maintenance, when planned, recognizes that not all faults can be identified prior to their onset, and that in these cases, it is critical that machines be brought back on line as quickly as possible. It involves development of a capability for rapid identification of machine failures when they occur by efficiently utilizing all the sources of information available, such as likelihood of failures of components, repair records, vibration data, and operator inspections/reports.

Preventive maintenance involves the periodic application of maintenance routines regardless of machinery condition. Typically, this entails the periodic lubrication, coating, and cleaning of machinery components. Some types of failures are effectively prevented in this fashion; for instance, bearing wear can be significantly reduced by periodic renewal of lubricating oil. Preventive maintenance aims to either slow down the deterioration process leading to faults, or prevent faults by replacing components at regular intervals shorter than their expected useful life. It is thus most suitable for time-dependent failure modes, although even in this case there is no guarantee that major disabling faults will not occur due to the probabilistic nature of machine failure. Figure 1.1 displays a typical life distribution for a component or system of components. The X-axis shows time, while the Y-axis indicates the probability of failure of the component by that time. Note that a component may fail immediately, and that there is no point at which the probability of failure is zero. Therefore, preventive maintenance can only reduce, but not eliminate, the occurrence of catastrophic failures.

A complete maintenance program, then, must integrate both prevention and prediction to achieve maximum effectiveness. Predictive maintenance utilizes periodic inspection, followed by overhaul or replacement of any components which have reached specified or subjective fault limits. It is well known that 99% of failures are preceded by trendable indicators (Bloch and Geitner 1983). This is particularly important when it is realized that a large percentage of the US industrial equipment has reached or even surpassed its useful life, moving into the region at the end of the failure curve where the failure rate begins to rise steeply, regardless of periodic maintenance (see Figure 1.1). Figure 1.2 depicts this aging process for the coal and utility industries. Through regular inspections, oncoming failures can be predicted before their occurrence, with the following benefits:

1. Replacement parts can be acquired in advance of the actual repair.
2. Maintenance may be scheduled for non-busy or off-hour time periods.
3. Unexpected catastrophic failures, and their inherent costs, are avoided.

Predictive maintenance holds that a machine will not be shut down for maintenance or repair unless it is necessary, that is, at the right time to do the right kind of maintenance – just-in-time maintenance. Therefore, determination of timing for maintenance becomes the major concern. Studies repeatedly show that predictive maintenance greatly reduces the overall amount of downtime. In addition, the failures are predicted in advance, and maintenance can be properly scheduled, whereas for preventive maintenance, the shutdowns cannot be predicted and are forced, resulting in higher repair costs and times.

While the importance of maintenance planning has been understood almost as long as there have been machines, the scientific analysis of such planning, primarily embodied under the heading of optimal maintenance control, has its roots in the 1960s, gaining strength with an article by Gerald Thompson, 'Optimal Maintenance Policy and Sale Date of a Machine' (Thompson 1968).

> The basic premise of optimal maintenance control is that maintenance actions act as signals to control the productive process of a machine subject to external uncontrollable deterioration. Maintenance actions may improve or simply slow the deterioration process, yielding profits from increased productivity. It is typically desired to apply maintenance actions in such a way as to maximize total gain; that is, productivity gains minus maintenance costs.

Existing work on maintenance control has almost exclusively been dedicated to the development of machine level maintenance policies. However, machines rarely operate independently in a production environment; the outputs of one machine are the inputs to another, and so their productivity, with respect to machine performance and availability, are related. Little work has been done to develop tools and models for

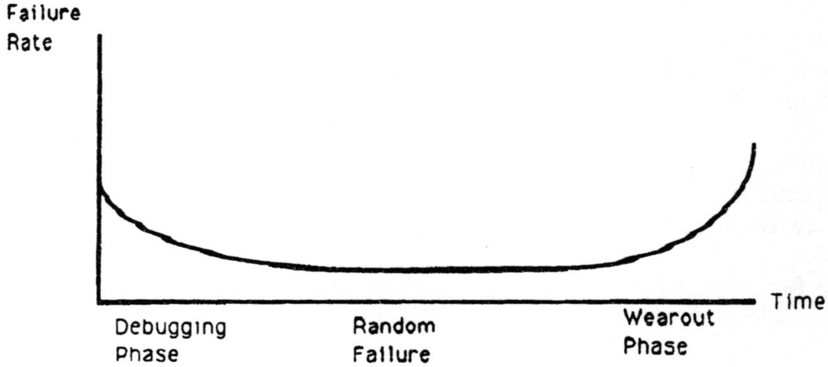

Fig. 1.1 Machine failure life distribution.

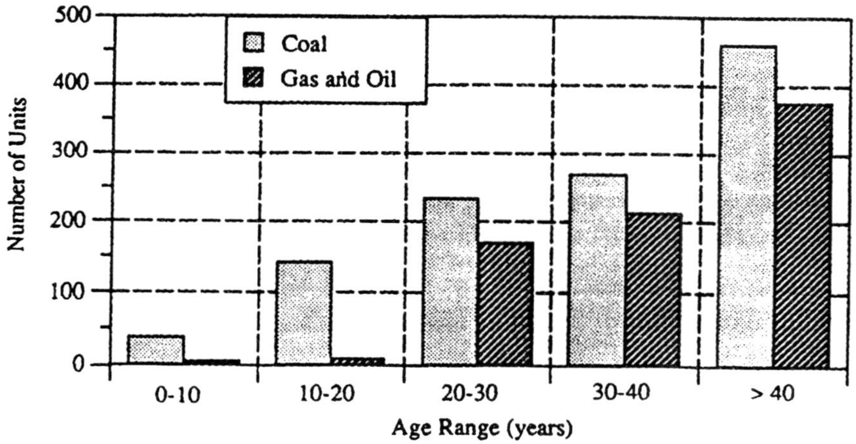

Fig. 1.2 Age distribution of US fossil fired units in the year 2000.

aiding the maintenance planner or designer to account for these relationships and how they effect the solution of optimal maintenance models.

In this research, we address the development of such tools for designing and implementing maintenance policy in a class of automated manufacturing systems called flexible manufacturing systems (FMS). FMS were chosen for their prevalence in industry and the high degree of production interdependence found between machines. Maintenance planning tools for FMS and automated manufacturing systems in general are required in three areas:

1. Design tools. The maintenance planner should have the capability to assess simply, rapidly, and accurately the maintenance costs on new FMS designs, or evolving existing systems.
2. In-process tools. Maintenance costs can be reduced by application of predictive maintenance techniques. Tools are required which consider the production relationships between machines, and which can efficiently automate the expensive task of manually analyzing huge qualities of sensor data.
3. Improvement. Tools are needed for automatically identifying and extracting predictive information from raw sensor data which can improve predictive monitoring performance.

Preventive maintenance, however, only statistically controls the system condition. Costs may often be further reduced by actively monitoring machine condition in an attempt to predict and head off unscheduled downtime. In such situations, it is often beneficial to have a predictive model which utilizes, for instance, proportional hazard modeling to analyze machine condition, based on sensor reading (such as vibration data). The predictive maintenance system determines whether a machine failure appears to be pending. The repair scheduling priority is then decided, using the results of the first section. Furthermore, if the approximate time distribution of the scheduling function is known, a cost model is developed which accounts for the reduction in failure costs in determining the optimal preventive maintenance cycle.

Total productive maintenance (TPM) (also called total maintenance control, or TMC), as in total quality control, is not only an approach to maintenance, but a philosophy which holds that we should continually seek to improve our control over the maintenance process. Thus, maintenance control cannot be simply a static process. Innovative systems which actively 'discover' trends in sensory data, and signal a follow-up investigation, are often needed. Upon confirmation of trend data, the network can adapt to incorporate this new knowledge. The information is then available for future predictive maintenance decisions.

1.5 MAINTENANCE CONTROL

The basic premise of optimal maintenance control is that maintenance actions act as signals to control the productive process of a machine subject to external uncontrollable deterioration. Maintenance actions may halt or simply slow the deterioration process, yielding profits from increased productivity. It is typically desired to apply maintenance actions in such a way as to maximize total gain; that is, productivity gains minus maintenance costs. It has been found that, under fairly general conditions, optimal static control policies (planned maintenance actions, either at fixed times or under fixed conditions) may be determined.

The history of maintenance control extends back to the 1950s and 60s. Researched models can be broadly classified according to approach, the two most common being those utilizing the Pontrjagin's Maximum principle (Arora and Lele 1970; Chikte and Deshmukh 1981; Fard and Zheng 1991; Gaimon and Thompson 1989; Haxtl 1983; Thompson 1968, Wong *et al.* 1991), and those based on Markovian assumptions (Eckles 1968; Hopp and Wu 1990; Murthy and Djamaludin 1990). A number of extensive survey papers have been published on this topic (Sheridf and Smith 1981; Valdez-Flores and Feldman 1989). This work may be categorized along a number of specific problem dimensions:

1. Deterioration process.
2. State space.
3. Maintenance actions.

We provide a brief introduction to the literature in each of these areas, and then return to examine more deeply works specifically relevant to this research.

1.5.1 Deterioration process

Maintenance models examined in the literature may be distinguished by the level of information available about the deterioration process. At one extreme, it is assumed that a failure distribution, specifying statistical information about the time between failures, is exactly known. Typically, Poisson failure processes are assumed, primarily because these have been shown to be applicable to a large array of machine failure problems, and also due to the wealth of theoretical knowledge and tools for solving such control problems. Because the failure information is statistically characterized, these models may be utilized in open-loop maintenance control; i.e. models in which no state feedback is available to the decision maker through inspection or other sensor measurement.

Many researchers have investigated systems in which deterioration can be modeled as a sequence of discrete 'shocks', the shocks occurring at random time intervals and each causing a random amount of cumulative damage. Failure is generally assumed to occur at shock points. Replacement (or repair) is generally assumed to be less costly prior to machine failure due to the ability to schedule downtime, etc. The replacement or repair of the system is assumed to renew the level of deterioration back to that of a new machine (Chikte and Deshmukh 1981). The distribution of shocks may be known (typically Poisson), as studied by Nakagawa and Kijima (1991) and Zuckerman (1986), or unknown (Bergman 1980).

A number of models have been developed for multiple component systems in which the repair of a single component does not renew the overall failure rate; only machine overhaul or replacement renews the overall failure rate (Barlow and Hunter 1960). Such models are an important characteristic problem in FMS environments, and will be the main focus of this research.

1.5.2 State space

The specification of the characteristics of the deterioration state space are important to the choice of formulation used to determine an optimal maintenance policy. Maintenance models may differ by whether the level of deterioration is considered discrete or continuous. The discrete case corresponds to instances in which deterioration may be classified into categories, the simplest classification being working/not working. The use of discrete states is generally conducive to solution via Markovian decision theory techniques, such as policy iteration, value analysis, stochastic dynamic programming, or linear programming. Discrete deterioration states have been assumed in, among others, the models of Eckles (1968), Hopp and Wu (1990), and Luss (1976). In the continuous case, deterioration is measured through continuous sensor variables which may directly or indirectly monitor the deterioration of the system. The optimal control policy then specifies the decision planes within the n-dimensional state space at which maintenance actions will be taken. Continuous state spaces have been assumed in Haxtl (1983), Thompson (1968), and Wong *et al.* (1991) among others, and are conducive to solution via application of the Pontrgagin Maximum principle and traditional control theory techniques.

Model assumptions about knowledge of the machine state vary. If the machine state is perfectly observable, knowledge of the state is complete and certain (Thompson 1968; Wong *et al.* 1991). In reality, the state is frequently deduced from imperfectly correlated and noisy sensor measurements. Thus, the true state may be only partially observable or known only with some certainty (Eckles 1968; Monahan 1982; Wang 1977).

1.5.3 Maintenance actions

Maintenance models may be distinguished by the level of detail of maintenance actions, the competency of repairs or inspections, the impact on overall failure rates of repair, and the timing of maintenance actions.

At the simplest level of detail are replacement models which consider at each decision point only whether to do nothing or replace the machine (Thompson 1968). At more detailed levels, specific repairs or replacement of system components may be considered. These repairs may be distinguished by their cost and effectiveness in changing the overall system failure rate (Chikte and Deshmukh 1981; Eckles 1968; Hopp and Wu 1990; Murthy and Djamaludin 1990).

Models may also be distinguished by the level of competency assumed for repair or inspection actions. Repairs may be assumed perfect (Thompson 1968), or to have random effectiveness (Eckles 1968; Nakagawa and Kijima 1991). Inspections may be assumed to yield either perfect or imperfect information about the state of the system.

1.5.4 System level maintenance control

Little work has been done on the study of maintenance control at the system level. Teo *et al.* (1989) develop a maintenance model for systems of machines where the return of the system is dominated by the machine with the worst condition, e.g. a bottleneck machine in a serial assembly line. The problem is formulated as a constrained optimal control problem. It is shown that the problem has an optimal solution, and a computational scheme for finding that solution, based on control parameterization, is developed.

Federgruen and So (1990) consider a single server queueing system with Poisson arrivals, general service times, and breakdowns according to a Poisson process. Upon breakdown, two repair actions are available. The objective is to find a maintenance policy which minimizes the long-run operating costs of the system. The model is formulated as a semi-Markov decision process, and it is shown, under several conditions, that a monotonically optimal solution space.

1.6 MACHINE DIAGNOSTICS

The quality of machined parts is a reflection of the machine's operating condition. Conditions sufficient to cause an unacceptable part may be classified into the following categories:

1. Unqualified materials.
2. Dulled/inadequate cutters.
3. Faults in the major rotating and transmission components of the machine system.

In this effort, the focus will be on the faults in the third category. There are three widely used 'evidence sources' for machine fault diagnostics: wear particle analysis, performance monitoring, and vibration signal analysis (Tranter 1989). These sources provide indicators for predicting and isolating fault occurrences. Within each source, numerous techniques are available to aid in the diagnostic process. These may be loosely distinguished by the following:

1. Signal processing techniques perform data transformations in an effort to enhance features, reduce noise, and make the data more amiable to analysis. These techniques typically serve as a front-end for the fault recognition and classification techniques.
2. Fault recognition techniques are typically Boolean indicators, which flag the presence or absence of a particular fault type based on a limit value of some function of the data.
3. Classification techniques that not only recognize the occurrence of a fault, but also make some decision about the identity of the fault.

In the following sections, we briefly describe techniques available for each of these evidence sources.

In wear particle analysis, particles in lubricating oils or greases are collected at periodic intervals and analyzed, using some form of spectro-chemical analysis. Particle size, size distribution, concentration (or the quantity) of particles, trace elements in suspension, and the shape of particles provide insight into machine wear conditions. For example, excessive amounts of copper in lubricating oil suggest that a bearing may have started to wear, because copper is the major constituent of bearing material. Likewise, an observation of an unusually large amount of iron particles in the lubricating oil indicates a worn gear.

This approach has been widely employed in the manufacturing industry.

Monitoring overall plant performance can also be an effective tool for detecting machine faults. In performance monitoring, data that relates to the overall performance of a plant may be used to detect specific machine or groups of machine faults. Typically, analysis consists simply of trending measurement data over time. A fault condition is recognized when certain limit values are exceeded. Examples of data that may provide insight into overall plant performance include electricity consumption, heat exchanger efficiency and pump differential pressure. In addition, total operating efficiency, number of machine failures and total time on-line can be used for this purpose. Thermal analysis is a frequently used performance measurement for monitoring bearing and coupling condition. Abnormal heat generation often accompanies bearing deterioration, and is typically proportional to such determination. In some instances, the relationship between machine component wear and part quality can be directly established. For example, bearing wear frequently causes play at the tool turret, resulting in part runout increases. By trending runout values, bearing wear can be indirectly inferred, and maintenance or replacement scheduled to correct the condition at the proper times to maintain specified part quality.

It is well known that the vibration characteristics of a machine will change when machine components start to deteriorate. Many studies were conducted on this subject in the past three decades (Li and Wu 1988; Lyon 1989; Reiter 1987; Serridge 1989; Tranter 1989). Generally, there are four popular approaches for vibration signal analysis: time-domain, phase-domain, frequency-domain, and parametric analysis techniques. Time-domain-based techniques include overall level (RMS) measurements, peak level detection, crest factor, shock pulse, spike energy, Kurtosis, demodulation, time waveform, and orbits (Tranter 1989). Parametric approaches utilize a two-step procedure. In the first step, a parametric model is fitted to the data. A commonly used model is the ARMA (auto-regressive moving average) model. Typically, assumptions are made about the underlying stochastic process, and then the parameters are estimated using techniques such as maximum likelihood estimators (MLE) or the Yule–Walker approach. In the second step, the parameters of the model are used to compute a power spectrum for the sampled data. While the parametric technique yields better data utilization than other techniques, such as Fourier analysis, the computational requirements are heavy and processing times are typically long. Phase-based approaches include Bode (rectangular) plots, and Polar (or Nyquist) plots. Frequency-based techniques include spectral analysis, waterfall plot, cepstral analysis, difference spectra, RMS of spectral difference, matched filter (Tranter 1989) and fuzzy set based methods (Zeng and Wang 1991). One of the most useful fault diagnostics techniques is that of spectral analysis. Each rotating element in a machine generates an identifiable frequency (signature), and any change within this frequency range can be measured and identified with a specific machine fault. Data is typically collected via accelerometer, velocity, or displacement sensors.

1.7 PREDICTIVE MAINTENANCE

Little work has been directed specifically at the problem of predictive maintenance planning, particularly at the system level. However, a great deal of work has been completed in closely related areas which is applicable to the current research. Work on component level and machine level diagnostics and monitoring has been an ongoing concern of industry since the earlier part of the century. Many tools have been developed for identifying and quantifying machine condition; some of these were discussed in the previous section on machine diagnostics. Condition information generated at the component and machine level forms the inputs necessary for successful predictive maintenance decisions. Some work has been done under maintenance control which pertains to predictive maintenance, although the term predictive is not specifically mentioned. Wan looked at a partially observable system whose deterioration can be represented by one of the integers O, . . . , L. Deterioration is known to move between states with known transition probabilities, but the current state cannot be observed directly; the only information available is the posterior probability that the system is in state i' (Wang 1977). Tapiero considers a system whose state can be indirectly observed from production quality. The number of units produced, rather than time, is used to measure system activity, and to decide replacement time (Tapiero 1986). Eckles considers a system whose condition is characterized at each period by the pair (i, t), where i is the vector of state variables, and t is the system age (Eckles 1968). At each period, the decision maker may choose from a finite set of actions k. Choosing an action leads to the system moving to a state j with some transition probability p_{ij}. Imperfect information about the state is derived by observing the outcome of an action and calculating the conditional probability of that outcome given that the state was i and the action taken k. Although these models consider some feedback information, none accurately represent the predictive maintenance process as implemented in many manufacturing systems.

Perhaps the most applicable work in maintenance control is that of Chu (1968). Chu assumes that the probability relationship between symptom and malfunction is not perfectly known. He then develops an optimal adaptive diagnostic procedure which yields the minimum expected diagnostic cost based on the present state of knowledge of p_{ij}. The procedure learns from past successful diagnosis, and is used in determining a simple optimal maintenance policy for a single machine.

Recently, research work has been on the statistical modeling of the effects of concomitant variables on time to failure. One method which has received considerable attention in this area has been proportional hazard modeling. Love and Guo utilized the historical correlation between several simple machine condition parameters and machine failure rates to determine optimal open-loop preventive maintenance cycles for a machine (Love 1991). An adaptive scheme was also presented to update the policy based on new feedback from the system. The approach was tested in an industrial setting and found to be quite successful, yielding improved maintenance plans. Jardine *et al.* (1987) applied a Weibull proportional hazards model to aircraft and marine engine failure data, and similarly found good results.

1.8 SUMMARY

Due to different priorities, maintenance and process, people often have different views of how a machine should run. Traditionally, maintenance has been wrongfully seen as a non value-adding operation. However, recent studies strongly indicate that a proactive maintenance approach always pays off. Machine failure occurs when a component, structure or system is unable to fulfill its intended purpose, resulting in its retirement from usable service. The ultimate goal of implementing a proactive maintenance program is to maintain a machine's operations at satisfactory levels, and possibly extend its useful life.

This book is intended to introduce recent research and development projects to the maintenance community. All projects introduced in this book have dual central themes: (1) computer-based technologies for 'intelligent' machine tool monitoring, and (2) fault modeling paradigms fundamental to those technologies. In order to make a connection between theory and real-life problem, case studies are collected and introduced throughout this book.

REFERENCES

Alban, L.E. (1985) *Systematic Analysis of Gear Failures*, American Society for Metals, Ohio.

Arora, S.R., and Lele, P.T. (1970) A note on optimal maintenance policy and sale date of a machine. *Management Science*, **17**(3), 170–3.

Barlow, R.E., and Hunter, L.C. (1960) Optimal preventive maintenance policies. *Operations Research*, **8**, 90–100.

Bergman, B. (1980) On the optimality of stationary replacement strategies. *Journal of Applied Probability*, **17**, 178–86.

Bloch, H.P., and Geitner, F.K. (1983) *Machinery Failure Analysis and Trouble-Shooting*, Gulf Publishing Co.

Chikte, S.D., and Deshmukh, S.D. (1981) Preventive maintenance and replacement under additive damage. *Naval Research Logistics Quarterly*, **28**(3), 33–46.

Chu, W.W. (1968) Adaptive diagnosis of faulty systems. *Operations Research*, **16**(5), September–October, 915–927.

Dale, B.G. (1990) Japanese manufacturing efficiency. A study in the electronics industry. *IEE Proceedings, Part A: Physical Science, Measurement, and Instrumentation, Management and Education, Reviews*, **137**(5), 293–301.

Eckles, J.E. (1968) Optimum maintenance with incomplete information. *Operations Research*, **16**(5), September–October, 1058–67.

Fard, N., and Zheng, Z. (1991) An approximate method for non-repairable systems based on opportunistic replacement policy. *Reliability Engineering and System Safety*, **33**, 277–88.

Federgruen, A., and So, K.T. (1990) Optimal maintenance policies for single-server queueing systems subject to breakdown. *Operations Research*, **38**(2), March–April, 330–48.

Gaimon, C., and Thompson, G.L. (1989) A real-time solution for preventive and repair maintenance. *Optimal Control Applications and Methods*, **10**, 211–28.

Hacker, M.E. (1991) Changing the maintenance culture in a traditional manufacturing environment. *1991 International Engineering Conference Proceedings*, Norcross, GA, May 19–22, 241–5.

Haxtl, R. (1983) Optimal maintenance and production rates for a machine: a nonlinear economic control problem. *Journal of Economic Dynamics and Control*, **6**, 281–306.

Hopp, W.J., and Wu, S-C. (1990) Machine maintenance with multiple maintenance actions. *IIE Transactions*, **22**(3), September, 226–33.

Jardine, A.K.S. (1973) *Maintenance, Replacement and Reliability*, Halsted Press, New York.

Jardine, A.K.S., Anderson, P.M., and Mann, D.S., (1987) Application of the Weibull proportional hazard model to aircraft and marine engine failure data. *Quality and Reliability Engineering International*, **3**, 77–82.

Li, P.G. and Wu, S.M. (1988) Monitoring drilling wear states by fuzzy pattern recognition techniques. *Journal of Engineering for Industry*, **110**, 297–300.

Love, C.E. (1991) Using proportional hazard modelling in plant maintenance. *Quality and Reliability Engineering International*, **7**, 7–17.

Luss, H. (1976) Maintenance policies when deterioration can be observed by inspection. *Operations Research*, **24**, 359–66.

Lyon, R.H. (1987) *Machinery Noise and Diagnostics*, Butterworths, Boston, MA.

Lyon, R.H. (1989) The monitoring and diagnostics of machines and process. *Proceedings of the 1st International Machinery Monitoring and Diagnostics Conference and Exhibit*, Las Vegas, Nevada, September, xi–xiii.

Maggard, B.N., Bailey, C.L., and Moss, D. (1989) Total productive maintenance; TPM that works. *Seventh IEEE/CHNT International Electronic Manufacturing Technology Symposium*, San Francisco, CA, September 25–27, 13–17.

Mobley, R.K. (1990) *An Introduction to Predictive Maintenance*, Van Nostrand Reinhold, New York.

Monahan, G.E. (1982) A survey of partially observable Markov decision process: theory, models, and algorithms. *Management Science*, **28**, 1–6.

Murthy, D.N.P., and Djamaludin, I. (1990) Quality control in a single state production system: open and closed loop policies. *International Journal of Production Research*, **28**(12), 2219–42.

Nakagawa, T., and Kijima, M. (1991) A cumulative damage shock model with imperfect preventive maintenance. *Naval Research Logistics*, **38**(2), April, 145–56.

Nisbet, T.S., and Mullett, G.W. (1978) *Roller Bearings in Service*, Hutchinson Benham, London.

Reiter, R. (1987) A theory of diagnosis from first principles. *Artificial Intelligence*, **32**, 57–95.

Serridge, M. (1989) Ten crucial concepts behind trustworthy fault detection in machine condition monitoring. *Proceedings of the 1st International Machinery Monitoring and Diagnostics Conference and Exhibit*, Las Vegas, Nevada, September 722–7.

Sheridf, Y.S., and Smith, M.L. (1981) Optimal maintenance models for systems subject to failure – a review. *Naval Research Logistics Quarterly*, **28**(3), 47–74.

Smith, J.D. (1983) *Gears and Their Vibration: A Basic Approach to Understanding Gear Noise*, Marcel Dekker, New York.

Tapiero, C.S. (1986) Continuous quality production and machine maintenance. *Naval Research Logistics Quarterly*, **33**, 489–99.

Teo, K.L., Wong, K.H., and Goh, C.J. (1989) Optimal maintenance of a system of machines with weakest-link-dependent performance. *Optimal Control Applications and Methods*, **10**, 113–27.

Thompson, G.L. (1968) Optimal maintenance policy and sale date of a machine. *Management Science*, **14**(9), May, 543–50.

Tranter, J. (1989) The fundamentals of, and the application of computer to, condition monitoring and predictive maintenance. *Proceedings of the 1st International Machinery Monitoring and Diagnostics Conference and Exhibit*, Las Vegas, Navada, September, 394–401.

Valdez-Flores, C., and Feldman, R.M. (1989) A survey of preventive maintenance models for stochastically deteriorating single-unit systems. *Naval Research Logistics*, **36**, 419–46.

Wang, R.C. (1977) Optimal replacement with unobservable states. *Journal of Applied Probability*, **14**, 340–8.

Wireman, T. (1990) *World Class Maintenance Management*, Industrial Press, New York.

Wong, K.H., Goh, C.J., and Teo, K.L. (1991) Optimal maintenance of stochastically deteriorating systems. *International Journal of Systems Science*, **22**(2), February, 375–81.

Zeng, L., and Wang, H.P. (1991) A fuzzy set approach for machine fault classification. Accepted for publication in *International Journal of Advanced Manufacturing Technology*, **6**(1), 83–94.

Zuckerman, D. (1986) Optimal replacement policy for stochastically failing equipment: a diffusion approximation. *Naval Research Logistics Quarterly*, **33**, 469–77.

CHAPTER 2

Fundamentals of sensory systems for maintenance engineering

Jay Lee, John Tsai and John Fildes

2.1 INTRODUCTION

Computer-aided maintenance depends on sensory systems to provide inputs for condition monitoring. Sensors can measure equipment performance and act like watchdogs to detect system faults and degradation.

Sensory systems play a critical role in computer-aided maintenance. Historically, maintenance engineering depended on visual and hand-on inspection to uncover potential problems. With the advance of system technologies, computers can now help maintenance engineering analyze trend lines, predict failures, and establish maintenance action. Sensor-based intelligent tools can diagnose the root cause of equipment failures, improve system availability and product quality. For these reasons, the National Research Council considers the development of sensory technologies as one of the six most critical needs for future manufacturing [1].

Technologies for sensory systems cover a broad and diversified field. Applications include supervisory control and data acquisition (SCADA), non-destructive evaluation, quality assurance, and condition monitoring for computer-aided maintenance [2,3,4]. Figure 2.1 shows a sensory system for advanced manufacturing. Since a comprehensive review of state-of-the-art sensor technologies is a major undertaking, this chapter provides only a brief overview of the technologies relevant to maintenance engineering.

With increasing emphasis on predictive maintenance, engineers and researchers are applying artificial intelligence to computer-aided maintenance. However, applications of advanced technologies must also consider human factors. Sensory systems can generate large amounts of data and overload the cognitive capabilities of users. Indiscriminate use of sensory systems can also be wasteful and counter-productive.

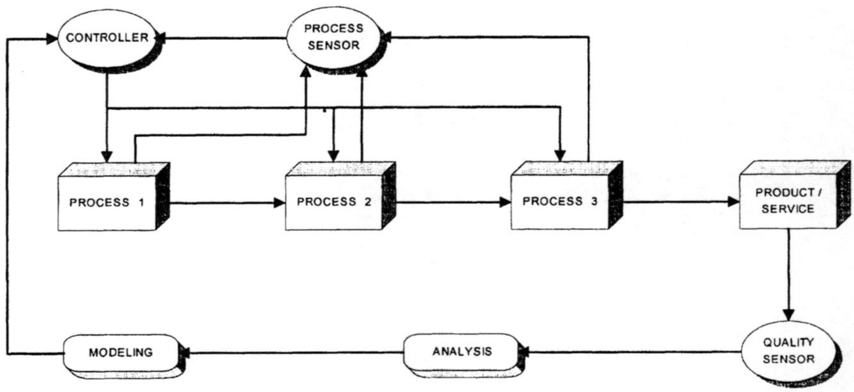

Fig. 2.1 Advanced manufacturing system.

2.2 SENSORY SYSTEM TECHNOLOGIES

Sensors are the key components in sensory systems. The word sensor comes from the Latin 'sentire', meaning 'to perceive.' A sensor is a physical device that responds to stimulus and produces a useful signal. Stimuli may include heat, light, sound, force, motion, and other phenomena. Since a transducer, by definition, is a device that converts one type of quantity into another type, a sensor is also a transducer. Figure 2.2 shows a block diagram of sensory systems.

2.2.1 System parameters

For typical applications, maintenance engineering would consider the following parameters in the design of sensory systems:

- High accuracy. High accuracy means low systematic errors about the true mean, minimum drift, linear output, low hysteresis, and stable calibration.
- High precision. High precision means low variability, low noise, good repeatability, and clean signals.
- Wide dynamic range. Wide dynamic range means sensors can respond to dynamic variables over a wide range of scales or values.
- High sensitivity. High sensitivity means sensors can respond to a small input level and produce a useful output signal.
- High resolution. High resolution means sensors can discriminate and detect small changes in measured variables.
- Fast response. Fast response means quick reaction and minimum time delay to change in measured variables.

Fig. 2.2 Sensory system.

- High reliability. High reliability means zero failures, zero defects, and robust operation in harsh industrial environments.
- Low cost. Low life-cycle cost means cost-effectiveness relative to performance, life, and other sensing alternatives.

2.2.2 Sensory systems

Sensors and sensory systems can measure a wide range of physical, chemical, and biological properties [5]. These include, for example, acceleration, acoustic emission, chemical composition, color, current, density, displacement, energy, flow rate, force, light, liquid level, mass, moisture, motion, opacity, pH, position, power, pressure, radiation, speed, strain, temperature, thickness, torque, velocity, vibration, viscosity, volume, voltage, weight, and others.

Sensors and sensory systems may use many different principles. These include, for example, capacitance, chromatography, conductivity, Doppler effect, eddy current, electro-magnetics, Hall effect, heat transfer, induction, infrared, laser, magnetic effect, microwave, optics, piezo-electric effect, pneumatics, radio frequency, resistance, semiconductor, strain gauge, thin film, ultrasound, vortex and others.

Products of sensory systems also include analog-to-digital converters, calibration equipment, data acquisition systems, data communication, data loggers, digital I/Os, encoders, interface boards, microprocessors, programmable controllers, recorders, signal analyzers, signal conditioners, and other hardware and their software drivers.

One related area to sensory systems is machine vision, which covers the acquisition, processing, and interpretation of visual images. In these systems, cameras replace sensors to measure a scene or an image. Since visual images, such as change in color or oil leakage, can reveal equipment problems, machine vision is a useful complementary technology for advanced maintenance.

2.2.3 Classification of sensors

A good taxonomy for sensors is provided by the requirements for different degrees of process control, which is shown in Figure 2.3. This taxonomy contains three classes: sensors used in regulatory feedback control loops, process analyzers, and product quality analyzers. In

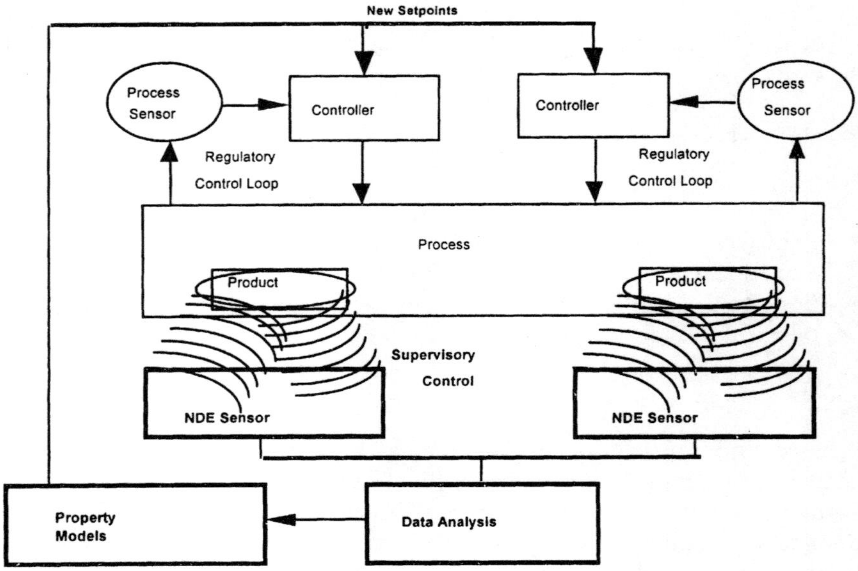

Fig. 2.3 Regulatory and intelligent control.

Figure 2.3, the basic element of control is the regulatory feedback loop that maintains controllable processing parameters at the desired values. These types of sensors, which usually provide a single value and are relatively generic in their applicability, are used for monitoring variables such as temperature, pressure, flow, level, displacement, proximity, and velocity. For the use of sensors in regulatory control, sensitivity, selectivity, simplicity, speed, reliability, and low-cost are the critical attributes.

Historically, sensors were almost solely used in this function, but this is no longer the case. Sensors, or more properly sensory systems, are now also used as process analyzers and product quality monitors. Sensors of this type, which are used for feedback in supervisory control, are more complex and application specific, and their output tends to be a matrix of values (e.g. a video image or a spectrum). Sensors of this type provide a representation of the process or product that has greater information content, but is more abstract than the representation provided by sensors used in regulatory control. Thus extensive computations and modeling are needed for process analyzers and product quality sensors, but there is also more ability to correct for deficiencies in sensitivity and selectivity through computational means. In this case, measurement speed is usually less demanding. These differences in the three types of sensors are summarized in Table 2.1.

Table 2.1 Characteristics of regulatory and intelligent process control

Characteristic	*Regulatory feedback sensors*	*Process and product sensors*
Speed of response	Very fast	Slow
Type of output	Single value	Matrix of values
Relationship of output to system parameter	Simple and direct	Abstract representation
Amount of computation	Little or none	Extensive
Sensitivity	Inherently high	Can be improved by computation and modeling
Selectivity	Inherently high	Can be improved by computation and modeling
Cost	Very low to moderate	High
Size	Small	Large
Applicability	Broad	Application specific

Table 2.2 Regulatory processing variables

Controllable variable	*Influenced processing factors*
Temperature	Rates of chemical reactions, degree of cure of polymers, degree of softening and viscosity, cooking times, drying times, annealing times, degree of carbonization
Pressure	Consolidation density, porosity, concentration of gaseous chemical reactants, forging, bending, injection times in molding
Flow	Mix ratios, quenching time, concentration of chemical reactants, color, pH, humidity
Level	Capacity, mix ratio, concentration of chemical reagents
Humidity	Drying of paper and grains, cooking
Position	Parts handling and placement, machining accuracy
Motion	Assembly, machining accuracy

The typical regulatory variables are temperature, pressure, flow, level, humidity, position, and motion. These variables form the basis for many process factors as summarized in Table 2.2 and the major sensor technologies for these variables are summarized in Table 2.3.

Integration refers to inclusion of signal conditioning on the same substrate as the transducer. One of the major advantages of integration is that the integrated sensor can operate with a very small transducer output, which results in significantly smaller sensors. With the advent in telecommunications and the microelectricalmechanical systems (MEMS) technologies, we can integrate software, hardware, power source, and communications on the same chip. Table 2.4 lists the advantages and disadvantages of integrated sensors.

Table 2.3 Major technologies for regulatory control

Variable	*Major sensor technologies*
Temperature	Thermocouples, thermistors, resistance thermometers, infrared pyrometers, acoustic pyrometers
Pressure	Manometers, bellows and diaphragms, strain gauges, piezoresistive, piezoelectric capacitive, thermocouple, ionization
Flow	Differential pressure (orifice plate, Venturi Pitot tube), velocity (turbine, vortex, electromagnetic, Doppler ultrasonic, time-of-flight ultrasonic), positive displacement (rotary sliding vane, gear, impeller), mass flow (thermal Coriolis), variable area (rotameters)
Level	Floats, pressure, radio frequency, ultrasonic, microwave, resistance tape, optical
Humidity	Dew point, length of hair, conductivity, capacitive, resonate
Position	Rotary encoder, linear variable differential transformer, potentiometers, magnetrostrictive, magnetic, inductive proximity, magnetic, ultrasonic
Motion	Tachometers, Pitot tube, anemometers

Table 2.4 Characteristics of integrated microsensors

Characteristic	*Advantage/Disadvantage*
Batch fabrication	Excellent control, low cost, sensor arrays
Loss of modularity	Difficult to package and limitations on materials
On-chip amplification	Better signal-to-noise ratio
On-chip compensation	Better accuracy through compensation for interferences
On-chip feedback	Better linearity
On-chip scaling and conversion	Standardized output
On-chip multiplexing	Sensor arrays, bus addressable

Smart sensors are defined as those that also include a logic function, rather than just signal conditioning electronics. Smart sensors may also include communications circuitry, diagnostics, and sometimes control outputs so that they can be directly connected to actuators, and documentation and trending functions. Conventional pressure sensors have rather broad accuracy limits, such as 0.25% of span. High-end smart pressure sensors have an accuracy of less than 0.1% of span, and mid-range smart pressure sensors approach this accuracy. Smart pressure transmitters are already available with built-in proportional–integral–derivative control functions, and fuzzy logic capabilities will soon be available. Multivariate measurement is not yet available, but will become so soon. Most likely, pressure and temperature will be the two measurements. Smart sensors are also available for presence detection, positioning, infrared photodetection with triangulation, and flowmeters, specifically magnetic meters, Coriolis mass flowmeters, and ultrasonic flowmeters. Integrated and smart sensor technology is also improving accelerometers, proximity detectors, and tactile sensors. The use of integrated and smart sensors will increase because of the overloading of shared resources and the use of distributed control schemes to solve this problem.

2.2.4 Maintenance parameters

Maintenance parameters are measurable attributes relating to system performance. For mechanical components these include, for example, force, lubrication characteristics, and physical dimensions. For electrical components these include, for example, heat, impedance, current, and voltage.

In mechanical systems, unwanted forces can come from misalignment, improper balance, vibration, clearance, or other sources. Force measurements typically include tension, compression, and impact. Common sensors for force measurement include, for example, piezo-electric devices and strain gauges.

In mechanical systems, heat can come from impact and friction. Heat can therefore become an indicator of wear. Excessive temperature rises in components can cause seizure, misalignment, or structure deformation. Heat can reduce the viscosity of lubricants and increase wear. The basic parameter to measure heat is temperature. Common sensors for such measurement include thermometers, pyrometers, thermal couples, and infrared devices.

Changes in lubrication usually come from contamination. Oil analysis determines the presence of a foreign substance such as water or wear particles that can degrade the lubricating properties of oil and cause premature bearing failure. The amount of metallic particles is also a direct indicator of wear. Lubrication analysis typically measures viscosity, particle size, concentration, or shape of wear particles.

Changes in geometry in mechanical components can increase wear and affect the overall performance of mechanical systems. Misalignment, imbalance, or thermal growth are the most common concerns. From dynamic or static measurement, maintenance engineering can determine displacement in X, Y, and Z axes, and rotational motion of pitch, yaw, and roll.

In electrical systems, excessive heat can come from restricted ventilation, high duty cycle, inadequate cooling, and power imbalance. Heat can cause premature insulation failure, component deterioration, and thermal noise. In addition to standard temperature sensors, current and voltage measurement can infer heat generation, such as from non-linear harmonics or poor power factors.

2.2.5 System engineering

The selection of sensors or sensory systems is application specific and technology dependent.

The first step in system engineering is to determine requirements – which component in which equipment or machine to monitor. Selection criteria may include failure rate, reliability, life, or mission importance. As a general guide, the failure rate of software is several orders of magnitude lower than hardware; electronic components with no moving parts are several orders of magnitude lower than mechanical components.

For mechanical components, life and failure rate can differ greatly. For example, ball bearings are relatively reliable components. The failure rate is approximately one per million hours of operation. Failure rates, by definition, assume only chance failures and do not include wear-out failures. The actual life expectancy of ball bearings is thus much lower, or approximately 6000 hours [6].

Experience from machine design suggests that unique parts designed by in-house engineers tend to fail more frequently. Off-the-shelf and commercially available components, such as ball bearings and electric motors, are usually more reliable and have a longer life.

The second step in system engineering is to select sensors. Sensor selection is subject to the following constraints: (1) not all variables are useful for maintenance purposes, and (2) not all variables are measurable (technology and cost constraints). For business decision making, the cost and benefit ratio must be favourable for the deployment of sensory systems. Cost justification is important.

At present, the most common technologies for maintenance include lubrication analysis (tribology), vibration monitoring, infrared thermography, and ultrasound monitoring [7].

2.2.6 Sensor system components

Because of the wide diversity of many different types of sensors and sensory systems, the present coverage can include only the most common components for computer-aided maintenance. These include:

(a) Light sensors

Light sensors are major components in factory automation, detecting the presence of objects and controlling physical movement. Such sensors usually use light interruption or reflection principles. In physical implementations, interruption schemes typically include the use of a vane or grid pattern to block the transmission of light.

A large number of photo-electric sensors are available commercially. For example, light-emitting diodes are very popular because of their relative immunity to ambient light and low cost. To reduce the effect of flash and reflection, detectors are often synchronized with source frequency to turn on and off for the proper duration. Light sensors may use fiber optic cables for light transmission. Some of these cables are only 1/16 inch in diameter and can fit into a tight space. They are rugged, explosion-proof, and do not radiate electro-magnetic interference.

Two divergent trends for light sensors are emerging that reflect the trend of system technologies. One group of sensors are becoming smaller, simpler, and of lower cost. These devices provide the basic on–off signals. Users must provide power supply, interface and conditioning devices. The other group of sensors are becoming more complex and integrated with self-contained, on-board logic and even signal processing capabilities. A number of new integrated sensory systems now have on-board 'intelligence' to perform self-diagnostics. For example, one major problem with light sensors is signal attenuation due to dirt, dust, moisture, and other contaminants from the environment. In such applications, the contrast or the differences in signal strength between light and dark conditions, can deteriorate over time. Smart or intelligent sensors can monitor themselves and compensate for changes in signal level.

(b) *Position sensors*

Position sensors include linear and rotational transducers that provide measurement of change in physical dimension or position. The rate of change of position is, by definition, velocity. These sensors can measure magnitude, direction, and the rate of mechanical movement. A number of different types and systems are available commercially.

For linear measurement, for example, encoders using glass scale and photo-sensing optics attached to the moving component can determine the relative motion between the scale and sensing head. In addition to optical sensors, magnetic encoders or potentiometric sensors can also provide similar measurement. Figure 2.4 shows a motion sensor that can measure linear translation. The direction of travel affects the output wave form (phase relationship) of the two sensors. A comparison can provide information on the direction of travel. The rate of change yields velocity.

For rotary motion, for example, synchros can operate as rotating transformers. With an AC reference signal passing through the rotor winding, the amplitude of voltage induced in the stator winding would vary in relation to the angular position of the rotor. The magnitude of the

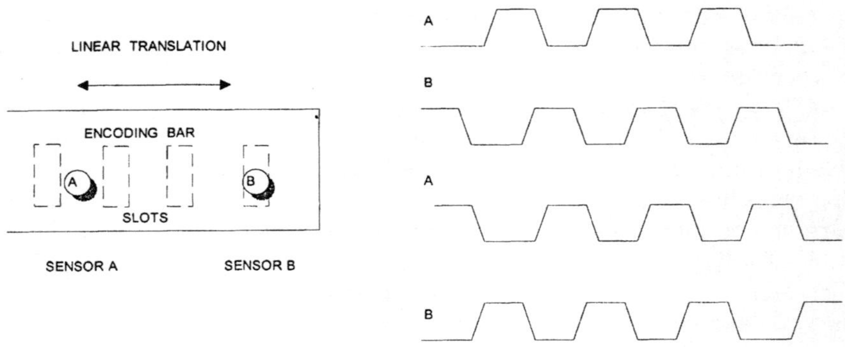

Fig. 2.4 Linear motion sensor.

induced voltage is proportional to the angle between the axis of rotor and the stator coil.

Another example of position sensor is a resolver that uses glass plates to generate sinusoidal wave form. The electrical phase differences between wave forms are functions of angular position. The frequency is fixed by the excitation circuit, typically about 2 kHz. To determine position, the resolver measures the relative phase relation between the reference and sensor output signals.

The cost of position sensors depends on the accuracy required. A high-accuracy optical encoder can resolve 1.2 arc seconds at a cost of $10 000 each. Resolvers that can resolve 7 arc seconds would cost $2000 each. Figure 2.5 shows a high-precision position sensor using a position sensitive detector. An LED illuminates the target and the resulting reflection passes through an optical system to reach a detector. Calculations can determine the distance of the object, using a triangular relationship and displacement from the center position (shown as 'x') on the detector.

Magnetic sensing can also provide position information. When a ferrous object approaches magnetic sensors, the magnetic flux changes and generates a measurable voltage. Magnetic sensors can measure both linear or rotational velocities. The most common types are eddy current and Hall-effect sensors. An eddy current device generates a high-frequency magnetic field that can induce eddy current in the metal target for sensing. A Hall-effect device generates voltages in the presence of an external magnetic field. Typical Hall-effect sensors use feedback to stabilize switching during flux reversals.

(c) Temperature sensors

In the industrial environment, temperature sensors are usually thermocouples, resistive temperature detectors, or thermistors. For maintenance

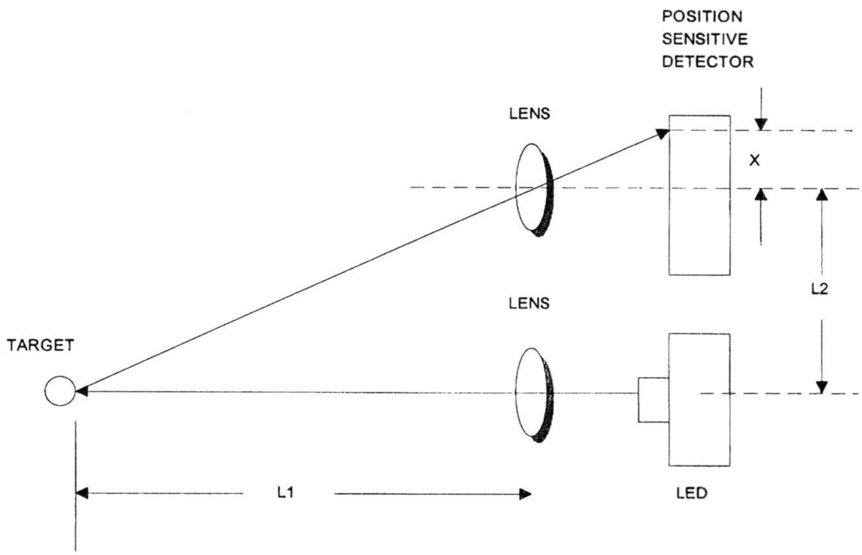

Fig. 2.5 Position sensing by triangulation.

engineering, changes in temperature may infer a number of potential operational problems, such as thermal growth in structure, bad bearing, shaft binding, or overheating of electric contact. For performance monitoring, a typical temperature sensor can have an accuracy of 0.1 °C and an operating range of –25 to +150 °C.

Thermocouples are by far the most numerous temperature sensors for process control; they are inexpensive and moderately accurate, despite low sensitivity. A thermocouple generates a very small voltage between the hot and cold (reference) end of two dissimilar metal wires. They can cover a wide temperature range, up to 1800 °C. They also come in different size (diameters) and response time. However, the voltage output is inherently non-linear and needs special circuits or software routines to provide proper amplification, compensation, or signal isolation.

Resistive temperature sensors are wire-wound resistors whose resistance changes with temperature. They have good linear response for short ranges. Copper, nickel, and platinum are the most common wire material. Users can connect sensors in different configurations using two, three or four wires. These sensors are more accurate than thermocouples, but require a current source for excitation and are sensitive to mechanical vibration. Cost is also higher. As an application example, resistive sensors are available in three standard ranges. The upper limit is typically 750 °F and the output is 4–20 mA, linear within the temperature range. Calibration and linearity errors are both less than 0.1% full scale. Response time is less than 0.2 seconds.

Thermistors exhibit large changes in resistance for relatively small changes in temperature. Their resistance drops exponentially with the increase in temperature. Made from a sintered mixture of oxide, they are relatively inexpensive. Their major shortcomings include poor interchange between sensors (even from the same manufacturing lot), and limited temperature range. Nevertheless, they provide a low cost alternative useful within their limitations.

(d) Infrared sensory systems

Infrared sensory systems measure or detect heat radiation, which reflects the temperature of a radiating body. This type of system can perform remote temperature measurement or thermal imaging (thermography). The basic components in such systems include lens, detector, and electronics, usually packaged as compact and lightweight cameras. Equipment suppliers also provide a wide range of software with trending and database capabilities for analysis and reporting purposes.

Infrared imaging can uncover a wide range of mechanical and electrical problems. In rotating devices, such as gear, shaft, belt, pulley, or bearing, excessive friction is an indication of wear, vibration, out-of-balance, or misalignment. The resulting heat rise can accelerate structural or material damage and causes premature failure. For electrical components, such as switch, fuse, or circuit breaker, overheating can cause insulation or contact breakdown.

Thermal imaging cameras are similar to video cameras in size and weight. A representative camera would have a 25 mm lens with a focal distance of 12 inch to infinity, spatial resolution of 1.2 mrad, spectral range of 3.6 to 5 microns, temperature range from 10°C to 1500°C, sensitivity of 0.1°C, and digital acquisition of 12 bits resolution. The camera may use 78 000 sensors in a focal plane of silicon array containing 320 × 244 pixels. Option devices include additional lens (close-up, wide-angle, and telephoto) filters, RS-232 interface, and others.

After image capturing, information processing is necessary to produce quantitative information. In a typical industrial application, a personal computer would perform statistical analysis. Users obtain temperature versus time trending for a selected temperature point of interest. However, users should develop an understanding of the physics of infrared radiation and characteristics of targets in relation to spectral response, emissivity, and reflection from the background. Such understanding is important for proper engineering evaluation and interpretation of thermal images.

Infrared thermography is becoming one of the major tools for inspection and predictive maintenance. Infrared measurement enables maintenance engineering to examine temperature distribution of plant equipment, detect abnormal temperatures and obtain early warning of potential breakdown.

(e) Lubrication sensory systems

Lubrication monitoring uses test instruments or sensory systems to measure a number of physical and chemical parameters relating to tribology. These systems include spectrometric analyzers and particle counters for determining wear, friction, cleanliness, and stability of fluid film. Since the depletion of oil additives and the presence of contaminants are both direct indicators of wear, off-line tests would measure, for example, viscosity, water, or acid content. In additional to chemical analysis, ferrography, or wear debris analysis can also determine the physical amount, size, shape, or appearance of wear particles or debris.

Lubrication analysis is a new tool for maintenance engineering. In the past, only specialist equipment laboratories had the capabilities and expertise to analyze lubricants. Such analysis may result in the reporting of a large amount of parameters. In recent years, however, diagnostic instruments and sensors have become less expensive and more computerized. Industrial plants can now set up in-plant laboratories. Although in-plant analysis cannot fully duplicate the accuracy and precision of laboratory analysis, maintenance engineering can perform limited diagnostics on a more frequent schedule and with quicker turnaround time.

(f) Vibration sensory systems

Mechanical vibration originates from a force whose magnitude and/or direction varies with time. Such forces may come from impact, imbalance of rotating element, misalignment, or excessive clearance. For maintenance, vibration is one of the most important parameters that determine the condition of the rotating element. Maintenance engineering can either listen to the sound of the vibration or use sensory systems for quantitative and objective measurements.

For trend analysis, maintenance engineering would monitor vibration signatures such as frequency versus amplitude over a period of time. The common procedure establishes a baseline or reference first, when the component is in good working condition. Additional measurements would then be made at periodic intervals during the remaining life of the component. Vibration analysis would compare the current measurement with the baseline to detect changes that may provide diagnostic information and early warning before component failure.

Vibration is directly related to force, which, in turn, is related to acceleration. For acceleration measurement, industrial sensors often measure force using piezo-electric devices. Piezo material changes electric resistance when subject to mechanical stress. In a simplified scalar representation, the relationship between specific resistance Q and the applied mechanical stress are:

$$\delta Q/Q = P * S$$

where δQ is the change in Q, P is the piezo-electric coefficient, and S is the mechanical stress. In a certain analytical representation, P is a tensor with 21 components.

In a typical piezo-resistive accelerometer, a pivoting arm would support a moving mass that can create compressive and tensile stresses on two pieces of piezo-electric material. The induced stresses are proportional to the magnitude of acceleration. The dynamic range of piezo-resistive accelerometers is in the order of 1000 g and can withstand shock of 10 000 g. Sensitivity is in the range of 0.1 to 2.5 mv g^{-1} and the resonant frequency is 50 Hz.

Low frequency vibration, such as 40 times the running speed, is useful for detecting component with imbalance, misalignment, looseness, or excessive play. Vibration from these sources has a frequency range from 0 to 1000 Hz. Within these frequencies, signals may contain different spectral lines from shaft-rate harmonics, mechanical resonance, and other causes. Detection and isolation of a specific vibration can be difficult. For this reason, sensors for low frequency measurement would have a minimum sensitivity of 500 mv g^{-1} to provide adequate signal-to-noise ratio.

Since signals from vibration sensors contain higher harmonics from different modes of vibration, vibration analysis often converts vibration signatures or wave forms from the time domain into the frequency domain. This conversion would clearly show the magnitude or amplitude or vibration at various discrete frequencies and helps to isolate the origin of vibration.

The most common method for data analysis of vibration is the Fourier analysis that provides the mathematical translation between time domain and frequency domain. To analyze a continuous spectrum, digital processing would sample waveform at discrete instants to obtain sampled-data signals. Mathematically, sampled-data signals consist of a set of weighted Dirac pulses and their frequency components. The corresponding Fourier transform becomes a Discrete Fourier Transform (DFT) that has a set of discrete values at various discrete frequencies. DFT divides the wave-form over a period of T into N samples. Each discrete spectral response is equivalent to the output of one of the N shifting filters, having a bandwidth of 1 T^{-1}.

The DFT signal thus contains N samples of harmonics with sine or cosine components. Evaluation of each sampled spectrum line involves the multiplication of each sampled signal by the entire set of sine or cosine terms. Calculations can become tedious. However, mathematical analysis shows that multiplication consists of repeated calculation. An efficient recipe or algorithm in the name of Fast Fourier Transform (FFT) is available to eliminate redundant and repeated calculations. Since many textbooks on FFT are available, this chapter will not cover the derivation of equations. Numerous softwave packages are also on the commercial market to perform the mathematical manipulation [8].

A number of other analytical techniques are also available for the analysis of vibration signals or noise, especially for non-stationary signals that change structurally during the observation period. For example, for transient signals, wavelet transformation can decompose vibration data into time-localized waveforms using a time-scale parameter. Parametric modeling, such as the autoregressive model, extracts characteristic parameters from observed data and then uses the parameters of the model to compute the power spectrums. In these approaches, signals are often treated as a time-series random process. Recursive computation would then attempt to minimize residual errors to obtain the best sets of coefficients for the models [9].

2.6.6 Signal processing

The first step in signal processing is signal conditioning that improves, enhances, or modifies signals from sensors. In an industrial environment, electrical noise, cross-talk, internal ground loop, and interference from AC power can all contaminate signals from sensors. Noise may come from many sources, including the sensors themselves. Signals can also leak between channels in multiple-channel systems. Signal conditioning typically performs restoration, enhancement, isolation or noise abatement.

Industrial signal conditioning systems come in many different forms ranging from single-channel I/O modules to multi-channel and multiple functional modules that contain data acquisition and communication boards. A typical single-channel module is a fixed-function, plug-in board designed for one type for sensor.

The second step in signal processing is data acquisition, which performs multiplexing and analog-to-digital conversion.

Multiplexing is the simultaneous transmission of two or more signals using the same media or channel. In signal processing, a multiplexer enables a multiple number of sensors to time-share a single module for analog-to-digital conversion. A typical multiplexer can accommodate 10 to 100 channels.

Analog-to-digital (A/D) conversion changes analog signals to digital form. The conversion process includes two steps: sampling and quantization. Sampling measures a continuous function at discrete time intervals periodically. An important step in sampling is the sample-and-hold operation. For fast-changing analog signals, a holding period is necessary to provide time for A/D conversion.

Quantization assigns a digital value to the analog signal sampled at a discrete interval. The assigned value corresponds to a finite level of voltage, typically between 0 to 10 volts. The number of quantization levels to cover the full range of sensor response is:

Number of quantization = 2 ^ n

The resolution is equivalent to the level of spacing, that is:

$$\text{Resolution} = \frac{\text{full-scale range}}{\text{no. of quantization}}$$

The hardware for data acquisition is typically I/O plug-in boards for PCs, workstations, or other computers with parallel ports. Detailed specifications would depend on input requirements, such as number of channels, sampling rate, resolution, and the input range.

In the industrial environment, personal computers or PCs are becoming the most common enterprise-wide platforms. For maintenance engineering, PCs interface with management information systems at the plant level and programmable logic controllers (PLCs) at the machine control level. PLCs can now plug into PC backplanes and PCs can plug into PLC backplanes.

2.3 SYSTEM APPLICATION

Sensor fusion is a major system application of sensory systems. Sensor fusion integrates output from (a) multiple dissimilar sensors, and/or (b) spatial and temporally separated similar sensors. The purpose of sensor fusion is to acquire more knowledge about the environment. Humans also use sensor fusion extensively by processing multiple inputs from sight, smell, sound, or touch. A number of references are available on the subject [10,11].

Sensor fusion creates redundant and complementary data, allowing cross-checking of sensor signals. Sensor fusion combines and correlates data from multiple sensors to make inference. The benefit of sensor fusion includes an increase in confidence and the reduction of ambiguities from measurement. Figure 2.6 shows an example of a multi-level sensor fusion using a combination of sensors to produce one single output. Each node in the diagram represents a fusion of multiple signals at different levels of combination.

2.3.1 Advantages

Sensor fusion offers the following key advantages:

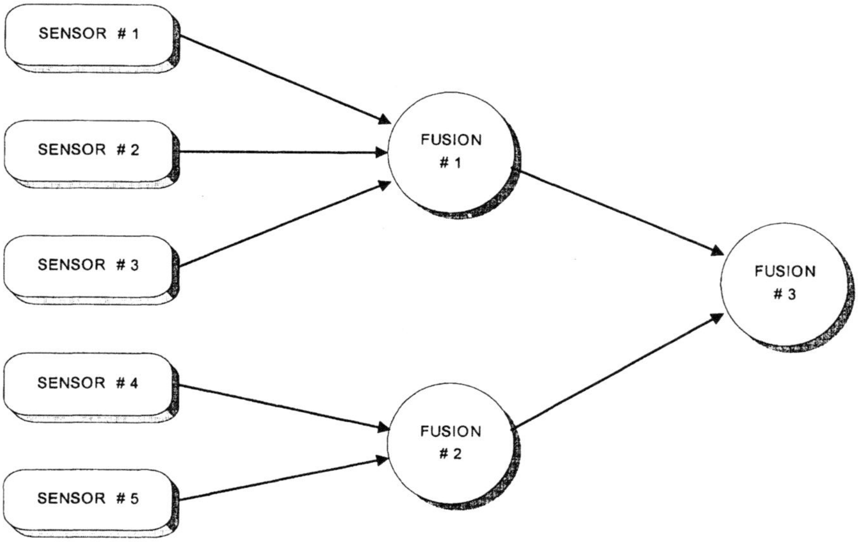

Fig. 2.6 Sensor fusion structure.

1. Redundancy. Multiple sensors measuring the same parameter can collaborate and cross-check one another to improve accuracy and reliability.
2. Complement. Multiple sensors measuring different parameters can correlate and complement one another to reinforce the output from individual measurement.
3. Spatial-temporal sensitivity. Multiple sensors placed in a different position or location can better capture the dynamics of disturbance or change.
4. Functional extension. Multiple sensors measuring the same parameter can overcome limitations in life, range, linearity, precision, uncertainties, signal-to-noise ratio, or response time.
5. Effectiveness. Depending on application, multiple sensors can be more efficient or cost effective than a single 'super' sensor, resulting in increased accuracy, resolved ambiguities, and improved reliability.

2.3.2 Principles

Functionally, sensor fusion operates on the following four principles:

1. Competition. Output from each sensor may conflict with other outputs. Competitive fusion typically uses multiple sensors for the same measurement and may ignore outputs that do not meet pre-established criteria.

2. Complement. Sensors measuring different parameters of the same phenomenon can provide comparable or complementary data to reconcile differences and use collateral data to support or verify the same measurement.
3. Validation. By comparing sensor output to process dynamics and reference response, sensor fusion can validate the accuracy and precision of measurements. When multiple sensors measure the same parameter, sensors can cross-check and benchmark against one another to detect drift, malfunction, or sensor malfunction.
4. Reconciliation. Multiple sensors can introduce inconsistencies and uncertainties. Reconciliation examines sensor output using current measurement, empirical and analytical modeling, or historical data to resolve discrepancies in measurement.

2.3.3 Algorithms

Fusion algorithms interpret and extract information from multiple sensors or sensory systems. Currently, over 75 fusion algorithms are available for various applications [12]. Most of the techniques perform pattern recognition and classification using the following general steps.

The first step in sensor fusion is preprocessing. This step performs feature extraction, removing extraneous information, and enhancing the detection of pertinent features. Ideally, features should represent the essence of a pattern. Simple features consist of a point or a line and complex features include circles, squares, or other identifiable shapes.

The second step is partitioning. Signal patterns are usually too large for processing as a single unit. Partitioning isolates the pattern into elements of workable size or 'windows.' This approach enables processing to concentrate or focus on samples of spatial or temporal patterns.

The third step is feature extraction. 'Feature' refers to a simple element of spatial or temporal arrangement of a pattern. Mathematically, a pattern contains a set of feature vectors that describe the state of the system. For extraction, fusion algorithms would establish feature vectors based on the presence or absence of specific or known features.

The fourth step is decision-making. Following the successful identification of features, the decision process determines the presence or absence of a particular pattern and declares the identity of the classification. This step uses certain decision criteria such as voting method, Bayesian inference, or other empirical methods.

For simple sensor fusion, physical or mathematical models can relate measured variables to system variables, based on relationship of physics, chemistry or biology. In these cases, signals from sensor fusion $f(X)$ are vector functions of components $(X_1, X_2 \ldots X_n)$. For time varying signals:

$$f(X) = f(X, t)$$

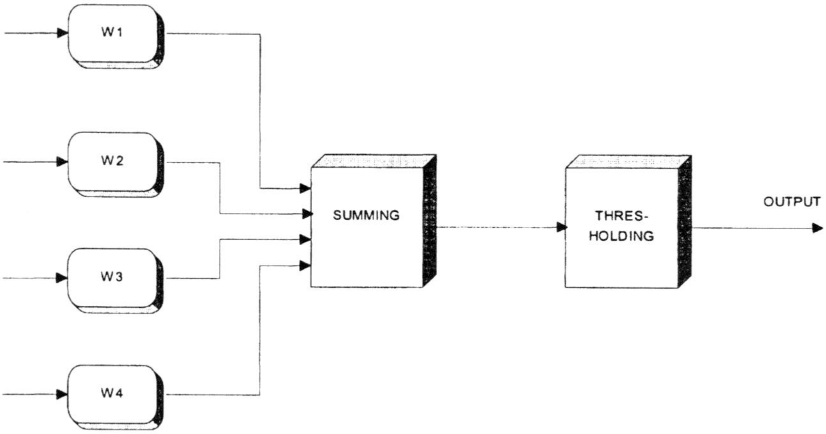

Fig. 2.7 Pattern classification steps.

Figure 2.7 shows a simple sensor fusion using threshold logic. Each input uses a weight whose value is adjustable. Outputs from the sensor pass through an operational amplifier with summing registers to add up all the sensor outputs. The resulting output goes to a threshold logic for comparison against a predetermined threshold established by an analytical or empirical relationship. Other fusion algorithms for simple cases include averaging, ranking, scaling, indexing, regression analysis, or time-series analysis.

For more complex cases, feature extraction uses more advanced techniques, including physical modeling, histogram analysis, spatial frequency analysis, parametric templates, cluster analysis, expert systems, fuzzy logic and neural networks.

Histograms show the relative frequency of occurrence of identifiable features, such as mean, variance, or peak-to-valley value. Histograms may represent probability density functions. The processing algorithm would compared the shape of the histogram of input patterns with known patterns. The algorithm then analyzes the significance of the differences or deviations and provide indices for the acceptance or rejection of a hypothesis.

Spatial frequency analysis converts patterns into frequency spectrums. While computers can perform the necessary conversion readily, coherent optics can also convert sensor patterns into spatial frequencies for feature extraction. In the optical approach, spherical waves illuminate the input plane and produce a corresponding Fourier transform in the plane where the spherical waves come to focus. Since optics can process spatial variables simultaneously, they are inherently fast and are true parallel processors. A number of references provide the technique of optical transformation [13].

Parametric templates identify feature boundaries of patterns from correlation with reference function, template, or mask. The location of each feature vector in relation to template boundaries determines the classification of the vector. A positive match is achieved when the reference function, stored in computer memory, and the input pattern or vector are nearly identical.

Cluster analysis uses heuristic methods to group data into natural clusters that represent object identity and classification. A cluster is a group of similar variables. The analysis determines if their feature vectors match the reference vectors, using criteria such as matching coefficients or distance measurement. Cluster analysis provides qualitative interpretation for sensor patterns. One example is the QI-Map technique that maps sensor data into clusters of normal, abnormal, and uncertain regions of operation [14].

Expert or knowledge-based systems make inferences on sensor output using symbolic abstraction or semantic rules. A number of different approaches are available for inferences, such as using rules, frames, cases, or models for reasoning. To perform pattern identification, domain knowledge can come from mapping of parametric features and symbolic representations as trend, landmark, or state description.

Fuzzy logic holds great potential for combining fuzzy reasoning with classical techniques of pattern recognition. Fuzzy logic provides an effective means of handling the approximate and inexact nature of data or patterns that are subject to noise and ambiguity. This reasoning framework enables the adaptation of empirical or heuristic knowledge, such as a 'rule of thumb' experience or engineering judgment, as the basis for logical inference. For this reason, fuzzy logic is emerging as a critical enabling technique for sensor fusion [15].

The neural network model is another common technique of sensor fusion. Using this paradigm, sensor signals become input vectors for neurons to process. Each neuron has a threshold of activation. The link between neurons has an adjustable weight to represent the strength of connection. The networks can store patterns from 'training,' that is from adjusting the weight by trial and error. Networks such as the classical Perceptron act like filters to perform pattern classification [16]. Conventional neural networks, however, require extensive training and a large amount of memory. To overcome this shortcoming, Albus developed the Cerebella Model Articulation Controller (CMAC) by modeling human information processing [17]. CMAC has a number of successful applications for pattern recognition [18].

A major limitation of pattern recognition is the need for 'training' patterns in order to establish the boundary of feature space. Reference patterns are the basis for identifying unknown vectors. For industrial applications, relevant historical or empirical data are often unreliable and/or unavailable. Data collection in maintenance is also seldom comprehensive or systematic because operating personnel are just too busy with daily assignments or 'fire-fighting'.

For degradation monitoring, Lee has therefore proposed a methodology to reduce the dependency on historical data. He uses CMAC and look-up tables (see Figure 2.8) to store the direction of change of sensor signals instead of the absolute magnitude of change. The direction can become a valuable trend indicator to provide an index or confidence number for describing the nature and extent of machine degradation [19].

2.4 CONCLUSIONS

In computer-aided maintenance, sensors and sensory systems are the major sources for providing important information for maintenance engineering. Using sensory outputs, condition monitoring can become fully automated. Sensory systems can supplement or replace human observations, providing objective and quantitative data for analysis and decision support. With the advance in computer technologies and microchips, smart sensory systems will emerge as the workhouse or predictive and proactive maintenance.

Facing growing competitive pressure in the market place, enterprises need to improve their manufacturing capabilities such as quality and system availability. The challenge is to achieve zero defects in products

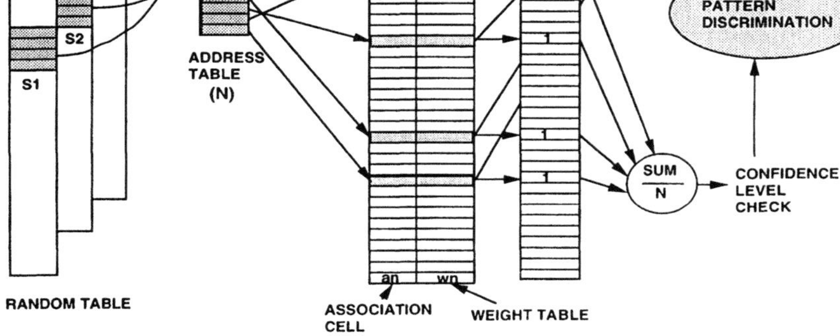

Fig 2.8 CMAC pattern discrimination model (PDM).

(or services) and zero breakdowns in operations. Sensory systems can provide the leverages to improve or re-engineer maintenance.

REFERENCES

[1] NRS, National Research Council (1995) *Unit Manufacturing Processes*, National Academy Press, Washington, DC, 1–10.

[2] Norton, H.N. (1992) *Sensor and Analyzer Handbook*, Prentice-Hall Englewood Cliffs, NJ.

[3] Fraden, J. (1993) *AIP Handbook of Modern Sensors*, American Institute of Physics, New York, NY.

[4] Carr, J.J. (1993) *Sensor and Circuits*, Prentice-Hall, New York, NY.

[5] Zuech, H. (ed.) (1992) *Handbook of Intelligent Sensors for Industrial Automation*, Addison-Wesley, Reading, MA.

[6] Levinson, C. (1964) System reliability analysis, in *Mechanical Design and Systems Handbook* (ed. H. Rothbart), McGraw-Hill, New York, NY, 1837–53.

[7] Mendelbaum, G. and Mizuno, R. (1992) Directions in maintenance. *Maintenance Technology*, June, 45–8.

[8] Proakis, J. and Manolakis, K. (1988) *Introduction to Digital Signal Processing*, Macmillan, NY.

[9] Gratlan, T. (ed.) (1993) *Sensors VI Technology, Systems and Applications*, Institute of Physics, Philadelphia, PA.

[10] Chiu, S.L., Morley, D.J., and Martin, J.L. (1987) Sensor fusion on a parallel processor. *Proceedings of IEEE International Conference on Robot and Automation*, 1629–33.

[11] Waltz, E. and Llinas, J. (1990) *Multisensor Data Fusion*, Artech House, Norwood, MA.

[12] Hall, D.L. (1992) *Mathematical Techniques in Multisensor Data Fusion*, Artech House, Norwood, MA, p. 35.

[13] Leith, E.N. and Upatnieks, J. (1967) Recent advances in holograph, in *Progress in Optics*, North-Holland, Amsterdam, **6**, 1–52.

[14] Whiteley, J.R. and Davis, J.F. (1993) Qualitative interpretation of sensor patterns. *IEEE Expert*, April, 54–63.

[15] Cox, E. (1993) Adaptive fuzzy systems. *IEEE Spectrum*, February, 27–31.

[16] Fausett, L. (1994) *Fundamentals of Neural Networks*, Prentice-Hall, Englewood Cliffs, NJ.

[17] Albus, J.S. (1995) Data storage in the cerebellar model articulation controller (CMAC). *Journal of Dynamic Systems, Measurement, and Control*, Trans. ASME, September, 228–33.

[18] Lee, J. and Tsai, J. (1991) Intelligent control, learning, reasoning methodology for intelligent material handling. *Issues in Design/Manufacturing Integration*, ASME De-Vol 39, 47–52.

[19] Lee, J. (1992) *In Process Machine Degradation Monitoring and Fault Detection Using a Neural Network Approach*; PhD thesis, Univ. of George Washington, Washington, DC.

[20] Fildes, J. (1997) Sensors in manufacturing, *CRC Mechanical Engineering Handbook*.

CHAPTER 3

Related work on machine monitoring and diagnostics

Hsin Hao (Tom) Huang and Ben Wang

This chapter reviews related research work on machine monitoring and diagnostics. Specifically, the following four categories of work are reviewed: signal processing techniques, artificial intelligence, artificial neural networks, and sensor fusion. Section 3.1 describes two of the most popular signal processing techniques, namely, vibration analysis and parametric modeling. Section 3.2 addresses the use of artificial intelligence approaches to fault diagnostics, particularly expert systems and model-based reasoning. Section 3.3 introduces a new technology, the artificial neural network, and its related studies in the area of machine monitoring and diagnostics. Finally, Section 3.4 provides an overview of the sensor fusion approach.

3.1 SIGNAL PROCESSING TECHNIQUES

Over the years, most machine monitoring and diagnostic systems have been performed by gathering the sensory data from the process, then analyzing the data off-line through a signal processing technique. One of the most widely used signal processing techniques is vibration analysis. This is because no other parameter can reveal as wide a range of machine fault types as vibration (Serridge 1989).

3.1.1 Vibration analysis

Vibration analysis deals with the extraction of information from measured vibration signals. It is well recognized that vibration characteristics will change as a machine's condition changes. Wear or damage to rotating elements, imbalance, and resonance can generate excessive vibration. For example, as wear increases, the clearance in the bearing housing increases and the shaft begins to vibrate excessively. Vibration analysis is able to reveal this excessive vibration. Generally, vibration data can be analyzed in two different domains: time and frequency (Tranter 1989, Li and Wu 1989). Time-domain analysis involves designing indices that are sensitive to the amount of impulsive vibrations observed. This technique includes overall level (RMS) measurements, peak level detection, crest factor, shock pulse, spike energy, kurtosis analysis, time waveform, and orbits. Frequency-domain analysis involves transforming the vibration waveform to show a train of impulses at different frequencies. This technique includes spectrum analysis, waterfall plot, difference spectra, RMS of spectral difference, envelope analysis, high frequency resonance analysis (HFRT), and matched filter.

Vibration analysis involves a number of steps, the first one being the recording of the baseline vibration signal which corresponds to a normal machine condition. If the frequency-domain analysis is applied, this baseline vibration signal is processed to show vibration levels (in amplitude) at different frequencies. Then, actual vibration signals during a machine's operating period are collected and processed in the same domain. Finally, the actual vibration signal is compared with the baseline signal to determine the machine condition and the source of machine failure if applicable.

One of the most powerful vibration analysis techniques is spectrum analysis, which estimates the spectrum or power spectral density (PSD) from a vibration signal by performing a Fast Fourier Transform (FFT). The reason for the popularity of this FFT-based technique is because of its high computational speed. In addition, analysis of a machine vibration spectrum can yield important information regarding the condition of machine components because each rotating component in a machine generates identifiable frequencies; thus, changes at a given frequency range can be related directly to a specific component failure. It is found that over 75% of the current commercial systems use spectrum analysis in their system (Carlson and Spiewak 1992). However, there are some problems with this FFT-based technique (Kay and Marple 1981, Braun 1986, Potts and Hakimmashhadi 1989):

- low frequency resolution,
- implicit windowing of the data,
- no significant data reduction.

3.1.2 Parametric modeling

In addition to spectrum analysis, the parametric modeling technique has been used for estimating the vibration spectrum. It is used in an attempt to alleviate the inherent limitations of the FFT approach. Two major advantages of using a parametric modeling technique are given below:

- improvement of frequency resolution over FFT by suppressing the noise from the real signal, and
- reduction of data by using few parameters to describe the signal globally.

A number of parametric modeling techniques have been reported to estimate the vibration spectrum, for example, the autoregressive (AR) method, the autoregressive and moving average (ARMA) method, Prony's method, the minimum variance method, and the covariance method. A detailed review of these techniques can be found in Kay and Marple (1981), and Braun (1986). In Chapter 4 of this book, a project utilizing parametric techniques for machine fault classification is presented. Normally, a parametric modeling technique is a two-step procedure:

1. Fit a model to the observed data.
2. Use the parameters of the model to compute the power spectrum.

One of the most widely used parametric methods is autoregressive (AR) modeling. The interest stems from two reasons. First, an AR spectrum can give a high resolution spectrum with sharp peaks. Second, estimates of AR parameters can be obtained as solutions to a linear equation. Vibration signals are treated as if they were generated from a time series random process x_n,

$$x_n,\ n = -\infty, \ldots, 0, \ldots, \infty$$

where the observed interval is from $n = 1, \ldots, N$. The autoregressive model of x_n is as follows:

$$x_n = -a_1 x_{n-1} - a_2 x_{n-2} - \ldots - a_p x_{n-p} + e_n$$

where e_n is the prediction error. The spectrum can be computed by plugging all p a_k parameters into the theoretical power spectral density (PSD) function defined below.

$$P_{AR}(f) = \frac{2\,\Delta t\,\sigma^2}{\left| 1 + \sum_{k=1}^{p} a_k \exp(-i2\pi f k \Delta t) \right|^2}$$

$$-\frac{1}{2} \leq f \leq \frac{1}{2}$$

$$\Delta t = \frac{1}{S}$$

where S is the sampling rate used in data acquisition, f is the fraction of the sampling rate, p is the prediction lags or order of the AR model, and σ^2 is the sum of square prediction errors. Therefore, if the prediction coefficients, a_k, can be estimated accurately, the spectrum of the random process can be calculated correctly from the above equation.

Another popular parametric modeling technique is ARMA modeling. The motivation for the use of an ARMA model, instead of an AR model, lies in the achieved parsimony, as such models often yield a much smaller set of necessary parameters (Braun 1986). The ARMA (p,q) model consists of a deterministic part (autoregressive) and a non-deterministic part (moving average).

The parametric methods described above have been applied in the area of fault detection. For example, Matsushima *et al.* (1982) applied an AR(28) model to detect tool breakage by monitoring the spindle current. The residual error of the model was monitored to determine if there was any tool breakage. The result showed that a large residual error was encountered even for a small tool breakage. Sidahmed (1989) applied an AR(10) model to detect the teeth spalling in a gearbox. His study showed that the AR model had the ability to detect small changes in the signal at the beginning of the spalling.

In the work of Liang and Dornfeld (1989), an AR(6) model was used to monitor tool wear condition. Acoustic emission generated during the cutting process was fitted by the model. The parameters of the AR model were monitored, but only the first and four parameters were shown to be important features in determining the amount of tool wear. They reported that the AR parameters of the acoustic emission signal showed good sensitivity to tool wear over the other machining conditions. Wu *et al.* (1980) used an ARMA model to monitor machine condition and their result indicated that the ARMA spectrum was able to catch a small variation of imbalance while the FFT spectrum could not. Their approach was to monitor the machine condition by examining the change in the residuals' term in the parametric model. If the variance of the residuals increased over a certain threshold value, the machine condition was then considered to be abnormal, and vice versa.

The disadvantage of parametric modeling is that it is not easy to find an optimal order for the model. The general guideline in the selection of the model order is based on the minimization of the sum of square errors. Akaike (1969, 1974) proposed two criteria, final prediction error (FPE) and Akaike information criteria (AIC), which can be used as the objective functions for order selection. In the resent work of Lin (1992), the order with the highest FPE and AIC level was selected as the optimal order.

Both parametric models mentioned above performed well in the early detection of machine failure. However, they were unable to identify the cause of the failure. This fault recognition task is usually done by the analyzer who identifies the cause of the fault by visual inspection of the spectrum. This is not an easy task because it requires experience and knowledge in order to make a correct diagnosis.

3.2 ARTIFICIAL INTELLIGENCE

Although vibration analysis and parametric modeling techniques have been proven to be very useful for machine monitoring and diagnostics, they are also knowledge-intensive techniques. In other words, they need to be performed by a highly trained and experienced engineer in order to identify the source of the machine fault correctly. To overcome this problem, a number of artificial intelligence approaches have been reported. In the past few years, the application of artificial intelligence to fault diagnostics has received much attention. Two of the most popular artificial intelligence approaches are expert systems and model-based reasoning.

3.2.1 Expert systems

One of the major successes in the field of artificial intelligence is expert systems. An expert system is a computer system which is programmed to exhibit expert knowledge in solving a particular domain problem. A typical expert system consists of the following components:

- knowledge base (which contains knowledge about the problem, i.e. rules and facts),
- inference engine (which is the method for combining rules and facts to reach conclusions),
- explanation component (which explains why and how the conclusions are reached),
- user interface (which includes knowledge and data acquisition).

Generally, the knowledge is represented in the form of an 'if-then' rule. This rule is based on problem-solving, heuristics generated by the expert. The inference engine controls the use of the knowledge base. Its control strategy can initiate from the facts or symptoms to reach a conclusion (forward chaining), or from a possible conclusion and search through the facts to verify the conclusion (backward chaining).

Early applications of expert systems were concentrated in medical diagnostics. Currently, efforts are expanding to the area of machine fault diagnostics. Many expert systems have been developed during the past several years. For example, Puetz and Eichhom (1987) used an expert system approach for diagnosing CNC (computer numerical control) machine faults. Malaureille (1987) developed an expert system to tackle the automatic transmission diagnosis problem. An expert system called 'Charley' was utilized by General Motors to diagnose machine condition using vibration signals (Marczewski 1988). Wang (1989) developed an expert system called MODEX (Model diagnostic expert system) for machinery diagnostics applications. Piety and Corley (1989) also developed an expert system called DXPERT to diagnose vibration problems on a wide variety of rotating machines. A typical rule used in DXPERT is shown below. Duggan and Ashley (1986) developed an expert system to evaluate the performance of a typical turbine generator from various machine parameters, such as bearing temperature, thrust bearing wear, shaft vibration, and lubricating oil quality. A typical expert system rule looks like this.

IF Bearing type is SLEEVE
  AND Vibration spectrum is complex
  AND Vibration change is rapid
  AND Frequency = 2 × RPM
  AND High axial vibration is YES
  AND Both bearings is NO
THEN Problem is cocked bearing

Ramamurthi *et al.* (1990) developed an expert system called IDES (influence diagram based expert system) for integrating multiple sensors for monitoring and control of a drilling operation in an automated manufacturing environment. A more detailed survey of fault-diagnostic expert systems can be found in Tzafestas (1989).

Although expert systems are easy to use and able to provide expert knowledge to solve a specific domain problem, there are many difficulties in using this approach (David and Krivine 1987), such as:

- difficulty in formalizing the problem,
- difficulty in obtaining knowledge,
- difficulty in validating the system.

In addition, there are many drawbacks with building an expert system for machine monitoring and diagnostics. One of the major drawbacks is its long execution time. This is particularly true when complex relations and a large knowledge base are involved in the reasoning process. Because expert systems must work through complex chains of reasoning in order to reach a conclusion, more processing time is required. Hence, the short response time required to perform on-line machine monitoring and diagnostics makes the application of expert systems in this area difficult (Pandelidis 1992).

3.2.2 Model-based reasoning

As an alternate approach to expert systems, model-based reasoning has been proposed to solve diagnostic reasoning problems. One of the most promising techniques of model-based reasoning is 'reasoning from structure and behaviour' (Davis 1984). This technique begins with a description of the system, together with the observation(s) of the system behaviour. If the observation conflicts with the way the system is meant to behave, then one concludes that a system failure has occurred. Given symptoms of misbehaviour, possible fault candidates are identified using the structural model by following a dependency chain back from a violated prediction to each component that contributed to that prediction.

Many of the notable applications of model-based reasoning to diagnostic problems have been in the digital electronics field. This is because the structure of digital circuits can be represented in a fairly obvious way, and the intended behaviour of the circuit is strongly implied by its structure (Price and Lee, 1988). For example, Rastogi and Sierzega (1990) used the model-based approach to diagnose the failure of mixed-signal circuits. Dvorak and Kuipers (1991) built a process monitoring system using the model-based approach. They demonstrated the system's capability of detecting faults by observing its dynamic behaviour with a water heater example.

Compared to the expert system approach, knowledge acquisition for the model-based system is easier. In addition, the model-based system is more robust and maintainable. It is able to diagnose multiple faults, avoiding exponential growth in the model size. However, it still poses problems for real-time diagnosis because the system has to look for all possible fault candidates and then has to classify them one by one according to likelihood, which means more reasoning time is needed.

3.3 ARTIFICIAL NEURAL NETWORK

The identification of a machine or component fault can be seen as a pattern recognition problem. In the past, a number of pattern recognition techniques, such as linear discriminant function and fuzzy sets, have been applied to solve this class of problems. Normally, these techniques classify machine or component condition into a two-state situation, that is normal or abnormal. Examples can be found in Li and Wu (1988), and Emel and Kannatey-Asibu (1988). Today, artificial neural networks are among the most popular approaches to solving pattern recognition problems. The following sections introduce this new technology and its related research work on machine monitoring and diagnostics.

An artificial neural network is a simulation of the biological neural network of the human brain. It has been studied in the hope of achieving human-like performance in solving problems with complex, incomplete, or seemingly unrelated data which cannot be solved by conventional programming techniques.

The power of artificial neural networks lies in their computational speed and ability to provide a high degree of robustness or fault tolerance (Lippmann 1987). Because artificial neural networks are formed and operate in a parallel fashion, they have been found to be able to perform complex computations simultaneously and rapidly. In addition, because a network consists of a number of neurons (processing elements), when even a few neurons or their interconnections are damaged, the network can still maintain its regular performance (Hopfield 1982, Hopfield 1984, Lippmann 1987).

A typical artificial neural network consists of the following components: a number of connected neurons or processing nodes, and a learning algorithm. A neuron is, in turn, composed of three elements (see Fig. 3.1):

- weighted connections,
- an integration function,
- an activation function,

Through weighted connections, a neuron receives inputs from those connected to it in the previous layer, and transfers output to those connected to it in the next layer. The integration function simply sums up the received inputs. The activation function, which usually is in the form of a non-linear sigmoid function, converts the integrated input into an output.

There are two primary connection types, inhibitory and excitatory. Inhibitory connections decrease the activation of processing elements to which they are connected, while excitatory connections increase the activation. Therefore, a portion of the connections to a particular neuron may have negative weights (inhibitory) and the remainder have positive weights (excitatory). The most important property of neural networks is their learning ability. A learning algorithm is used to train the network to learn a mapping, sometimes arbitrary, between the input space and the output space by adjusting the interconnection weights. Two types of learning algorithms are generally used in training the network: supervised learning and unsupervised learning. Supervised learning requires an external feedback of error signals after each mapping. Examples are back propagation (BP) (Rumelhart and McClelland 1986), ARTMAP (Carpenter *et al.* 1991), Cerebellar Model Arithmetic Computer (CMAC)

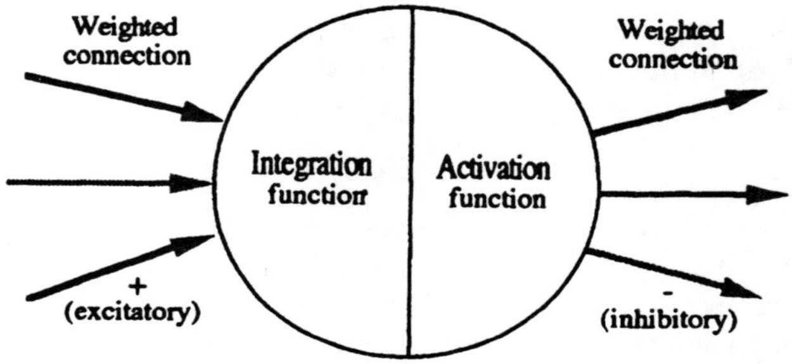

Fig. 3.1 Components of a neuron.

(Miller *et al.* 1990), and Brain-State-in-a-Box (BSB) (Anderson 1977). Unsupervised learning does not require any external feedback during the training process. Examples are Adaptive Resonance Theory (ART) (Carpenter and Grossburg 1987a, 1987b) and the Hopfield network (Hopfield 1982).

There are six different types of neural networks suitable for pattern classification (Lippmann 1987, Mangrulkar 1990). Among them, multi-layer feedforward networks are the most popular paradigms because they can solve non-linear problems which are unable to be solved by a linear single-layer network, known as perceptions. The multi-layer feed-forward network has one or more layers of processing elements between the input and output layers. These layers are called hidden layers.

One of the most powerful and popular multi-layer feedforward networks is trained with backpropagation. Backpropagation was proposed by Rumelhart and McClelland (1986) as an algorithm for finding the optimal assignment of weights of network connections. It employs an iterative gradient descent algorithm to minimize the error measure between the actual output of the network and the desired output.

Recently, artificial neural networks have been applied successfully in the area of machine monitoring and diagnostics. Dietz *et al.* (1989) used a neural network approach to diagnose jet and rocket engine faults. The proposed network utilized backpropagation, and was divided into many subnetworks; each evaluated an individual engine fault. Some of the networks also evaluated severity of the faults as well as the fault type. Their study showed that the system performed with a high degree of accuracy.

Marko *et al.* (1989) used neural networks to diagnose automotive control systems. A back propagation network was used to classify engine faults automatically based on the data taken from sensors in idling car engines. A total of 26 different engine faults (shorted plug, open plug, plugged injector, broken manifold pressure sensor, etc.) were used to train the network. Results indicated that the trained neural network was able to match the accuracy of the expert diagnostician and, in addition, diagnose the problem faster.

Sunil *et al.* (1990) applied the neural network approach to estimate tool wear and surface finish from three components of force signals. Two three-layer backpropagation networks were used in their study. A 2-3-3 network was used to predict tool wear, while another 3-3-2 network was used to predict surface finish. Their result showed that the networks were able to predict flank wear, crater wear, and surface finish quite accurately.

Hoskins *et al.* (1990) proposed a backpropagation network for fault detection and diagnosis in a chemical plant. Fourteen essential sensory measurements in the absorber section of the plant were used as the inputs. Their result indicated that the neural network was able to classify the input measurements perfectly even with noise in the sensor data. Liu and Ko (1990) utilized a backpropagation network to recognize drill wear from thrust and vertical acceleration. Their result showed that the network could achieve a success rate of over 85% for on-line recognition of drill wear. Guo and Dooley (1990) also utilized a backpropagation neural network approach to assist in a diagnostic problem in quality control. Simulation experiments showed that the network correctly classified the problem 94% of the time.

Liu and Mengel (1991) used an 8-12-1 backpropagation neural network to predict ball bearing conditions. Peak amplitude in the frequency domain, peak RMS, and the power spectrum of vibration signals were used as the inputs to the network. The output of the network indicated the state of the bearing. The network could distinguish between the normal bearings and defective bearing with 100% success rate. In addition, it could also categorize the bearing conditions into six different states with a success rate of 97%. In the recent work of Knapp and Wang (1992), a backpropagation network was employed to determine fault classification from the vibration signals. Four fault conditions were investigated in the study. Results showed that the network was able to classify the fault conditions with 100% accuracy.

One of the major problems with artificial neural networks is that neural networks are not designed with an explanatory capability (Caudill 1990). In order to eliminate this so-called 'black box' approach, it may be necessary to build such an explanatory capability into a neural network system. An apparent approach is to combine expert systems and neural networks into a hybrid system. Examples of combining expert systems and neural networks can be found in Caudill (1990) and Hillman (1990). Much research has been done in investigating this hybrid approach. For instance, Kraft *et al.* (1991) used the hybrid neural network–expert system approach for monitoring and diagnosing a boiler system. Rabelo *et al.* (1990) studied the capability of a hybrid system for real-time scheduling of flexible manufacturing systems.

3.4 SENSOR FUSION

Sensor fusion, sometimes referred to as multisensor integration, is a process of integrating the information obtained from a variety of sensors. It is utilized with the hope of achieving human-like performance (i.e. the ability to effectively combine information from his/her senses) in decision making, especially in applications of image or signal processing where the information from individual sensors is generally noisy, uncertain, and insufficient.

There are four key advantages of using sensor fusion (Fildes 1992, Luo and Kay 1988):

1. Redundancy. Fusion of redundant information obtained from a group of sensors (or a single sensor over time) concerning the same feature can increase accuracy as well as enhance reliability in the case of sensor error or failure.
2. Complementarity. The complementary information can be yielded by using multiple sensors to measure different aspects of the feature if the required information could not be obtained by individual sensors acting alone.
3. Timeliness. Multiple sensors can provide more timely information, as compared to the speed at which it could be provided by a single sensor, particularly when parallelism is involved in the integration process.
4. Cost effectiveness. Multiple sensors can provide required information at a lower cost when compared to the equivalent information obtained from individual sensors.

The objective of sensor fusion is to combine individual information into a representative pattern that provides a higher level of information than the sum of the information from individual sensors (Fildes 1992). The information from individual sensors can be raw data or processed data. The processed data is normally generated by a preprocessing procedure which performs pattern recognition, noise faltering, or data reduction. It can be in the form of either estimates of parameters (such as parameters of an autoregressive model), or evidence supporting certain propositions, or decisions favoring certain hypotheses (Thomopoulos 1989).

Determining a method to integrate different types of sensors in order to provide reliable and consistent information is the most challenging task in sensor fusion. However, a large number of methods are available to achieve this task. These methods extend from low-level probability distributions for statistical inference to high-level production rules for logical inference. Luo and Kay (1988) reviewed six general methods for sensor fusion:

- statistical pattern recognition,
- Bayesian estimation,
- multi-Bayesian,
- Shafer–Dempster evidential reasoning,
- fuzzy logic,
- production rules.

In their work, no comparison between these methods was made.

In the work of Chryssolouris and Domroese (1988, 1989), four different methods for sensor fusion were reviewed and compared. These methods include

- neural network,
- multiple regression,

- inverse least-squares approach,
- group method of data handling (GMDH).

Two conclusions were made by the authors:

- the neural network approach is more effective in learning a relationship for providing parameter estimates, particularly when the relationship between the sensor-based information and the actual parameter is nonlinear, and
- the neural network approach is less sensitive to deterministic errors in the sensor-based information than the other three approaches.

In recent years, neural networks have been used as alternatives to quantitative and statistical techniques for sensor fusion. Ajjimarangsee and Huntsberger (1988) used a neural network model for the fusion of visible and thermal infrared sensor outputs. The result of their studies indicated that more detailed information could be obtained by using the neural network approach. Furthermore, Eggers and Khuon (1990) introduced neural network data fusion concepts and applications for automatic decision-making. The system they developed consisted of a statistically based adaptive preprocessor for each sensor, followed by a neural network for associating the preprocessor outputs to the appropriate decisions. Smotroff *et al.* (1990) also developed a neural network data fusion system for meteorological classification of satellite imagery. The system included a neural network for classifying image data and an expert system for controlling the outputs of the neural network. Their results showed a promising potential for using neural networks in sensor fusion.

In the past few years, the applications of sensor fusion have been focused on robot control and military applications. Examples are the work of Luo and Lin (1987), Chen (1987), Holm (1987), and Kishel (1990). Recently, there has been a growing interest in the use of sensor fusion for machine monitoring and diagnostics.

Rege and Agogino (1986) developed an expert system for the hierarchical integration of sensors and diagnostic reasoning in automated manufacturing and process control. Alexander *et al.* (1989) proposed an architecture for sensor fusion in intelligent process monitoring. In this architecture, the Bayesian method and Dempster–Shafer's evidential reasoning were used for fusion processor and a knowledge base was used to control the fusion processor.

Dornfeld and Pan (1985) used linear discriminant functions (or perceptrons) to integrate acoustic emission signals with machining parameters (cutting speed and feedrate) for the detection of continuous or discontinuous chip formation in turning. Liu and Wu (1988) used a linear discriminant function to integrate acceleration and thrust signals in drill wear monitoring. Emel and Kannatey-Asibu (1989) also applied a linear discriminant function to combine acoustic emission signal and cutting force to detect tool breakage, chip form, and a threshold level of tool flank wear in turning.

In the earlier work of Rangwala and Dornfeld (1987), the feasibility of using neural networks for sensor fusion was investigated because most of the above approaches suffered from a time-consuming training process and were inefficient for use in real-time process monitoring. A backpropagation neural network was used to integrate information from acoustic emission and force sensors to monitor tool wear state in a turning operation. The network achieved 95% recognition accuracy. Further, Chryssolouris and Domroese (1988, 1989), as mentioned earlier, performed comparisons between four different sensor fusion techniques and proposed the use of neural networks as the decision-making component in tool wear monitoring.

Liu and Ko (1990) used a backpropagation neural network to recognize drill wear by integrating vertical acceleration and thrust. In their study, the backpropagation network was further compared with a linear discriminant function. It was concluded that the neural network approach had better performance over a linear discriminant function. In a more recent work of Dornfeld (1990), a backpropagation neural network was developed to combine acoustic emission, force, and spindle motor current for tool wear monitoring. Their approach used AR model and power spectrum amplitudes as inputs to the network. If the network's output value was greater than 0.5, then the state of the cutting tool would be 'worn tool.' The result showed that the neural network system was able to perform on-line tool wear monitoring. Okafor *et al.* (1991) proposed a backpropagation network to integrate multi-sensor time domain signals (acoustic emission, spindle vibration, and cutting force) to estimate surface roughness and bore tolerance in circular end-milling. The result showed that most estimation errors were below 10%.

3.5 SUMMARY

Several popular approaches in the area of machine monitoring and diagnostics have been reviewed. Each approach has its strengths and weaknesses. A significant amount of research has been conducted in the development and application of each individual approach. However, little research has been done in incorporating these different approaches into an intelligent system.

Efficiency, accuracy, and knowledge are the most important properties in an intelligent system. It has been shown that signal processing techniques and artificial intelligence approaches are unable to meet the speed requirements of the intelligent system. On the other hand, sensor fusion using the neural network technique has been demonstrated to be the most effective tool in providing robustness and on-line monitoring capability for machine monitoring. Therefore, it is clear that sensor fusion using the neural network approach is the potential tool to build and enhance the properties of intelligent machine monitoring and diagnostic system.

Most of the applications in sensor fusion or neural networks have been focused on machine tool monitoring. Not many studies have been carried out to monitor the condition of machine components. Additionally, there are a number of different neural network paradigms available which can be applied to solve pattern classification or pattern recognition problems in machine fault-diagnostic fields. Most of the current studies reviewed use backpropagation networks for their applications. However, backpropagation has the drawbacks of slow learning speed and being easy to fall into a local minimum. Therefore, it is necessary to investigate the performance of different neural network technologies in order to determine a suitable one for future applications.

Finally, current machine monitoring and diagnostic systems do not have deep reasoning capability. Therefore, it is necessary to embrace this capability into an intelligent machine monitoring and diagnostic system in order to improve its reliability.

REFERENCES

Ajjimarangsee, P. and Huntsberger, T.L. (1988) Neural network model for fusion of visible and infrared sensor outputs. *Sensor Fusion: Spatial Reasoning and Scene Interpretation: Proceedings of SPIE*, **1003**, 153–60.

Akaike, H. (1969) Power spectrum estimation through autoregression model fitting. *Ann. Inst. Stat. Math.*, **21**, 407–19.

Akaike, H. (1974) A new look at the statistical model identification. *IEEE Trans. Autom. Control*, **AC-19**, December, 716–23.

Alexander, S.M., Vaidya, C.M., and Kamel, K.A. (1989) An architecture for sensor fusion in intelligent process monitoring. *Computers Industrial Engineering*, **16**(2), 307–11.

Anderson, J.A. (1977) Neural models with cognitive implications. In *Basic Processes in Reading Perception and Comprehension*, D. LaBerge and S.J. Samuels (eds), Erlbaum, NJ, 27–90.

Braun, S. (1986) *Mechanical Signature Analysis*: Theory and Applications, Academic Press, London, 1986.

Carlson, J. and Spiewak, S.A. (1992) *Vibration based machine condition monitoring: the state of practice. Proceedings of the 16th Annual Meeting on Vibration Institute*. Williamsburg, June 9–11, 87–95.

Carpenter, G.A. and Grossberg, S. (1987a) A massively parallel architecture for a self-organizing neural pattern recognition machine. *Computer Vision, Graphics, and Image Processing*, **37**, 54–115.

Carpenter, G.A. and Grossberg, S. (1987b) ART 2: Self-organization of stable category recognition codes for analog input patterns. *Applied Optics*, **26**(23), December, 4919–30.

Carpenter, G.A., Grossberg, S., and Reynolds, J. (1991) ARTMAP: Supervised real-time learning and classification of nonstationary data by a self-organizing neural network. *Neural Networks*, **4**, 569–88.

Caudill, M. (1990) Using neural nets: hybrid expert networks. *AI Expert*, November, 49–54.

Chen, S. (1987) Multisensor fusion and navigation of mobile robots. *International Journal of Intelligent Systems*, **2**(2), 227–51.

Chryssolouris, G. and Domroese, M. (1988) Sensor integration for tool wear estimation in machining. *Sensors and Controls for Manufacturing*: presented at The Winter Annual Meeting of the ASME, Chicago, IL, November 27–December 2, 115–23.

Chryssolouris, G. and Domroese, M. (1989) An experimental study of strategies for integrating sensor information in machining. *Annals of the CIRP*, **38**(1), 425–28.

David, J.M. and Krivine, J.P. (1987) Three artificial intelligence issues in fault diagnosis: declarative programming, expert systems, and model-based reasoning. *Proceedings of the Second European Workshop on Fault Diagnostics, Reliability and Related Knowledge Based Approaches*, UMIST, Manchester, April 6–8, 19–196.

Davis, R. (1984) Diagnostic reasoning based on structure and behaviour. *Artificial Intelligence*, **24**, 347–410.

Dietz, W.E., Kiech, E.I., and Ali, M. (1989) Jet and rocket engine fault diagnosis in real time. *Journal of Neural Network Computing*, 5–18.

Dornfeld, D.A. (1990) Neural network sensor fusion for tool condition monitoring. *Annals of the CIRP*, **39**(1), 101–5.

Duggan, P. and Ashley, G.K., II (1986) Artificial intelligence (expert systems): a tool for predictive maintenance. *Proceedings of the American Power Conference*, Chicago, IL, April 14–16, 562–6.

Dvorak, D. and Kuipers, B. (1991) Process monitoring and diagnosis – a model-based approach. *IEEE Expert*, June, 67–74.

Eggers, M. and Khuon, T. (1990) Neural network data fusion concepts and application. *Proceedings of the International Joint Conference on Neural Networks (IJCNN)*, **2**, 7–16.

Emel, E. and Kannatey-Asibu, E., Jr. (1988) Tool failure monitoring in tuning by pattern recognition analysis of AE signals. *Journal of Engineering for Industry*, **110**, May, 137–45.

Emel, E. and Kannatey-Asibu, E., Jr. (1989) Acoustic emission and force sensor fusion for monitoring the cutting process. *International Journal Mechanical Science*, **31**(11/12), 795–809.

Fildes, J.M. (1992) Sensor fusion for manufacturing. *Sensors*, January, 11–15.

Guo, Y. and Dooley, K.J. (1990) The application of neural networks to a diagnostic problem in quality control. *Monitoring and Control for Manufacturing Processes*: presented at the Winter Annual Meeting of the ASME, Dallas, TX, November 25–30, 111–19.

Hillman, D.V. (1990) Integrating neural nets and expert systems. *AI Expert*, June, 54–9.

Holm, W.A. (1987) Air-to-ground dual-mode MMW/IR sensor scene registration. *Infrared Sensors and Sensor Fusion: Proceedings of SPIE*, Orlando, FL, **782**, May 19–21, 20–27.

Hopfield, J.J. (1982) Neural networks and physical systems with emergent collective computational abilities. *Proceedings of the National Academy of Sciences*, **74**, 2554–8.

Hopfield, J.J. (1984) Neurons with graded response have collective computational properties like those of two-state neurons. *Proceedings of the National Academy of Sciences*, **81**, 3088–92.

Hoskins, J.C., Kaliyur, K.M., and Mimmelblau, D.M. (1990) Insipient fault detection and diagnosis using neural networks. *Proceedings of the International Joint Conference on Neural Networks (IJCNN)*, **1**, 81–6.

Kay, S.M. and Marple, S.L. (1991) Spectrum analysis – a modern perspective. *Proceedings of the IEEE*, **69**(11), November, 1380–1419.

Kishel, J. (1990) Survey of multi-function and advanced sensors for smart weapons, *Sensor fusion III – Proceedings of SPIE*, **1306**, 220–4.

Knapp, G.M. and Wang, H.-P. (1992) Machine fault classification: a neural network approach. *International Journal of Production Research*, **30**(4), 811–23.

Kraft, T., Okagaki, K., Ishii, R., Surko, P., Brandon, A., DeWeese, A., Peterson, S., and Bjordal, R. (1991) Hybrid neural net and rule based system for boiler monitoring and diagnosis. *Proceedings of the 53th Annual Meeting of the American Power Conference*, Chicago, IL, April 29–May 1, 952–7.

Li, P.G. and Wu, S.M. (1988) Monitoring drilling wear states by a fuzzy pattern recognition technique. *Journal of Engineering for Industry*, 110, August, 297–300.

Li, C.J. and Wu, S.M. (1989) On-line detection of localized defects in bearings by pattern recognition analysis. *Journal of Engineering for Industry*, **111**, November, 331–6.

Liang, S.Y. and Dornfeld, D.A. (1989) Tool wear detection using time series analysis of acoustic emission. Journal of Engineering for Industry, **111**, August, 199–205.

Lin, C.C., (1992) *Classification of Autoregressive Spectral Estimated Signal Patterns Using an Adaptive Resonance Theory (ART)*, Master's Thesis, Department of Industrial Engineering, The University of Iowa, Iowa City, Iowa, August.

Lippmann, R.P. (1987) An introduction to computing with neural nets. *IEEE ASSP Magazine*, April, 4–22.

Liu, T.I. and Ko, E.J. (1990) On-line recognition of drill wear via artificial neural networks. *Monitoring and Control for Manufacturing Processes*: presented at the Winter Annual Meeting of the ASME, Dallas, TX, November 25–30, 101–10.

Liu, T.I. and Mengel, J.M. (1991) Detection of ball bearing conditions by an AI Approach. *Proceedings of the Winter Annual Meeting of the ASME*, Atlanta, GA, December 1–6, 13–21.

Liu, T.I. and Wu, S.M. (1988) On-line drill wear monitoring. *Sensors and Controls for Manufacturing*: presented at the Winter Annual Meeting of the ASME, Chicago, IL, November 27–December 2, 99–104.

Luo, R.C. and Lin, M.-H. (1987) Multi-sensor integrated intelligent robot for automated assembly. *Spatial Reasoning and Multi-Sensor Fusion: Proceedings of the 1987 Workshop*, St. Charles, IL, October 5–7, 351–60.

Luo, R.C. and Kay, M.G. (1988) Multisensor integration and fusion: issues and approaches. *Sensor Fusion: Proceedings of the SPIE*, **931**, 42–49.

Lyon, R.H. (1987) *Machinery Noise and Diagnostics*, Butterworths, Boston, MA.

Malaureille, P. (1987) MULTIDIAG: A Parameterizable Expert System for the Diagnosis of Mechanical Failure. *Knowledge Based Expert Systems for Engineering: Classification, Education and Control*, 239–46.

Mangrulkar, S. (1990) Artificial neural systems. *ISA Transactions*, **29**(1), 5–7.

Marczewski, R. (1988) An expert system for machine tool diagnosis. *Proceedings of 2nd Engineering Society of Detroit Conference on Expert Systems*, Detroit, MI, April, 161–72.

Marko, K., James, J., Dosdall, J., and Murphy, J. (1989) Automotive control system diagnostics using neural nets for rapid pattern classification of large data sets. *Proceedings of International Joint Conference on Neural Networks (IJCNN)*, **II**, 13–15.

Matsushima, K., Bertok, P., and Sata, T. (1982) In-process detection of tool breakage by monitoring the spindle current of a machine tool. *Proceedings of ASME Winter Annual Meeting*, Phoenix, AZ, 145–54.

Miller, W.T., III, Glanz, F.H., and Kraft, L.G., III. (1990) CMAC: an associative neural network alternative to backpropagation. *Proceedings of the IEEE*, **78**(10), October, 1561–7.

Okafor, A.C., Marcus, M., and Tipirneni, R. (1991) Multiple sensor integration via neural networks for estimating surface roughness and bore tolerance in circular end milling - Part 1: time domain. *Condition Monitoring and Diagnostic Technology*, **2**(2), October, 49–57.

Pandelidis, I.O. (1992) Machine diagnostics, in *Intelligent Design and Manufacturing*, A. Kusiak (ed), John Wiley & Sons, Inc., 523–43.

Piety, P. and Corley, J.E. (1989) Development of an expert system to diagnose machinery vibration problems. *Proceedings of the 1st International Machinery Monitoring and Diagnostics Conference and Exhibit*, Las Vegas, NV, September, 523–7.

Potts, S.L. and Hakimmashhadi, H. (1989) Application of parametric spectral estimation to machine vibration analysis. *Proceedings of the 1st International Machinery Monitoring and Diagnostics Conference and Exhibit*, Las Vegas, NV, September, 161–7.

Price, C. and Lee, M. (1988) Applications for deep knowledge. *Artificial Intelligence in Engineering*, **3**(1), 12–17.

Puetz, R.D. and Eichhorn, R. (1987) Expert systems for fault diagnosis on CNC machines. *Knowledge Based Expert Systems for Engineering*: Classification, Education and Control, 387–96.

Rabelo, L.G., Alptekin, S., and Kiran, A.S. (1990) Synergy of artificial neural networks and knowledge-based expert systems for intelligent FMS scheduling. *Proceedings of the International Joint Conference on Neural Networks (IJCNN)*, **1**, 359–66.

Ramamurthi, K., Shaver, D.P., and Agogino, A.M. (1990) Real time expert system for predictive diagnostics and control of drilling operation. *Proceedings of the 6th Conference on Artificial Intelligence Applications*, Santa Barbara, CA, March 5–9, 62–9.

Rangwala, S.S. and Dornfeld, D.A. (1987) Integration of sensors via neural networks for detection of tool wear states. *Intelligent and Integrated Manufacturing Analysis and Synthesis*: presented at the Winter Meeting of the ASME, Boston, MA, 13–18 December, 109–120.

Rastogi, R. and Sierzega, K. (1990) A new approach to mixed-signal diagnosis. *Proceedings of IEEE International Test Conference*, 591–7.

Rege, A. and Agogino, A.M. (1986) Sensor-integrated expert system for manufacturing and process diagnostics. *Knowledge-Based Expert Systems for Manufacturing*: presented at the Winter Annual Meeting of the ASME, Anaheim, CA, 7–12 December, 63–83.

Rumelhart, D.E. and McClelland, J.L., (eds) (1986) *Parallel Distributed Processing*, **Vol. 1**, MIT Press, Cambridge, MA.

Serridge, M. (1989) Ten crucial concepts behind trustworthy fault detection in machine condition monitoring. *Proceedings of the 1st International Machinery Monitoring and Diagnostics Conference and Exhibit*, Las Vegas, NV, September, 722–7.

Sidahmed, M. (1989) Contribution of parametric signal processing techniques to machinery condition monitoring. *Proceedings of the 1st International Machinery Monitoring and Diagnostics Conference and Exhibit*, Las Vegas, NV, September,

Smotroff, I.G., Howells, T.P., and Lehar, S. (1990) Meteorological classification of satellite imagery using neural network data fusion. *Proceedings of the International Joint Conference on Neural Networks (IJCNN)*, **2**, 23–8.

Sunil, E.V.T., Shin, Y.C., and Kumara, S. (1990) Machining condition monitoring for automation using neural networks. *Monitoring and Control for Manufacturing Processes*: presented at the Winter Annual Meeting of the ASME, Dallas, TX, 25–30 November, 85–100.

Thomopoulos, S.C.A. (1989) Sensor integration and data fusion. *Sensor Fusion II: Human and Machine Strategies: Proceedings of the SPIE*, **1198**, 178–91.

Tranter, J. (1989) The fundamentals of, and the application of computer to, condition monitoring and predictive maintenance. *Proceedings of the 1st International Machinery Monitoring and Diagnostics Conference and Exhibit*, Las Vegas, NV, September, 394–401.

Tzafestas, S.G. (1989) System fault diagnosis using the knowledge-based methodology. In *Fault Diagnosis in Dynamic Systems: Theory and Applications*, R. Patton, P. Frank, and R. Clark (eds), Prentice-Hall, New York.

Wang, S.S. (1989) AI and expert systems for diagnostics. *Proceedings of the 1st International Machinery Monitoring and Diagnostics Conference and Exhibit*, Las Vegas, NV, September, 516–22.

Wu, S.M., Tobin, T.H., Jr., and Chow, M.C. (1980) Signature analysis for mechanical systems via dynamic data system (DDS) monitoring technique. *Journal of Mechanical Design*, **102**, April, 217–21.

CHAPTER 4

Parametric modeling methods: theory and a case study

Julie Spoerre and Ben Wang

4.1 BASICS OF PARAMETRIC MODELING

It has been proven that as a machine condition changes, the vibration characteristics will also change [26]. Renwick and Bubson have proven that the vibrations that exist in rotating machinery can be used to indicate many defect modes before catastrophic failure occurs, whether the defect serves as a source for periodic energy or merely modifies the transmission path [20]. Jianfang, Yueming and Qi developed an effective method for decomposing bearing vibration signals and detecting faults using adaptive noise canceling and the kurtosis analysis technique [12]. Peczely used vibration analysis to characterize the condition of gas-engine units and detect different types of failure [17]. Piety, Piety, and Scheibel [18] applied predictive maintenance techniques to centrifugal fans and found that several faults were detectable through the application of spectral frequency analysis of vibration signals. These recognizable faults include imbalance, misalignment, bent or bowed shafts, bearing faults of all types, and structural degradation, to name a few.

The vibration characteristics of a ball bearing with corrosive pitting were studied in detail by Wang, Shi and Xu. They concluded that the vibration signal in the time domain has distinct features; the impulse energy, peak and background value can directly reflect the working state of the bearing [27].

Seth [23] also discussed using time domain analysis in detecting abnormalities in rotating machines and found that the time domain analysis of signatures, which preserves the history of events, is better suited for monitoring the status of rotating machinery.

Chen, Li, and Song used time series analysis in the condition monitoring and fault diagnosis of the aero-engine. Through the establishment of an autoregressive (AR) model and estimation of autoregressive coefficients, the characteristic parameters were extracted and a classification function determined to recognize some special faults in the engine [7].

Jixi and Yujun modeled and analyzed vibration signals using the technique of time series analysis to detect the main factors of the surface waviness which cause the vibration of a ball bearing [13].

A quantitative (parametric) monitoring technique of mechanical systems was introduced and developed by Wu, Tobin, and Chow. This monitoring technique utilized the concepts of statistical quality control to identify the occurrence of system abnormalities, reflected in the system's operational signature [31].

Sidahmed, through preliminary testing on an industrial gear test bench, showed that spalling can be detected early using autoregressive (AR) modeling and automatic linear prediction error (LPE) monitoring [25].

Potts and Hakimmashhadi determined the suitability (in terms of performance and computational speed) of parametric spectral analysis, in particular, AR models and modified covariance method, in the early prediction of machine component failure. The modified covariance was found to give excellent performance in terms of frequency resolution, line splitting, and frequency bias [19].

Iserman presented a general structure of process fault detection methods that rely on the determination of changes in the process in comparison to the normal status based on a model of the observed process, as well as models of the faulty process. Fault detection and diagnosis based on parameter monitoring was shown to be effective in the fault detection of an electrically driven centrifugal pump [10].

A new statistical method was investigated by McElroy and Fruchtman [15] to predict equipment reliability with the following proven applications:

1. Determine whether or not to repair or replace large drive motors.
2. Estimate probability of future tube failures in a nuclear steam generator.
3. Determine when to replace a feedwater heater.
4. Evaluate cost-effectiveness of a generator rewind.

In this new technique, historical failure data for a specific component are collected and advanced statistical analysis is used to define the common fault pattern.

Chen [6] proposed a method of machine diagnostic monitoring using a model-based prediction error analysis algorithm to extract reliable indices based on the system model. After data collection, a model for the normal condition was established and an AR(5) model was found to be adequate. This method was applied to the monitoring of a spindle system and achieved an overall success rate of 94.5%.

Applying statistical methods to maintenance functions was the topic of a paper by Brick, Michael, and Morganstein [5]. The authors listed several questions that can be answered by applying statistical thinking:

1. Are equipment failures in statistical 'control,' in the sense that they are only random failures and not due to some identifiable factor (local fault)?
2. Are factors that are predictive of failures, such as oil consumption, oil pressure drops, vibration amplitudes, or consumption of electrical power within normal (not necessarily constant) limits?
3. Can failure be predicted on the basis of factors such as those in the preceding questions?
4. How many bearings, fans, refractories, and other parts should be kept in stock to meet maintenance demands?
5. How can the average lifetime of a part or component be estimated on the basis of the lives of the first few that wear out?
6. How should a maintenance program be chosen that minimizes total cost, that is the cost of both maintenance and production lost due to maintenance problems? In particular, should a component be allowed to remain in service until it fails or should it be replaced on a defined schedule? If it should be routinely replaced, what is the best replacement frequency?

Crowder and Hamilton introduced an exponentially weighted moving average (EWMA) for monitoring a process standard deviation and showed that the EWMA has the capability of quickly detecting small percentage increases in the process standard deviation of a normal process [8].

Parametric modeling methods operate on incoming data or autocorrelation estimates to compute a set of parameters which correspond to an a priori model of the data statistic. This concept can be shown in Fig. 4.1 which gives a parametric modeling schematic where the linear prediction of y_t is $\hat{y}_t = a_1 y_{t-1} + a_2 y_{t-2} + \ldots + a_p y_{t-p}$ and p is the number of autoregressive parameters in the model.

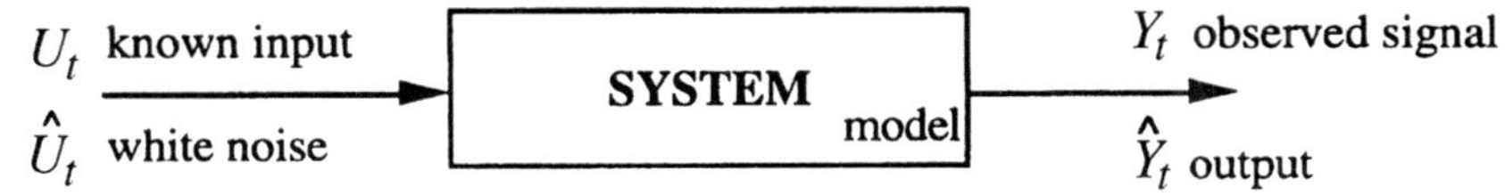

Fig. 4.1 Parametric modeling schematic.

An autoregressive process is represented by a difference equation of the form:

$$X(n) = \sum_{i=1}^{\rho} \phi_i \quad X(n-i) + e(n) \tag{4.1}$$

where $X(n)$ is the real random sequence, ϕ_i, $i = 1, \ldots, \rho$ are parameters, and $e(n)$ is a sequence of independent and identically distributed zero-mean Gaussian random variables with constant variance, that is,

$$E\{e(n)\} = 0 \tag{4.2}$$

$$E\{e(n)e(j)\} = \begin{cases} \sigma^2 N, & \text{for } n = j \\ 0 & \text{for } n \neq j \end{cases} \tag{4.3}$$

$$f_{e(n)}(1) = \frac{1}{\sqrt{2\pi}\,\sigma_N} exp\left\{\frac{-\lambda^2}{2\sigma_N^2}\right\} \tag{4.4}$$

The sequence $e(n)$ is called white Gaussian noise. Thus, an auto-regressive process is a linear difference equation model when the input or forcing function is white Gaussian noise [11].

Thus, for a machine operating under normal conditions, the vibration condition of the machine can be described by an AR process where the values of $e(n)$ are white Gaussian noise.

In order to validate the model representing the normal condition, the error values, or residuals (calculated from the normal model) for each of the normal data sets, were determined and analyzed using MINITAB, a statistical software package.

If the model is suitable, the vibration signal will be defined to a large degree by the model, and the deviations, or residuals, of the predicted signal from the actual signal for each point in time are distributed as white Gaussian noise and, therefore, randomly distributed about the mean value zero. If the stochastic component is white noise and the trend in the vibration signal is adequately modeled, a plot of the residuals over time is expected to exhibit a rectangular scatter plot with no discernable pattern, meaning the variance is constant. This anticipated random distribution of the residuals is apparent in the plot of the residuals over time for the first data set representing a normal machine condition (see Fig. 4.2). Similar plots were obtained when plotting the residuals versus time for the other seven data sets used in describing the normal model.

Gross non-normality can be evaluated by plotting a histogram of the residuals. Since the errors are expected to be normally distributed, the histogram should closely resemble a normal distribution. The presence of normality in the standardized residuals from the first data set representing a normal machine condition is evident in Fig. 4.3. Similar histograms were found when testing the other seven normal data sets for normality in the residuals.

A final check for normality was conducted by calculating the normal scores of the residuals. The tth normal score is defined to be the $(t - 3/8)/(n + 1/4)$ percentage point of the standard normal distribution. With normally distributed data, a plot of the tth ordered data value (residual) versus the corresponding normal score should fall approximately on a straight line. This phenomenon occurs in the normal plot of a normal data set (see Fig. 4.4). Likewise, each of the other seven data

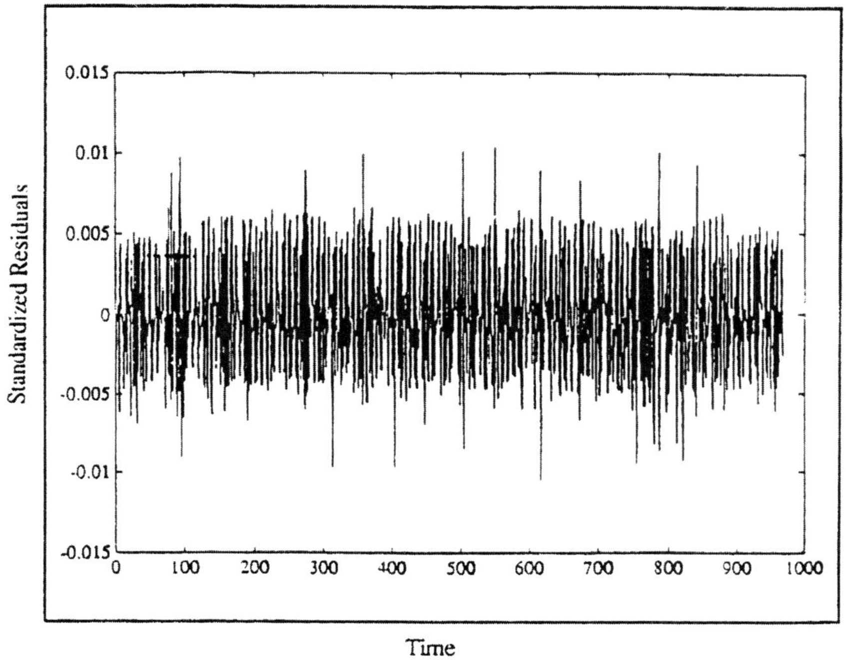

Fig. 4.2 Time series plot of standardized residuals of 'normal' AR model.

sets showed an approximate straight line in the plot of the standardized residuals versus normal scores.

To determine that the residuals are independent, a runs test was performed on the normal data sets. The runs test examines the standardized residuals in sequence to search for patterns. For example, patterns will occur if the residuals are correlated or if the model misrepresents the data in a systematic way. Since the standardized residuals are being analyzed, the runs test will display the number of observations above and below the mean value of zero and the number of runs in the observed residuals. For the first normal data set, the runs test resulted in the following information:

Normal vibration data set:
Observed number of runs = 488
Expected number of runs = 482.9359
Observations above mean = 511
Observations below mean = 456
Significant at 0.7437
Cannot reject at $\alpha = 0.05$

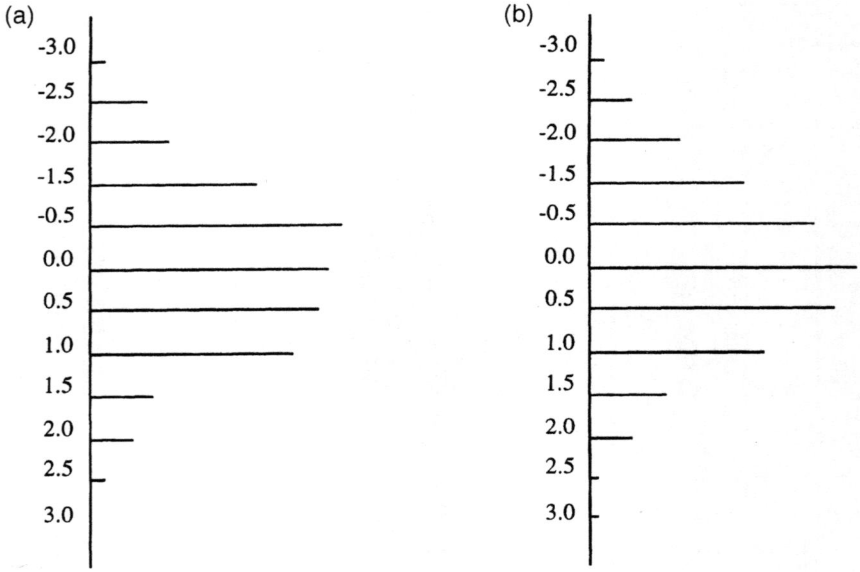

Fig. 4.3 Test for normality in the standardized residuals. Histograms of (a) standardized residuals from normal vibration data; (b) data from a random normal distribution ($\mu = 0$, $\sigma = 1$).

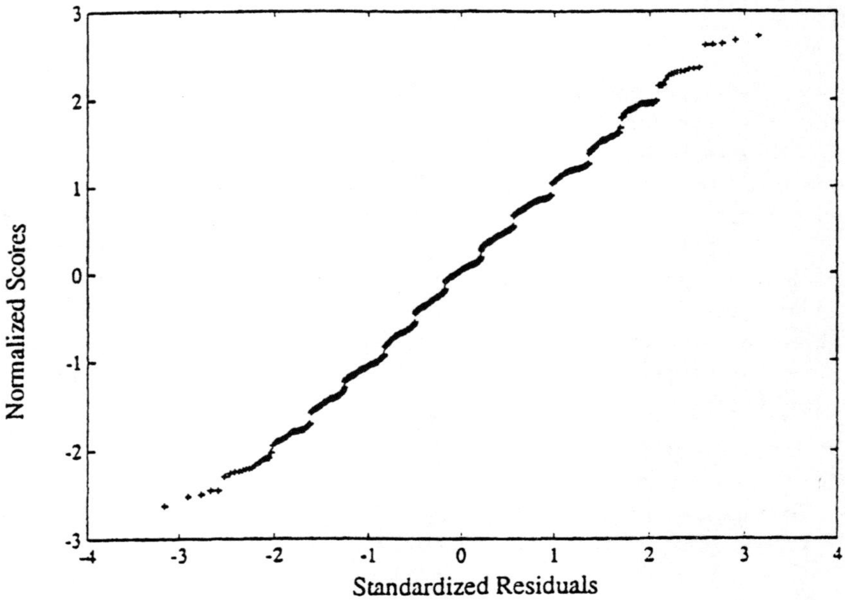

Fig. 4.4 Plot of normalized scores vs standardized residuals.

The statistics for the same test on the data generated from a normal distribution with mean 0 and standard deviation 1 is given below:

Random normal data set:
Observed number of runs = 486
Expected number of runs = 484.1230
Observations above mean = 497
Observations below mean = 470
Significant at 0.9036
Cannot reject at $\alpha = 0.05$

From the information on the normal vibration data, it is shown that the number of observations above and below the mean are nearly equal, which means that the residuals are evenly distributed about the mean level. In addition, there is no trending in the data because the observed number (488) of runs is large and thus proves independence in the residuals. In addition, a runs test performed on a randomly generated normal data set yields highly similar results. Once again, the test for independence was performed and validated for the other seven data sets.

Determination of the AR parameters can be achieved by one of several techniques: Yule-Walker method, Burg method, covariance method, modified covariance method, etc. [14]. The modified covariance method was chosen since it eliminates problems encountered by using the other methods – frequency resolution, spectral line splitting, and bias of the frequency estimate [11].

1. Frequency bias describes a non-zero average shift in the detected frequency of a peak in a spectral estimate from its true value.
2. Line splitting occurs when a single peak in the true spectrum is incorrectly modeled as two peaks in a spectral estimate.
3. Frequency resolution is the ability to distinguish closely spaced spectral peaks.

Criteria used for the selection of the AR model order were final prediction error (FPE), Akaike information criteria (AIC), and criterion autoregressive transfer (CAT) function.

FPE selects the order of the AR process so that the average error variance for a one-step prediction is minimized, where the error variance is the sum of the power in the unpredictable part of the process and a quantity representing the inaccuracies in estimating the AR parameters. The FPE for an AR process is defined as follows:

$$FPE\,[p] = \hat{\rho}_p \left(\frac{N + (p+1)}{N - (p+1)} \right) \tag{4.5}$$

where N is the number of data samples, p is the order, and $\hat{\rho}_p$ is the estimated white noise variance.

The AIC determines the model order by minimizing an information theoretic function. Assuming the process has Gaussian statistics, the AIC for an AR process has the following form:

$$AIC[p] = N\, ln\, (\hat{\rho}_p) + 2p \tag{4.6}$$

A final criterion, CAT, selects the order p as that which minimizes the estimate of the difference between mean square errors of the true prediction error filter and the estimated filter. This difference is calculated from the following equation:

$$CAT[p] = \left(\frac{1}{n} \sum_{j=1}^{p} \overline{\rho_j^{-1}} \right) - \bar{\rho}_p^{-1} \tag{4.7}$$

where $\bar{\rho}j = [N/(N-j)]\, \hat{\rho}_p$ and p is chosen to minimize CAT[p].

The pth-order forward and backward linear prediction errors for the modified covariance method may be represented as the vector inner products.

$$\mathbf{e}_p^f\ [n] = x_\rho^T\ [n]\ \mathbf{a}_p^{fb}\ [n] \tag{4.8}$$

$$\mathbf{e}_\rho^b\ [n] = x_\rho^T\ [n]\ \mathbf{J}\mathbf{a}_p^{fb}\ , \tag{4.9}$$

where the data vector $\mathbf{x}_\rho[n]$ and linear prediction coefficient vector $\mathbf{a}_p^{fb}$ are defined as follows:

$$\mathbf{x}_\rho[n] = \begin{pmatrix} x[n] \\ x[n-1] \\ \cdot \\ \cdot \\ \cdot \\ \cdot \\ x[n-p] \end{pmatrix}, \quad \mathbf{a}_p^{fb} = \begin{pmatrix} 1 \\ a_p[1] \\ \cdot \\ \cdot \\ \cdot \\ \cdot \\ a_p[p] \end{pmatrix} \tag{4.10}$$

and $\mathbf{J}$ is a $(p + 1) \times (p + 1)$ reflection matrix. Based on measured data samples $x[1], \ldots, x[N]$, the modified covariance method minimizes the average of the forward and backward linear prediction squared errors.

$$\rho^{fb} = \frac{1}{2} \left[\sum_{n=p+1}^{N} \left[\left| e_p^f[n] \right|^2 + \left| e_p^b[n] \right|^2 \right] \right] \tag{4.11}$$

The exponentially weighted moving average (EWMA) control statistic is defined as follows:

$$EWMA_t = \max\{(1-\lambda)EWMA_{t-1} + \lambda\, ln\, [\rho^{fb}_{\text{normalized}}], 0\} \tag{4.12}$$

where

$EWMA_0 = 0$,
$EWMA_t$ = predicted EWMA value at time t (new EWMA),
$EWMA_{t-1}$ = predicted EWMA value at time $t-1$ (old EWMA),
$ln[\rho^{fb}_{normalized}]$ is the sample variance of observed values at time t,
λ is a smoothing constant satisfying $0 < \lambda \leq 1$ that determines the depth of memory of the EWMA.

It is shown that the EWMA can be written as:

$$y_{t+1} = \sum_{i=0}^{t} w_i y_i$$

where the w_i are weights and $w_i = \lambda(1-\lambda)^{t-i}$.

The sum of the weights $\sum_{i=0}^{t} w_i = 1$. The constant λ determines the 'memory' of the EWMA statistic. That is, λ determines the rate of decay of the weights and, in turn, the amount of information recollected from the past data. As λ approaches 1, w_1 approaches 1 and $\hat{y}_{t+1}$ is nearly equivalent to the most recent observation y_t. On the other hand, as λ approaches 0, the most recent observation has small weight and previous observations nearly equal weights.

The upper control limit for the EWMA statistic is

$$UCL = \mu_{\rho^{fb}} + 4\sigma_{\rho^{fb}} \tag{4.13}$$

where

$$\mu_{\rho^{fb}} = \frac{1}{n} \sum_{i=1}^{n} \mu_{\rho_i^{fb}} ,$$

$$\sigma_{\rho^{fb}} = \sqrt{\frac{1}{n} \sum_{i=1}^{n} \rho_i^{fb} - \mu_{\rho_i^{fb}}} ,$$

and n = number of data sets initially collected under a normal machine condition.

The EWMA is a statistic with the characteristic that it gives less and less weight to data as it becomes older and older [3]. This concept is shown in Fig. 4.5.

The EWMA chart was chosen since it has been proven superior to the range chart or s^2 in terms of its ability to quickly detect small percentage increases in the process standard deviation [8]. In addition, the EWMA is easy to plot, easy to interpret, and its control limits are easy to obtain. A

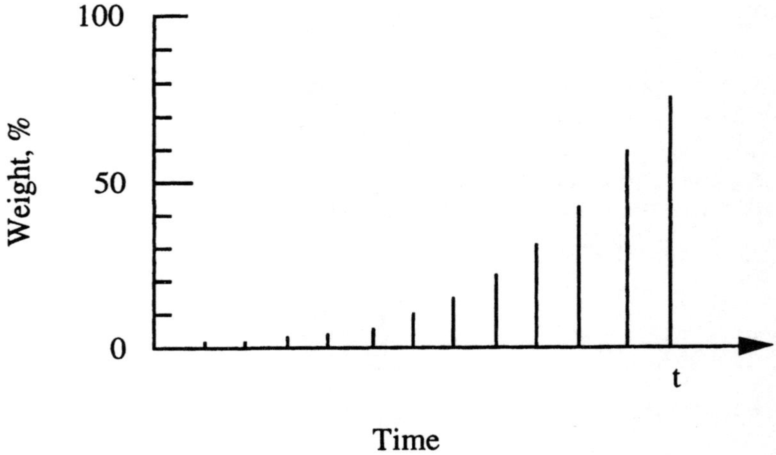

Fig. 4.5 Data weighting for the EWMA chart.

major advantage of employing EWMA is that it provides a mechanism for dynamic process control.

To control a process it is convenient to forecast where the process will be in the next instance of time. Then, if the forecast shows a future deviation from target that is too large, some electro-mechanical control system or process operator can take remedial action to compel the forecast to equal the target. In manufacturing, a forecast based on the unfolding historical record can be used to initiate a feedback control loop to adjust the process [3].

Lambda (λ) determines the 'memory' of the EWMA statistic; that is, λ determines the rate of decay of the weight and hence the amount of information secured from the historical data. The choice of λ is somewhat arbitrary and was experimentally chosen to provide the smallest predicted variance (error) with a value of 0.7.

It can be seen that the logarithmic scale was used. The meaningful presentation of vibration data is essential in order to enable a diagnostician to accurately determine the true condition of a machine. The use of a logarithmic scale provides a representation closer to the vibrational behaviour of machines [2]. Box, Hunter, and Hunter [3] and others have recommended using the log of the sample variances when making inferences about variances of normally distributed data. One reason is that the logs of the sample variances will be much more normally distributed than the sample variances themselves. Also, the variance of ln (ρ^{fb}) is independent of s^2 and depends only on the sample size n.

An increase in s^2 corresponds to an increase in the location parameter of the log-gamma distribution (the distribution of ln (ρ^{fb})). Thus, an increase in the underlying process standard deviation should cause an increase in the mean level of the plotted EWMA values. Because of its simplicity and the properties listed above, the log transformation was considered to be an appropriate transformation for this research. Since vibration signals collected on rotating machinery can be very complex, as the vibration signal measured at a given point on the machine is the sum of all the internal forces applied to the machine modified by their respective transmission paths, the logarithmic scale provides a large range on which to display all the relevant data [2].

Lemma:
Assume that the random variable X has a gamma distribution with shape parameter $a > 0$ and scale parameter $b > 0$ i.e X Gamma (a,b). Then the pdf of X is

$$f(x) = \frac{x^{\alpha-1} \exp(-x/\beta)}{\Gamma(\alpha)\, \beta^{\alpha}} \quad \text{for } x > 0$$

If $W = ln\,(X)$, then W has a log-gamma distribution with pdf

$$f_w(w) = \frac{\exp\{\alpha w - \exp(w)/\beta\}}{\Gamma(\alpha)\, \beta^{\alpha}} \quad \text{for } -\infty < w < \infty$$

Re-expressing $f_w(w)$ as

$$f_w(w) = \frac{\exp\{\alpha(w - \ln(\beta)) - \exp(w - \ln(\beta))\}}{\Gamma(\alpha)}$$

it can be seen that while the parameter b associated with the gamma distribution is a scale parameter, $\ln(b)$ is a location parameter of the corresponding log-gamma distribution. Since the EWMA is designed primarily to track location changes in the process, it is reasonable to use the logarithmic transformation to monitor and detect changes in the original process scale parameter.

4.2 EXPERIMENTATION

The methodology used in this research is summarized in the flowchart given below (Figure 4.6). A more detailed explanation of the figure is given below.

1. Initially, time series data (vibration signals) were collected for the normal machine condition. Each of the data sets collected consisted of 1000 data points.
2. An appropriate AR model (one that adequately describes the vibration

data being collected) was selected using the criteria defined earlier. A suitable AR order was chosen.

3. After choosing a suitable AR order for the normal condition, an AR model was calculated for each of the data sets collected under a normal machine condition, i.e. first order parameters (ϕ_{i1}), second order parameters (ϕ_{i2}), up to p order parameters (ϕ_{ip}), for $i = 1, 2, \ldots, n$ data sets.
4. From step 3, an average value was calculated for the first order AR parameter through the pth order AR parameter. This calculation was performed in order to define a model that would be representative of a normal machine condition under the conditions defined in the lab setup.
5. Once the model was established for the normal machine condition, new data was collected for an abnormal machine condition. Then the abnormal vibration signals were fitted to the 'normal' model found in step 4 as an indicator of how closely the normal model fits the data set collected under the current condition.
6. In order to measure the fit of the data to the normal model, forward and backward prediction errors were calculated to determine the $\rho^{fb}_{normalized}$ value of the data. The normalization is based on the normal machine condition since the purpose of this study is to be able to detect any deviations from the normal machine condition. Normalization is utilized to uncover all possible collections or sets containing the same data and allows current and previous data for a machine to be superimposed, regardless of the operating speed. In addition, normalization allows the creation of an average data file for each specific machine type. Vibration signatures for many identical machines taken at different times at slightly different operating speeds can be accumulated statistically and represented by a single set of averaged narrow band spectra [28]. To determine whether the vibration signature of a machine is significant, the current condition is compared to the normal condition through the ρ^{fb} values.
7. Given the ρ^{fb} value found in step 6, an exponentially weighted moving average (EWMA) statistic was calculated, with $\lambda = 0.7$.
8. The calculated EWMA statistic is an indicator of the overall machine condition and is compared to the upper control limit (UCL) to determine if the machine is in a state of control or if it is out-of-control. If the EWMA value exceeds the UCL, this is a signal that an abnormal machine condition exists and action should be taken.

Data collection was conducted using the following setup: a DC motor connected to a shaft by a drive belt, two cylindrical pillow block bearings mounted on each end of the shaft and secured to a steel plate, an oscilloscope to display the raw vibration signal collected, an amplifier to

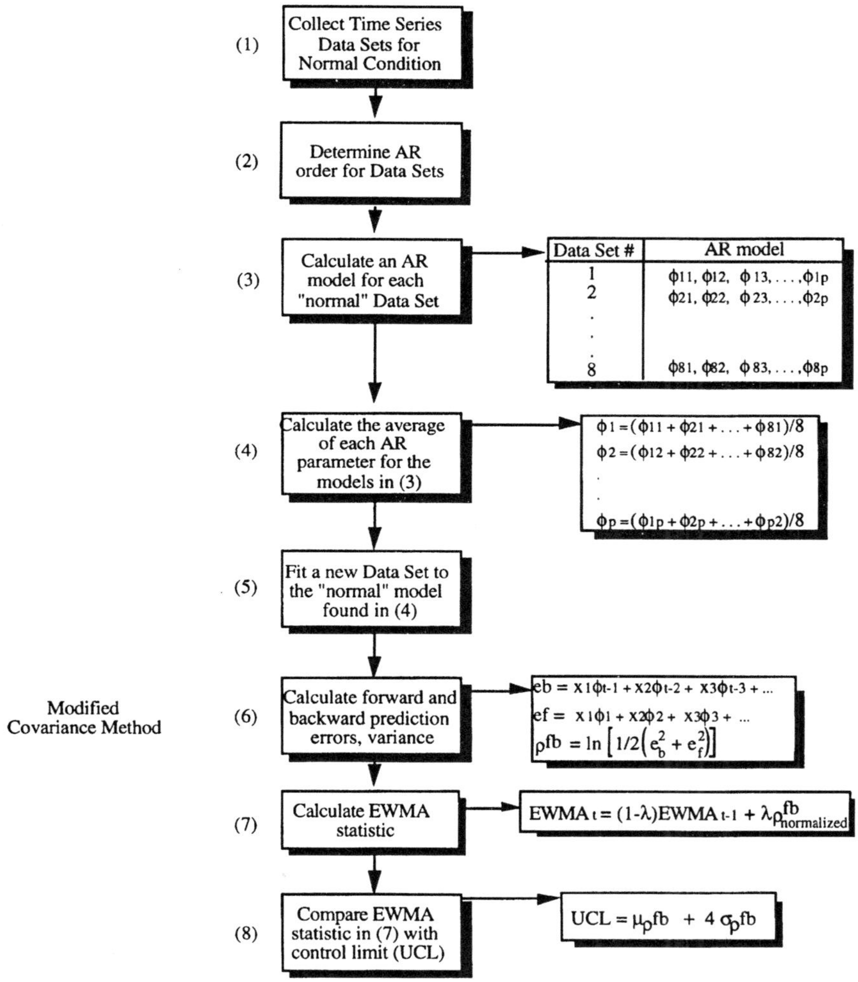

Fig. 4.6 Methodology for monitoring and diagnosing a machine condition.

magnify the signal, and a DT2821-G-8DI data acquisition board. Vibration signals were collected from the bearing using 328C04 PCB accelerometers mounted on the bearing housing.

Accelerometers are transducers whose voltage output is proportional to acceleration or, as a more useful description, the internal forces in the machine. If the acceleration level is high, then the internal forces are high. Forces are the cause of oscillation, in addition to excessive wear and premature failure. The sensitivity response (amplitude) of an accelerometer versus frequencies is displayed in Fig. 4.7.

Accelerometers are the preferred transducers in machine condition monitoring due to the following reasons:

1. Extreme ruggedness.
2. Large frequency response.
3. Large dynamic range – accelerometers can detect very small vibrations without being damaged by large vibrations.
4. Output is proportional to forces which are the cause of internal damage.
5. High-frequency sensitivity for detecting bearing faults.

The initial machine condition was normal. Eight sets of data were collected under this condition, each containing 1000 data points. The motor was operating at a speed of 1420 Hz. For all machine conditions imitated in the lab, the motor speed was kept constant.

4.3 ANALYSIS AND RESULTS

After collecting data, the modified covariance method and the three criteria methods described above were used to determine the appropriate AR model that would adequately describe the normal machine condition. The most suitable order for the AR model was found to be 33, resulting in a ratio of AR order-to-sequence length (1000) of 0.033. This

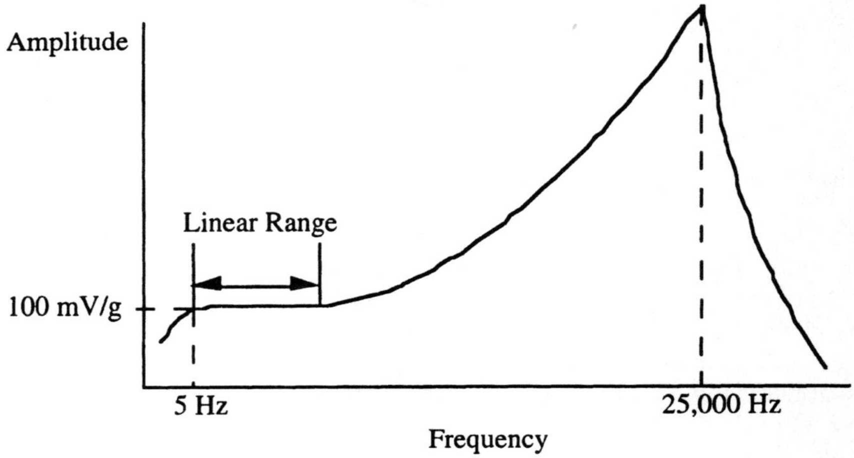

Fig. 4.7 Sensitivity vs frequency for typical, general-purpose accelerometer used for machinery monitoring.

value is preferably small since frequency bias and line splitting increase with an increasing ratio of AR order-to-sequence length [19].

Once the model was established for the normal condition, vibration data were collected under three abnormal operating conditions:

1. Misaligned shaft.
2. Loose bearing.
3. Contaminated bearing.

For each data set collected under one of the four conditions, the minimum and maximum variances were determined (see Table 4.1). These variances were normalized by dividing the variance by the average variance under a normal condition (0.0000048582) and taking the natural log of the variance for both the minimum and maximum variances for each machine condition. The variance values were used to determine the upper and lower bounds of the EWMA (see Table 4.2).

The numbers in the upper portion of each cell in Table 4.1 represent the minimum and maximum observed EWMA statistic based on actual data collected in the lab. The numbers in the lower portion of the cell are based on statistics of the collected data and were calculated using $\mu_{\rho^{fb}} - 4\sigma_{\rho^{fb}}$ for the lower bound and $\mu_{\rho^{fb}} - 4\sigma_{\rho^{fb}}$ for the upper bound, where $\mu_{\rho^{fb}}$

Table 4.1 Calculated 'normalized' variances for a normal machine condition and three abnormal machine conditions

Condition	*Average value* $\mu_{\rho^{fb}}$	*Standard deviation* $\sigma_{\rho^{fb}}$
Normal	0.100081	0.124959
Misalignment	2.561171	0.116305
Contamination	5.002439	0.302079
Looseness	3.449739	0.125359

Table 4.2 Calculated EWMA statistics for a normal machine condition and three abnormal machine conditions

Condition	*EWMA statistic*	
	Lower bound	*Upper bound*
Normal	0.000000	0.134323
	(0.000000)	(0.496657)
Misalignment	1.165905	1.315764
	(1.072859)	(1.5011256)
Contamination	2.283083	2.487441
	(2.201966)	(2.709934)
Looseness	1.835429	1.975226
	(1.740603)	(2.162600)

is the average value for the EWMA statistic for each machine condition, i.e. $\mu_{\rho^{fb}} = \frac{1}{n}\left[\sum_{i=1}^{n} \rho_i^{fb}\right]$ where n is the number of samples collected.

The standard deviation, as well as $\mu_{\rho^{fb}}$, was determined for each bearing condition. Standard deviations are calculated by the following formula:

$$\sigma_{\rho^{fb}} = \sqrt{\frac{1}{n}\left[\sum_{i=1}^{n} \left(\rho_{i^{fb}} - \mu_{\rho^{fb}}\right)^2\right]}$$

In Table 4.2, the upper and lower bound of EWMA values for the abnormal conditions (misalignment, contamination, and looseness) are calculated based on the minimum and maximum EWMA statistic calculated for the normal condition and the variances of the abnormal conditions in Table 4.1. For example, under the normal condition, the minimum and maximum EWMA values for all eight data sets collected was 0.000000 and 0.126535, consecutively. Referring to Equation 4.6, the EWMA statistic is calculated by $\text{EWMA}_t = \max\{(1 - \lambda)\text{EWMA}_{t-1} + \lambda \ln \rho^{fb}_{\text{normalized}}, 0\}$.

Under a misalignment condition, for example, the minimum EWMA statistic that could occur (based on initial data) is calculated using the minimum variance found under a misalignment condition and the minimum EWMA statistic for the normal condition. Similarly, the maximum EWMA statistic that could occur (based on initial data) is calculated using the maximum variance found under a misalignment condition and the maximum EWMA statistic for the normal condition. In the following calculations below, it is shown how the values in Table 4.2 were determined using the statistics from Table 4.1.

For a normal condition

The lower bound for the EWMA statistic is as follows:

$$LB_{\text{normal}} = [ln\ (\min \rho^{fb}_{\text{normal}})][0.7] = 0.000000$$

The upper bound for the EWMA statistic is

$$UB_{\text{normal}} = [ln\ (\max \rho^{fb}_{\text{normal}})][0.7] = 0.134323$$

For a misaligned condition

The lower bound for the EWMA statistic is

$$\begin{aligned} LB_{\text{misalignment}} &= (1 - \lambda)EWMA_{LB(\text{normal})} + \lambda[ln\ (\min \rho^{fb}_{\text{misalignment}})] \\ &= (0.3)\ (0.000000) + (0.7)\ (1.665578) \\ &= 1.165905 \end{aligned}$$

The upper bound for the EWMA statistic is

$$
\begin{aligned}
UB_{\text{misalignment}} &= (1-\lambda)EWMA_{UB(\text{normal})} + \lambda[ln\ (\max \rho^{\text{fb}}_{\text{misalignment}})] \\
&= (0.3)\ (0.134323) + (0.7)\ (1.822096) \\
&= 1.315764
\end{aligned}
$$

For a contaminated condition

The lower bound for the EWMA statistic is

$$
\begin{aligned}
LB_{\text{contaminated}} &= (1-\lambda)EWMA_{LB(\text{normal})} + \lambda[ln\ (\min \rho^{\text{fb}}_{\text{contaminated}})] \\
&= (0.3)\ (0.000000) + (0.7)\ (3.261547) \\
&= 2.283083
\end{aligned}
$$

The upper bound for the EWMA statistic is

$$
\begin{aligned}
UB_{\text{contaminated}} &= (1-\lambda)EWMA_{UB(\text{normal})} + \lambda[ln\ (\max \rho^{\text{fb}}_{\text{contaminated}})] \\
&= (0.3)\ (0.134323) + (0.7)\ (3.495920) \\
&= 2.487441
\end{aligned}
$$

For a loose condition

The lower bound for the EWMA statistic is

$$
\begin{aligned}
\text{LB}_{\text{loose}} &= (1-\lambda)EWMA_{LB(\text{normal})} + \lambda[\ln\ (\min \rho^{\text{fb}}_{\text{loose}})] \\
&= (0.3)\ (0.000000) + (0.7)\ (2.622042) \\
&= 1.835429
\end{aligned}
$$

The upper bound for the EWMA statistic is

$$
\begin{aligned}
\text{UB}_{\text{loose}} &= (1-\lambda)EWMA_{UB(\text{normal})} + \lambda[\ln\ (\max \rho^{\text{fb}}_{\text{loose}})] \\
&= (0.3)\ (0.134323) + (0.7)\ (2.764185) \\
&= 1.975226
\end{aligned}
$$

In Table 4.2, the numbers in parentheses are determined based on the equations given below:

(1) Minimum normalized variance = $\mu_{\rho^{\text{fb}}} - 4\sigma_{\rho^{\text{fb}}}$
(2) Maximum normalized variance = $\mu_{\rho^{\text{fb}}} - 4\sigma_{\rho^{\text{fb}}}$

where the minimum and maximum normalized variances, average normalized variances and standard deviation of normalized variances for each machine condition are found in Table 4.1.

To determine the upper and lower bounds of the EWMA values for each machine condition under the 4σ limits, the following equation was used for each machine condition:

$$
\begin{aligned}
\text{EWMA}_{\min} &= (0.3)(\text{EWMA}_{\min(\text{norm})}) + (0.7)(\text{lower } 4\sigma \text{ normalized variance}) \\
\text{EWMA}_{\max} &= (0.3)(\text{EWMA}_{\max(\text{norm})}) + (0.7)(\text{upper } 4\sigma \text{ normalized variance})
\end{aligned}
$$

The second set of upper and lower bounds, based on 4σ limits, were calculated as follows (see Table 4.3):

For a normal condition

The 4σ lower bound for the EWMA statistic is as follows:

$$\begin{aligned} LB_{\text{normal}} &= [\text{Min } \rho^{\text{fb}}_{\text{normal}}]\,[0.3] \\ &= \mu_{\rho^{\text{fb}}_{\text{normal}}} + 4\sigma_{\rho^{\text{fb}}_{\text{normal}}} \\ &= (0.00)(0.3) = 0.000000 \end{aligned}$$

The 4σ upper bound for the EWMA statistic is

$$\begin{aligned} UB_{\text{normal}} &= [\text{Min } \rho^{\text{fb}}_{\text{normal}}]\,[0.3] \\ &= [\mu_{\rho^{\text{fb}}_{\text{normal}}} + 4\sigma_{\rho^{\text{fb}}_{\text{normal}}}] \quad (0.7) \\ &= [-0.00357 + 4(0.17827)](0.7) = 0.496657 \end{aligned}$$

For a misaligned condition

The 4σ lower bound for the EWMA statistic is

$$\begin{aligned} LB_{\text{misalignment}} &= (1-\lambda)(\text{Min } \rho^{\text{fb}}_{\text{normal}}) + \lambda(\text{Min } \rho^{\text{fb}}_{\text{misalignment}}) \\ &= (0.3)(0.00) + (0.7)\left(\mu_{\rho^{\text{fb}}_{\text{misalignment}}} - 4\sigma_{\rho^{\text{fb}}_{\text{misalignment}}}\right) \\ &= (0.00) + (0.7)\,(1.732227 - 4(0.049893)) \\ &= 1.072859 \end{aligned}$$

The 4σ upper bound for the EWMA statistic is

$$\begin{aligned} UB_{\text{misalignment}} &= (1-\lambda)(\text{Max } \rho^{\text{fb}}_{\text{normal}}) + \lambda(\text{Max } \rho^{\text{fb}}_{\text{misalignment}}) \\ &= (0.3)(0.496657) + (0.7)\left(\mu_{\rho^{\text{fb}}_{\text{misalignment}}} - 4\sigma_{\rho^{\text{fb}}_{\text{misalignment}}}\right) \\ &= (0.3)(0.496657) + (0.7)(1.732227 + 4(0.049893)) \\ &= 1.501256 \end{aligned}$$

For a contaminated condition

The 4σ lower bound for the EWMA statistic is

$$\begin{aligned} LB_{\text{contaminated}} &= (1-\lambda)(\text{Min } \rho^{\text{fb}}_{\text{normal}}) + \lambda(\text{Min } \rho^{\text{fb}}_{\text{contaminated}}) \\ &= (0.3)(0.00) + (0.7)\left(\mu_{\rho^{\text{fb}}_{\text{contaminated}}} - 4\sigma_{\rho^{\text{fb}}_{\text{contaminated}}}\right) \\ &= (0.3)(0.00) + (0.7)(3.402073 - 4(0.064102)) \\ &= 2.201966 \end{aligned}$$

The 4σ upper bound for the EWMA statistic is

$$\begin{aligned} UB_{\text{contaminated}} &= (1-\lambda)(\text{Max } \rho^{\text{fb}}_{\text{normal}}) + \lambda(\text{Max } \rho^{\text{fb}}_{\text{contaminated}}) \\ &= (0.3)(0.496657) + (0.7)\left(\mu_{\rho^{\text{fb}}_{\text{contaminated}}} - 4\sigma_{\rho^{\text{fb}}_{\text{contaminated}}}\right) \\ &= (0.3)(0.496657) + (0.7)(3.402073 + 4(0.064102)) \\ &= 2.709934 \end{aligned}$$

For a loose condition

The 4σ lower bound for the EWMA statistic is

$$\begin{aligned} LB_{\text{loose}} &= (1-\lambda)(\text{Min } \rho^{\text{fb}}_{\text{normal}}) + \lambda(\text{Min } \rho^{\text{fb}}_{\text{loose}}) \\ &= (0.3)(0.000000) + (0.7)\left(\mu_{\rho^{\text{fb}}_{\text{loose}}} - 4\sigma_{\rho^{\text{fb}}_{\text{loose}}}\right) \\ &= (0.3)(0.000000) + (0.7)(2.681575 - 4(0.048750)) \\ &= 1.740603 \end{aligned}$$

The $4\,\sigma$ upper bound for the EWMA statistic is

$$\begin{aligned} UB_{\text{loose}} &= (1-\lambda)(\text{Max } \rho^{\text{fb}}_{\text{normal}}) + \lambda(\text{Max } \rho^{\text{fb}}_{\text{loose}}) \\ &= (0.3)(0.496657) + (0.7)\left(\mu_{\rho^{\text{fb}}_{\text{loose}}} - 4\sigma_{\rho^{\text{fb}}_{\text{loose}}}\right) \\ &= (0.3)(0.496657) + (0.7)(2.681575 + 4(0.048750)) \\ &= 2.162600 \end{aligned}$$

The percentage of the data sets that will have a calculated normalized variance within the range of normalized variances found under initial experimentation is determined by calculating the probability that the normalized values for each machine condition will lie with the range (maximum normalized variance – minimum normalized variance). This probability is calculated by:

$$P(a < X < b) = \phi\left(\frac{b-\bar{x}}{\sigma}\right) - \phi\left(\frac{a-\bar{x}}{\sigma}\right)$$

where a is the minimum normalized variance (min $\rho^{\text{fb}}_{\text{normalized}}$), b is the maximum normalized variance (max $\rho^{\text{fb}}_{\text{normalized}}$), $\bar{x}$ is the average normalized variance (normalized $\mu_{\rho^{\text{fb}}}$), and s is the standard deviation of the normalized variances (normalized $\sigma_{\rho^{\text{fb}}}$) (see Table 4.1).

For a normal condition

$$\begin{aligned} P(\bar{x} \le 0.19189) &= \phi\left(\frac{0.19189 + 0.00357}{0.17827}\right) \\ &= \phi(1.096) = 0.86355 \end{aligned}$$

For a misaligned condition

$$P(1.665578 \le \bar{x} \ge 1.822096)$$

$$= \phi\left(\frac{1.822096 - 1.732227}{0.049893}\right) - \phi\left(\frac{1.665578 - 1.732227}{0.049893}\right)$$

$$= 0.87337$$

For a contaminated condition

$$P(3.261547 \leq \overline{x} \geq 3.495920)$$

$$= \phi\left(\frac{3.495920 - 3.402073}{0.064102}\right) - \phi\left(\frac{3.261547 - 3.402073}{0.064102}\right)$$

$$=0.91423$$

For a loose condition

$$P(2.622042 \leq \overline{x} \geq 2.764185)$$

$$= \phi\left(\frac{2.764185 - 2.681575}{0.048750}\right) - \phi\left(\frac{2.622042 - 2.681575}{0.048750}\right)$$

$$=0.84391$$

Replacing the minimum and maximum normalized variances by the 4*s* limits will increase the probability that the normalized variance for a particular machine condition will lie within the calculated variance range. Limits of 4*s* give a 99% probability that the data will lie within those 4*s* boundaries. Although we are not using the lower control limits in the exponentially weighted moving average statistic, an assumption is made that the variance will never be less than 0 when an abnormal condition occurs. Using the 4*s* range for each machine condition, with the average normalized variances and standard deviation of normalized variances shown in Table 4.1, the 99% probabilities are calculated below:

For the normal condition

$$P(\overline{x} \leq 0.70951) = \phi\left(\frac{0.70951 + 0.00357}{0.17827}\right)$$

$$= \phi(4) = 0.9997$$

Table 4.3 Calculated statistics of EWMA values for a normal machine condition and three abnormal machine conditions

Condition	*Minimum EWMA*	*Maximum EWMA*
Normal	0.0000	0.4967
Misalignment	1.0729	1.5013
Contamination	2.2020	2.7099
Looseness	1.7406	2.1626

For a misaligned condition

$$P(1.532655 \le \bar{x} \ge 1.931799)$$

$$= \phi\left(\frac{1.931799 - 1.732227}{0.049893}\right) - \phi\left(\frac{1.532655 - 1.732227}{0.049893}\right)$$

$$= 0.99994$$

For a contaminated condition

$$P(3.145665 \le \bar{x} \ge 3.658481) =$$

$$= \phi\left(\frac{3.658481 - 3.42073}{0.064102}\right) - \phi\left(\frac{3.145665 - 3.402073}{0.064102}\right)$$

$$= 0.99994$$

For a loose condition

$$P(2.486575 \le \bar{x} \ge 2.876575)$$

$$= \phi\left(\frac{2.876575 - 2.681575}{0.048750}\right) - \phi\left(\frac{2.486575 - 2.681575}{0.048750}\right)$$

$$= 0.99994$$

The range of values for each of the operating conditions investigated were well defined in the preliminary studies with no overlap among different machine conditions. Consequently, the diagnosis and monitoring procedure was established (see Fig. 4.8).

At the bottom of Table 4.4, the average value of all data sets tested is shown. A comparison of the average values of initial and testing conditions is shown in Table 4.5.

Table 4.4 displays the results of testing for an unknown bearing condition using the EWMA procedure described earlier. Under the testing procedure, each of the four bearing conditions (normal, contaminated, misaligned, and loose) was imitated in the lab. For each condition, eight data sets were collected containing 1000 points each. Out of a total of 32 data sets, the EWMA was able to correctly identify all 32 of the data sets. The testing was conducted to prove the validity of using the EWMA statistic for identifying a specific bearing condition based on predetermined limits for that condition.

In Fig. 4.9, the time series signals of the four machine conditions – normal, contaminated, misalignment, and loose bearings – are shown. The x-axis represents time and the y-axis represents the amplitude of the vibration signal (in g).

Two time series plots are displayed for each machine condition and the average amplitude of each time series sample is given below.

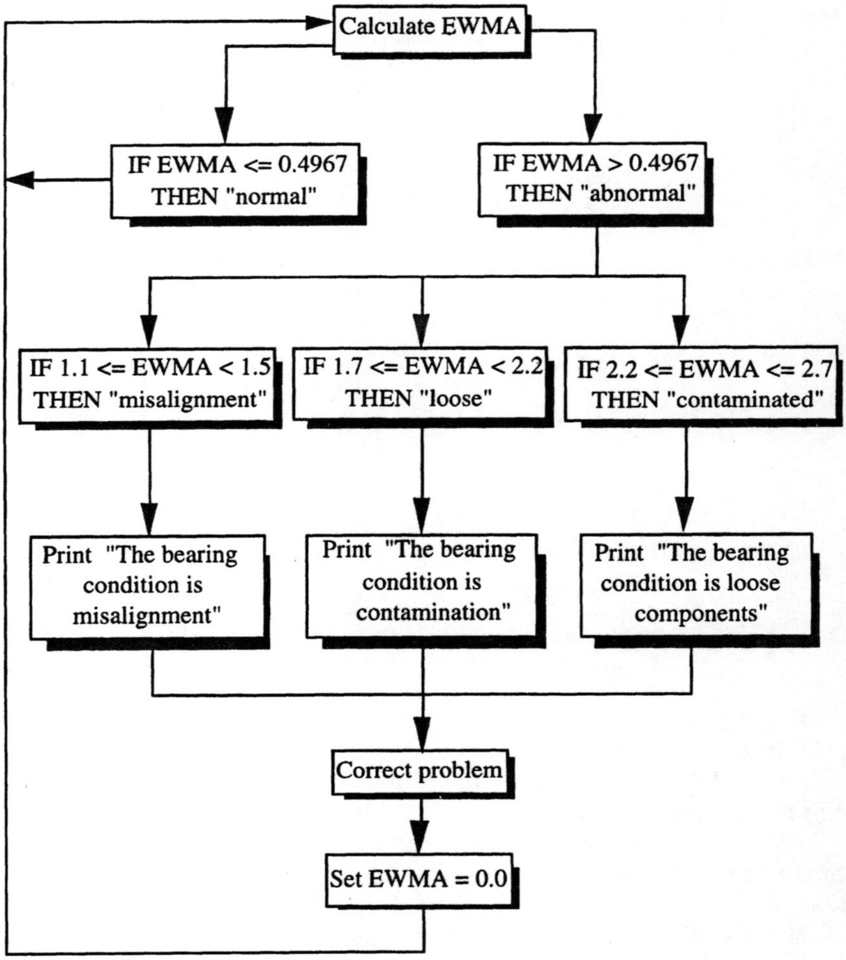

Fig. 4.8 Machine diagnostics and process monitoring technique.

• Contaminated:	0.0084615234
	0.0091725977
• Normal:	0.0025399219
	0.0024788086
• Misalignment:	0.0051446973
	0.004910332
• Looseness:	0.0061960254
	0.0064852148

The average of the amplitudes for the two time series signals for each machine condition is:

Table 4.4 Testing results based on EWMA statistic

Results of testing for unknown bearing condition based on EWMA statistic							
Normal	*Analysis*	*Contaminated*	*Analysis*	*Misaligned*	*Analysis*	*Loose*	*Analysis*
0.097145	Normal	2.577619	Contaminated	1.289229	Misaligned	1.892884	Loose
0.092735	Normal	2.429190	Contaminated	1.296169	Misaligned	1.816893	Loose
0.074031	Normal	2.391453	Contaminated	1.190132	Misaligned	1.949157	Loose
0.095964	Normal	2.504000	Contaminated	1.221219	Misaligned	2.026461	Loose
0.065993	Normal	2.476401	Contaminated	1.374886	Misaligned	2.039054	Loose
0.000000	Normal	2.367333	Contaminated	1.374886	Misaligned	2.039054	Loose
0.077476	Normal	2.547304	Contaminated	1.271684	Misaligned	2.052888	Loose
0.000000	Normal	2.354327	Contaminated	1.375385	Misaligned	1.843072	Loose
0.062918		2.455953		1.272073		1.957843	

Table 4.5 Comparison of calculated statistics of EWMA values for initial and test results under a normal machine condition and three abnormal machine conditions

Condition	*Average value* $\bar{x}_{EWMA}$		*Standard deviation* σ_{EWMA}	
	Initial	*Test*	*Initial*	*Test*
Normal	0.0580	0.0629	0.0379	0.0378
Misalignment	1.2126	1.2721	0.0349	0.0745
Contamination	2.3815	2.4560	0.0449	0.0449

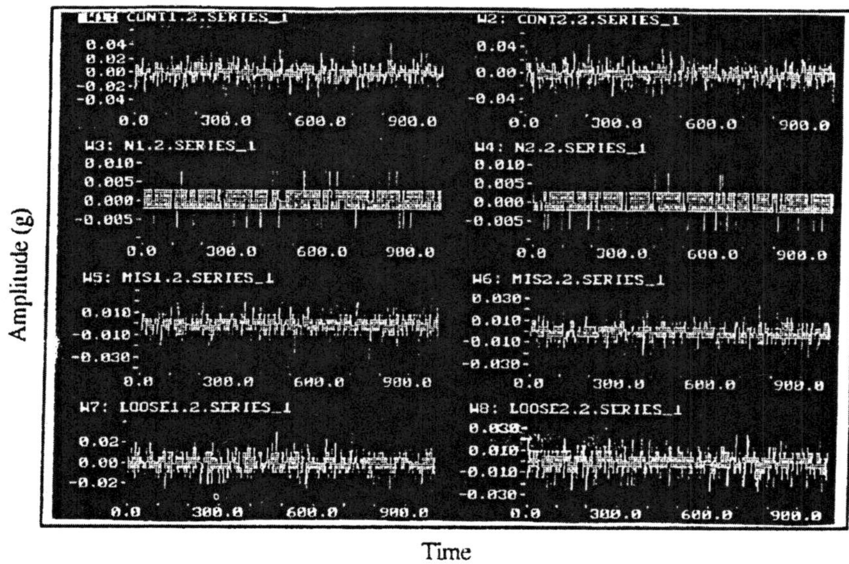

Fig. 4.9 Plot of vibration amplitudes vs time for four machine conditions.

- Contaminated: 0.00881706
- Normal: 0.002509365
- Misalignment: 0.005027514
- Looseness: 0.00634062

Thus, there is a positive correlation between the average amplitude of the signal and the average EWMA statistic for a given machine condition. As the average amplitude increases, the EWMA value increases. This implies that the EWMA statistic is sensitive to changes in amplitude. Since an increase in amplitude occurs when a bearing or other machine component undergoes the initial stages of failure, the EWMA technique has the ability to detect small changes in amplitude once a defect occurs. Table 4.6 shows the relationship between amplitude of the vibration signal and the corresponding EWMA statistic.

4.4 SUMMARY

The objective of the research presented was to develop an accurate machine monitoring and diagnostic technique based on vibration analysis. By developing an AR parametric model to characterize the normal machine condition, the EWMA control statistic was able to identify whether the machine was in a normal state ('in control') or in an abnormal state ('out-of-control').

Through experimentation, the applicability of employing AR parametric modeling and EWMA techniques concurrently in the accurate diagnosis of the condition of a machine was established. AR parametric modeling and EWMA process control techniques, in joint operation, were able to identify a normal machine condition from an abnormal machine condition 100% of the time. This is particularly significant, especially when viewed as a monitoring subsystem of a larger machine monitoring and fault diagnostics system.

In addition, the four machine conditions under study were easily distinguishable from one another in terms of their EWMA values. In Fig. 4.10, the abnormal machine conditions (contamination, looseness, and

Table 4.6 Comparison of average amplitude and average EWMA values for four machine conditions

Condition	*Average amplitude*	*Average EWMA*
Normal	0.002509365	0.066722
Misalignment	0.005027514	1.212559
Contamination	0.00881706	2.061447
Looseness	0.00634062	1.592654

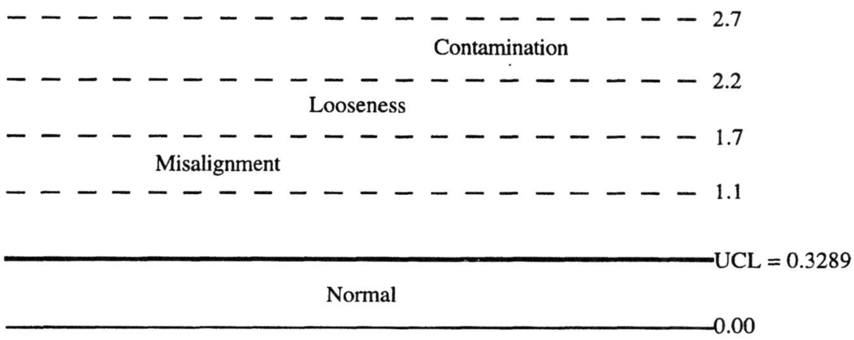

Fig. 4.10 EWMA control chart with upper control limit of normal machine condition and EWMA statistical ranges for three abnormal machine conditions.

misalignment) were easily identifiable from the normal condition, as well as from each other.

Since this research focused only on four different machine conditions, it cannot be stated that the EWMA technique is robust across all machine conditions. It is unlikely that the EWMA value would be able to distinguish all faults that occur; however, it would likely have the capability of distinguishing fault classes for those faults that were typically found within a range of EWMA values.

REFERENCES

[1] Angels, M. (1990) Choosing accelerometers for machinery health monitoring. *Sound and Vibration*, December, 20–4.

[2] Archambault, R. (1989) Getting more out of vibration signals: using the logarithmic scale. *Proceedings of the 1st International Machinery Monitoring and Diagnostics Conference*, Las Vegas, Nevada, 567–71.

[3] Box, G., Hunter, W., and Hunter, J. (1978) *Statistics for Experimenters*, John Wiley & Sons, New York, NY.

[4] Boyce, M., Bhargava, R., and Chinoy, R. (1989) On-line condition monitoring of turbomachinery. *Proceedings of the 1st International Machinery Monitoring and Diagnostics Conference*, Las Vegas, Nevada, 419–27.

[5] Brick, J., Michael, J., and Morganstein, D. (1989) Use statistical thinking to solve maintenance problems. *Quality Progress*, May, 55–60.

[6] Chen, Y. (1989) Machine diagnostic monitoring by prediction error analysis and information measures. *Proceedings of the 1st International Machinery Monitoring and Diagnostics Conference*, Las Vegas, Nevada, 553–9.

[7] Chen, Z., Li, Q., and Song, Z. (1989) Condition monitoring and fault diagnosis of the aero-engine using time series analysis method. *Proceedings of the 1st International Machinery Monitoring and Diagnostics Conference*, Las Vegas, Nevada, 415–18.

[8] Crowder, S. and Hamilton, M. (1992) An EWMA for monitoring a process standard deviation. *Journal of Quality Technology*, **24**(1), 12–21.

[9] Cryer, J. (1986) *Time Series Analysis*, PWS-Kent Publishing Company, Boston, 59–66.

[10] Isermann, R. (1984) Process fault detection based on modeling and estimation methods – a survey. *Automatica*, **20**, 387–404.

[11] Jangi, S. and Jain, Y. (1991) Embedding spectral analysis in equipment. *IEEE Spectrum*, February, 40–3.

[12] Jianfang, J., Yueming, S., and Qi, Z. (1989) A method of bearing vibration signal extraction and fault detection. *Proceedings of the 1st International Machinery Monitoring and Diagnostics Conference*, Las Vegas, Nevada, 149–54.

[13] Jixi, C. and Yiyun, H. (1989) Vibration analysis and quality detection of ball bearings. *Proceedings of the 1st International Machinery Monitoring and Diagnostics Conference*, Las Vegas, Nevada, 572–75.

[14] Marple, S. (1987) *Digital Spectral Analysis with Applications*, Prentice-Hall, Inc., 224–31, 251.

[15] McElroy, A. and Fruchtman, I. (1991) Use statistical analysis to predict equipment reliability. *Power*, October, 39.

[16] Pardue, E., Piety, K. and Moore, R. (1992) Elements of reliability-based machinery maintenance. *Sound and Vibration*, May, 14–20.

[17] Peczely, G. (1989) Vibration analysis of gas-engine units. *Proceedings of the 1st International Machinery Monitoring and Diagnostics Conference*, Las Vegas, Nevada, 306–12.

[18] Piety, K., Piety, R., and Scheibel, J. (1991) Re-evaluating fan standards and how they are applied. *Power*, December, 59–62.

[19] Potts, S. and Hakimmashhadi, H. (1989) Application of parametric spectral estimation to machine vibration analysis. *Proceedings of the 1st International Machinery Monitoring and Diagnostics Conference*, Las Vegas, Nevada, 190–5.

[20] Renwick, J. and Bubson, P. (1985) Vibration analysis – a proven technique as a predictive maintenance tool. *IEEE Trans. on Industry Applications*, **IA-21**(2), 324–32.

[21] Sachs, N. (1991) What does 'PM' really mean? *Journal of the Society of Tribologists and Lubrication Engineers*, November, 889–91.

[22] Serridge, M. (1989) Ten crucial concepts behind trustworthy fault detection in machine condition monitoring. *Proceedings of the 1st International Machinery Monitoring and Diagnostics Conference*, Las Vegas, Nevada, 722–3.

[23] Seth, B. (1989) Isolation of abnormalities in rotating machines using signature analysis. *Proceedings of the 1st International Machinery Monitoring and Diagnostics Conference*, Las Vegas, Nevada, 35–9.

[24] Shanmugan, K. and Breipohl, A. (1988) *Random Signals: Detection, Estimation and Data Analysis*, John Wiley & Sons, Inc., 250.

[25] Sidahmed, M. (1989) Contribution of parametric signal processing techniques to machinery condition monitoring. *Proceedings of the 1st International Machinery Monitoring and Diagnostics Conference*, Las Vegas, Nevada, 190–5.

[26] Tranter, J. (1990) The application of computers to machinery predictive maintenance. *Sound and Vibration*, December, 15.

[27] Wang, X., Shi, X., and Xu, M. (1989) The fault diagnosis and quality evaluation of ball bearing by vibration signal processing. *Proceedings of the 1st International Machinery Monitoring and Diagnostics Conference*, Las Vegas, Nevada, 318–26.

[28] Watts, W. and Van Dyke, D. (1989) A portable, automated machine condition diagnostics program using order normalized vibration data. *Proceedings of the 1st International Machinery Monitoring and Diagnostics Conference*, Las Vegas, Nevada, 637–43.

[29] Wei, W. (1990) Time series analysis: univariate and multivariate methods. Addison-Wesley Publishing Company, 57.

[30] Wowk, V. (1991) *Machinery Vibration: Measurement and Analysis*, McGraw-Hill, Inc., 69–73.

[31] Wu, S., Tobin, T., and Chow, M. (1980) Signature analysis for mechanical systems via dynamic data system (DDS) monitoring technique. *Journal of Mechanical Design*, April. 217–21.

CHAPTER 5

Machine performance estimation and reliability modeling

Chang-Ching Lin and Ben Wang

A prerequisite to effective maintenance planning is accurate information about current machine conditions. In this chapter, a new method of obtaining machine maintenance information is proposed. We combine reliability modeling methods and vibration-based condition monitoring techniques into an integrated system, CMAC-PEM. With sensor readings CMAC-PEM fuses the information and estimates machine performance. Then, based on the CMAC-PEM outputs, reliability analysis is applied to build a probabilistic model which is then used to derive important maintenance data, such as failure rates and mean time between failures (MTBF).

This chapter is divided into six sections. Section 5.1 introduces the theoretical background of reliability analysis. In section 5.2 the CMAC-PEM methodology is presented. Section 5.3 gives an example to illustrate the implementation procedure of CMAC-PEM. In section 5.4 a bearing deterioration experiment used to verify the methodology was conducted. Section 5.5 addresses the proportional hazards model (PHM), which has been used as a validation tool to verify the results from CMAC-PEM. The summary and conclusions of this chapter are given in section 5.6.

5.1 MACHINE RELIABILITY ANALYSIS

5.1.1 Hazard rate and reliability modeling

Reliability is the probability that a component, equipment, or a system will perform a required function under a certain operating condition for a stated period of time. The random variable used to quantify reliability is the 'time to failure' denoted by T. The reliability function $R(t)$ of a system is defined in Equation (5.1).

$$\begin{aligned} R(t) &= \Pr[T > t] \\ &= 1 - \Pr[T \leq t] \\ &= 1 - F_T(t) \end{aligned} \tag{5.1}$$

where $F_T(t)$ is the failure distribution function. Other related functions are given as follows:

$$\begin{aligned} F_T(t) &= \int_0^t f_T(x)dx \\ f_T(t) &= \frac{d(F_T(t))}{dt} \\ &= h(t)\, exp\left[-\int_0^t (x)dx\right] \\ h(t) &= \frac{f_T(t)}{R(t)} \\ E[T] &= \int_0^\infty t f_T(t)dt \\ &= \text{MTBF} \end{aligned} \tag{5.2}$$

where $f_T(t)$ is the failure probability density function of T, and $h(t)$ is the conditional failure rate or hazard rate. Hazard rate is one of the criteria used to select a specific model in lifetime studies. This number indicates what the probability of failure is in the next instant of time, given that the system is presently operating. Equipment can have an increasing, decreasing, or constant hazard rate corresponding to situations where the probability of failure in the next instant of time increases, decreases, or does not change with time. Equation (5.2) shows that given $h(t)$, the density function, $f_T(t)$, can be derived; hence, knowing the shape of the hazard rate can assist in the selection of a failure model. The shape of the hazard rate for most components or systems follows the 'bath-tub curve.'

The expected value $E[T]$ is also called 'mean time between failures' (MTBF) or 'mean time to failure' (MTTF) depending on the type of system. If the system is irreparable, MTTF is used; if it is a repairable system, MTBF is used.

5.1.2 Weibull model verification

Many parametric failure models, such as Poisson, exponential, Weibull, and log-normal distributions, have been proposed to describe the behaviour of machine degradation. Among them the Weibull failure model is most popular because it can accommodate several types of behavior including infant mortality and various aging defects in the 'bath-tub' curve (Archambault and Fillion 1991). Furthermore, it belongs to a special class of log-linear model possessing an interesting property which makes it particularly suitable for two other types of higher level probabilistic models: the proportional hazards model and the accelerated time model (ibid.).

The failure probability density function of the Weibull distribution is defined in Equation (5.3).

$$f_T(t) = \frac{\beta}{\theta}\left(\frac{t-c}{\theta}\right)^{\beta-1} exp\left[-\left(\frac{t-c}{\theta}\right)^{\beta}\right] \tag{5.3}$$

where β is called the shape or slope parameter, θ is called the characteristic life or scale parameter, and c is the location constant which indicates the minimum time when a failure starts.

Once the underlying failure model of a system has been determined the following reliability equations can be easily derived.

$$f_T(t) = 1 - exp\left[-\left(\frac{t-c}{\theta}\right)^{\beta}\right]$$

$$R(t) = exp\left[-\left(\frac{t-c}{\theta}\right)^{\beta}\right]$$

$$h(t) = \frac{\beta}{\theta}\left(\frac{t-c}{\theta}\right)^{\beta-1} \tag{5.4}$$

$$E[T] = \text{MTBF} = c + \theta\Gamma\left(1 + \frac{1}{\beta}\right)$$

$$Var\,[T] = \theta^2\left[\Gamma\left(1 + \frac{2}{\beta}\right) - \left(\Gamma\left(1 + \frac{1}{\beta}\right)\right)^2\right]$$

$$\Gamma[x] = \int_0^{\infty} t^{x-1}e^{-t}dt\,,\ x > 1$$

$\Gamma(x)$ is the well known gamma function. $R(t)$ can be rearranged in the form shown below:

$$ln\left(ln\left(\frac{1}{R(t)}\right)\right) = \beta\, ln(t-c) - \beta\, ln\,(\theta) \tag{5.5}$$

If the estimates of $R(t)$ can be obtained, validation of the assumed model may be accomplished by computing the linear regression line of $ln(ln(1/R(t)))$ with respect to $ln(t-c)$ as shown in Equation (5.5). By plotting

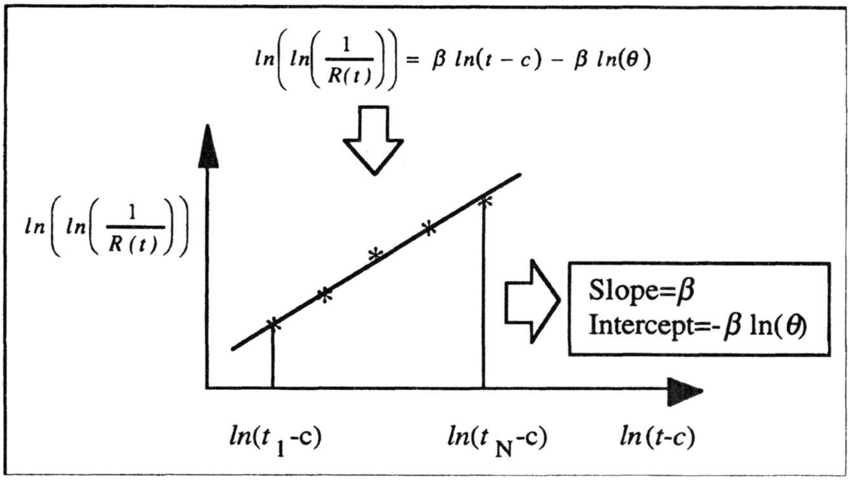

Fig. 5.1 Verification of Weibull model.

the estimates of $R(t)$ on a special scaled coordinate system and looking for a straight line, the slope of the linear regression line gives β and the intercept on the Y-axis gives θ. The value of c is usually zero, but could take different values if the plotted values exist on a curvature rather than a straight line. In such a case, a suitable value for c must be chosen in order to fit a straight line. Figure 5.1 illustrates the verification procedure of the Weibull model. The Chi-Square test may be used to validate the model fit for a given confidence level.

Once β, θ and c are known the failure rate, $h(t)$, MTBF, and $Var[T]$ can be calculated from Equation (5.4). These values may provide useful guidelines for maintenance scheduling.

5.2 CMAC PERFORMANCE ESTIMATION MODEL (CMAC-PEM)

5.2.1 Assumption and definition of CMAC-PEM

The problem at hand is how to calculate the reliability measure, $R(t)$, without looking at historical data or performing life tests on machines. Since CMAC can provide a very good estimate of machine performance, one can assume the machine performance levels estimated from CMAC are equivalent to the machine's reliability measures. This assumption is justified by a unique CMAC feature, the CMAC Confidence Level, explained in the next section. By taking advantage of the well-defined reliability equations described previously, useful information, such as

$h(t)$, MTBF, and $Var[T]$, can be computed. This is named the CMAC Performance Estimation Model (CMAC-PEM).

5.2.2 CMAC confidence level

To justify the equivalence between CMAC outputs and reliability measures a special feature of the CMAC algorithm must be investigated and implemented. According to the research by Lee and Kramer, one can calculate the percentage of shared weight cells in a trained CMAC when a test vector is presented to the CMAC. This percentage value, called the 'confidence level', represents the closeness between the test vector and the training vector (Lee and Kramer 1993). The confidence value is regarded as the level of behavioral change of the system being monitored. It is computed as follows:

$$\text{Confidence Level} = (\text{Repeatable\_Weight\_Location\_Count})/K \qquad (5.6)$$

where K is the number of quantizing functions or is actually the total number of cells used for storing the distributed weights of any input vector in a CMAC network. The Repeatable_Weight_Location_Count is the number of shared cells between the trained CMAC and a test vector.

In the context of machine condition monitoring the input vectors to CMAC are readings from sensors (which are formed by vibration trending indices in the research). If we train a CMCA network, using the readings from the normal condition of a machine, then in the CMCA test stage the confidence value may be seen as the machine's performance level compared to its normal condition. The performance level of the machine actually depicts the similarity between the current input vector and the input vectors used in training. In order words, a low performance level means the machine has deviated from the normal condition and hence requires close monitoring, but a high performance level ensures the machine stays in good condition.

5.2.3 Implementation procedure

Figure 5.2 shows the CMAC-PEM flow chart. The CMAC-PEM procedure is listed step-by-step as follows.

1. Select and define a normal operating condition of the monitored machine. The normal condition is usually defined by the operator's experience or from empirical data.
2. Train a CMAC network by using the normal condition vibration trending indices collected from the machine. The target value of the training set can be arbitrarily chosen since only one machine condition, the normal condition, is to be trained.

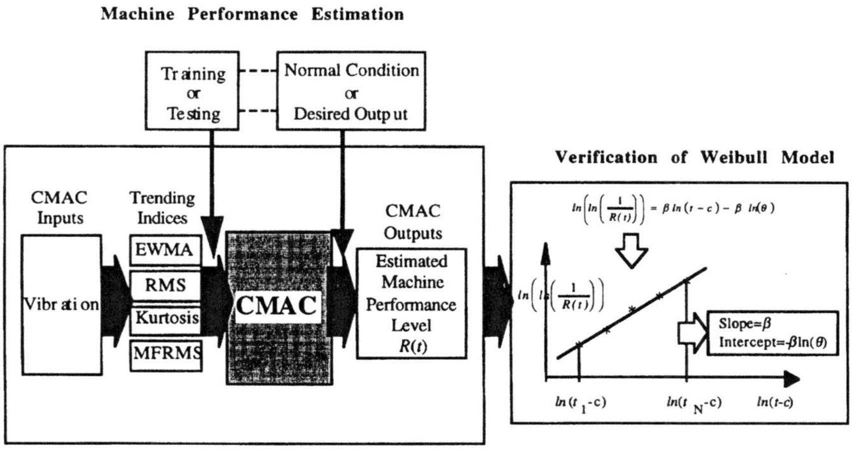

Fig. 5.2 CMAC-PEM flow chart.

3. Use the trained CMAC to estimate the machine reliability measure, $R(t)$, by feeding the current vibration trending indices to the network at the predetermined time intervals.
4. Use the computed reliability measures to verify the underlying Weibull model and calculate the model parameters β, θ, and c.
5. Derive the other reliability statistics for the current machine condition based on these three parameters.
6. Apply the derived information to devise a machine maintenance plan.

The idea behind the CMAC-PEM methodology comes from the special CMAC data store-retrieval system. CMAC is taught the normal condition data to direct the CMAC internal weight connections to build a baseline profile of the condition. In the CMAC testing stage the trained CMAC calculates the confidence level, which is, by the definition, the percentage deviation of the current input data to the baseline profile stored in the CMAC. Hence using this percentage deviation to represent the machine's reliability measure is a logical consequence.

The major advantages of CMAC-PEM are: (1) the only task is to have the CMAC network learn the baseline information of a system. It makes the CMAC learning extremely simple and easy because one does not have to collect data in various operating conditions; (2) the output of CMAC is an overall performance index of the monitored system. The index can be used in reliability calculation without further processing; and (3) CMAC contains an outstanding information fusion capability.

5.3 CMAC-PEM ANALYSIS ON MACHINE IMBALANCE DATA: AN ILLUSTRATIVE EXAMPLE

To illustrate the proposed methodology a simulated machine degradation process was created using the data from a machine imbalance experiment. A total of seven different imbalance conditions labeled 1 to 7 were created. Level 1 was assumed to be the normal condition of the text rig. It was the only training data used in this example. For each imbalance level, several signals were collected. Each signal was used to calculate one set of trending indices. Each set of trending indices included EWMA, Kurtosis, RMS, and MFRMS. The CMAC architecture used was a four-input, one-output model. In addition, time indices were assigned to the trending indices vectors to simulate a machine degradation process. The time sequence of the vectors was arranged in the order shown in Fig. 5.3. Each small cube in Fig. 5.3 is a data collection point of the hypothesized machine degradation process.

As mentioned earlier, the first CMAC configuration parameter we should determine is the physical memory size available on the computer. For the trending indices case, the following CMAC parameters were chosen.

$N = 4$
$R = 600$

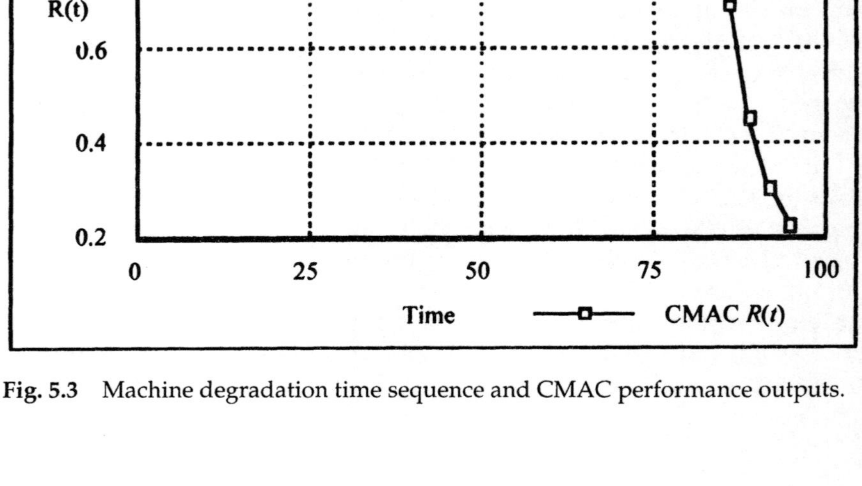

Fig. 5.3 Machine degradation time sequence and CMAC performance outputs.

$|\mathbf{Ap}| = 4096$
$\beta \quad = 1$
$K \quad = 128$

Following the CMAC-PEM implementation procedure, the level 1 condition was defined as the normal condition of the test rig. Then a CMAC network was trained using the trending indices vectors from the level 1 condition, which allowed the network to learn and store 'the normal profile' of the test rig. After the training stage the other trending indices vectors were used to test the network. The results are displayed in Fig. 5.3, where the Y-axis is the CMAC estimated reliability measure and the X-axis is the assigned time index for each test vector. The first 15 points show the test results of using the test vectors acquired from the normal condition. One can see their CMAC confidence levels are close to 1, meaning these test vectors represent conditions which are close to the baseline information learned by CMAC. When test vectors, other than level 1, were used the reliability estimates from CMAC-PEM decreased. Figure 5.3 denotes this decrease for output levels 2 to 7 from time units 75 to 100.

Although the level 1 condition was the only 'knowledge' the CMAC learned, the decreasing reliability measures following the increasing imbalance levels shown in Fig. 5.3 reveal a very consistent match between the estimated reliability values and the test rig physical conditions. By comparing the number of weight table cells shared between the learned condition and the test vectors CMAC-PEM could precisely calculate the distance between them and generate overall machine performance levels to reflect their physical situations.

In this experiment four vibration trending indices formed an input vector. The figure basically shows a very complicated input space. However, under such circumstances, CMAC could still perform a good generalization (information fusion) from the data to reveal machine conditions. Hence, it is clear CMAC comprises an excellent information fusion capability.

Using the data presented in Fig. 5.3 and Equations (5.4) and (5.5), one can actually derive the reliability model of the test rig. First, we transformed the data in Fig. 5.3 into Fig. 5.4 by applying Equation (5.5). Since the purpose was to calculate the failure model of the test rig, only the data from the imbalance conditions were needed. Therefore, the last six points in Fig. 5.4 were plotted in another graph (see Fig. 5.5). The location constant c was first left at zero. A linear regression model shown in the top of Fig. 5.5 was calculated. Since the plotted data did not fall on a straight line, c needed to be adjusted. It is known that when the line is concave downward, one must increase c to make it a straight line. The result, as shown in Fig. 5.6, yielded a reasonable straight line when c=79. Then the two Weibull parameters, β and θ, were calculated. Finally,

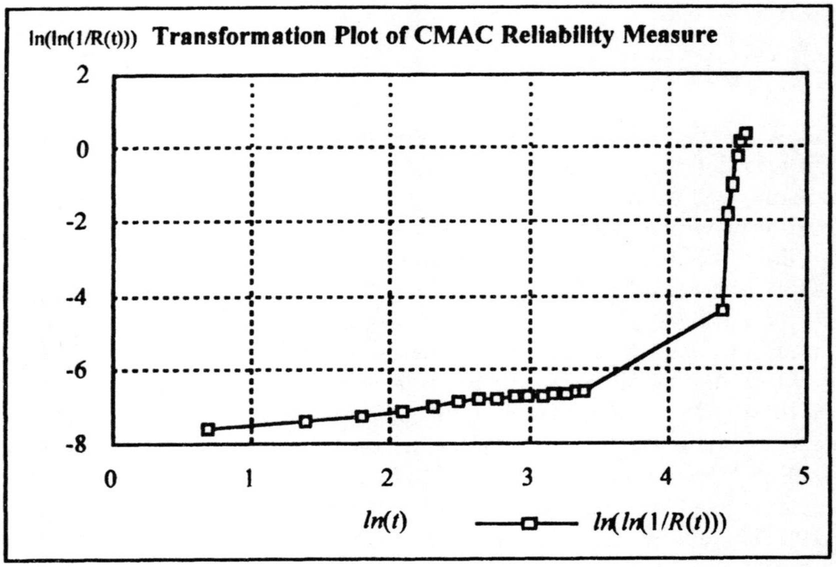

Fig. 5.4 Transformation plot of the CMAC reliability measures.

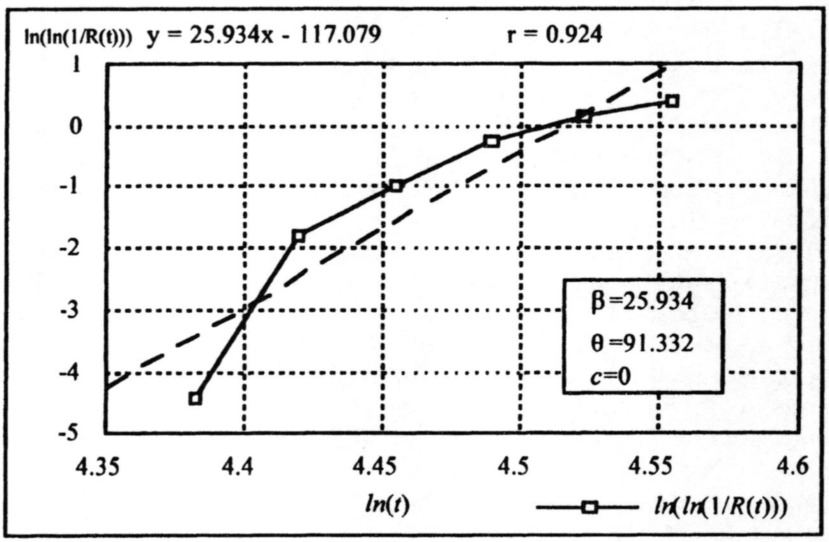

Fig. 5.5 Linear regression model of the test rig imbalance data (with $c = 0$).

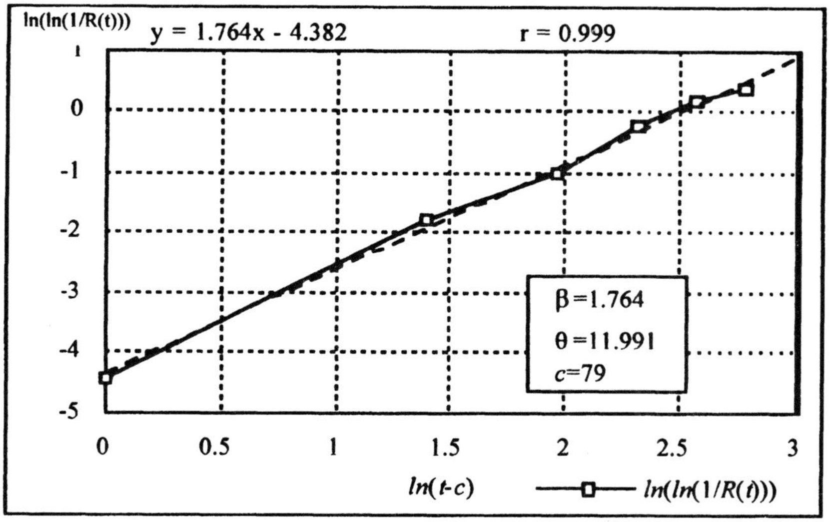

Fig. 5.6 Linear regression model of the test rig imbalance data (with $c = 79$).

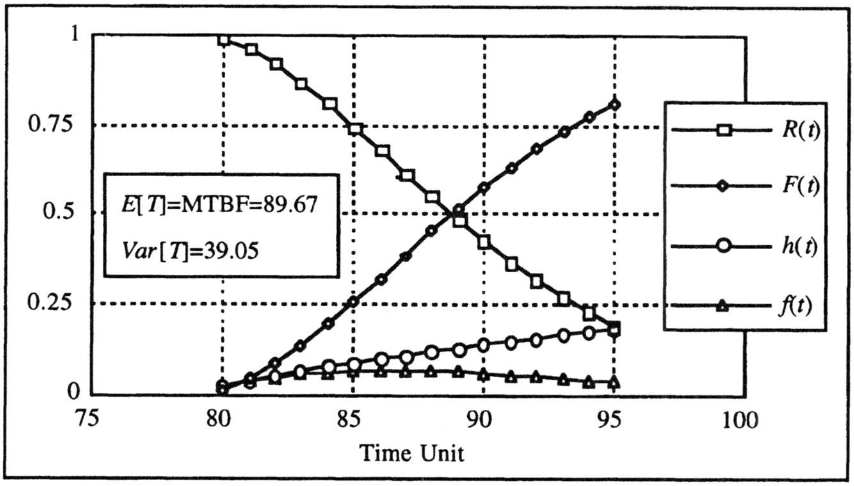

Fig. 5.7 Reliability functions plot of the test rig imbalance data.

using the functions in Equation (5.4), the reliability function $R(t)$, the failure distribution function $F_T(t)$, the probability density function $f_T(t)$, and the hazard rate $h(t)$ were prepared with respect to the specified time points. This graph is given in Fig. 5.7. MTBF and the variance, $Var[T]$, were calculated to be 89.67 time units and 39.05 time units, respectively.

Although the time indices were artificially assigned this still helps to give a clear illustration of the proposed CMAC-PEM methodology. For a real-world situation, the calculation can be carried out in real-time. Then we can obtain the updated machine reliability information over time and use this information to plan a better maintenance schedule.

5.4 ANALYSIS OF BEARING DETERIORATION PROCESS

5.4.1 Experimental setting

A bearing deteriorating experiment was conducted to test the CMAC-PEM performance in a real-world situation. The experimental test stand is shown in Fig. 5.8. During the experiment an extra constant load was added to a test bearing by using a turn buckle, driving the bearing housing downward. This created a constant stress on the top of the test bearing. The purpose of adding the extra force on the bearing was to accelerate the bearing deterioration process. The concept of acceleration occurs when a component or equipment, operating under a higher level of stress, will have exactly the same failure mechanisms as seen when operated at the normal stress level. The only difference is the process proceeds faster. The failure state occurs exactly as it had at lower stress – only the time scale is changed. The acceleration failure model has been widely used in industry for component reliability analysis.

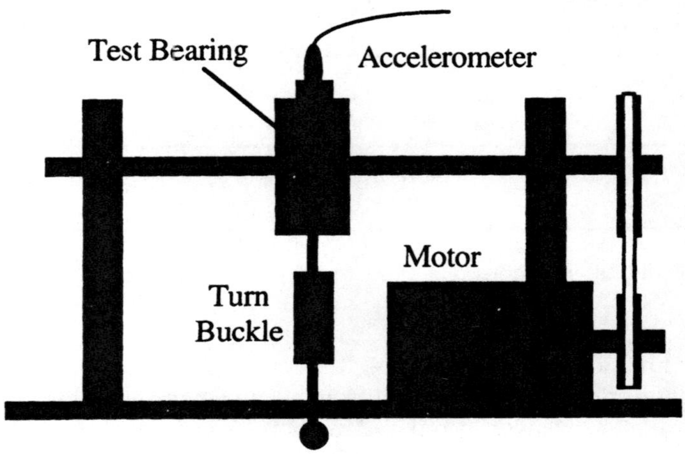

Fig. 5.8 Bearing test stand.

5.4.2 Experimental procedure and data acquisition

A bearing with a single point inner race defect was created and used as the test bearing in the experiment. It was inserted into the bearing housing located in the middle of a turning shaft (see Fig. 5.8). The motor was turned on and ran continuously while the experiment was proceeding. Vibration signals, collected in the time wave form from the test bearing, were saved into a personal computer for later analysis. Data sets were collected four to five times a day. The entire running time of the experiment was about a week. Table 5.1 lists the test stand running times and their normalized RMS, Kurtosis, and MFRMS vibration trending indices calculated at that time. Figure 5.9 plots the three trending indices corresponded with running times. The EMWA index was excluded from this experiment because its performance was similar to the RMS index.

5.4.3 Deterioration process of a single point inner race defect bearing

In order to investigate the mechanism changes inside the defective bearing, the vibration signals collected from the test stand were transformed

Table 5.1 Test stand running times and vibration trending indices

Condition	*RMS*	*Kurtosis*	*MFRMS*
2.5	0.000	0.069	0.000
4.0	0.025	0.005	0.388
6.3	0.410	0.047	0.897
10.5	0.550	0.001	0.527
15.8	0.220	0.011	0.615
25.5	0.810	0.059	0.671
28.5	1.000	0.029	0.935
31.0	0.776	0.028	0.744
34.0	0.962	0.020	0.828
39.8	0.651	0.087	1.000
49.5	0.438	0.000	0.786
52.5	0.304	0.045	0.677
58.0	0.473	0.055	0.823
64.0	0.495	0.086	0.787
74.0	0.924	0.116	0.589
78.5	0.711	0.149	0.988
81.5	0.725	0.108	0.912
88.0	0.638	1.000	0.689
98.0	0.528	0.035	0.758
103.0	0.899	0.095	0.712
108.0	0.724	0.213	0.823
124.0	0.881	0.063	0.889
133.0	0.548	0.085	0.579

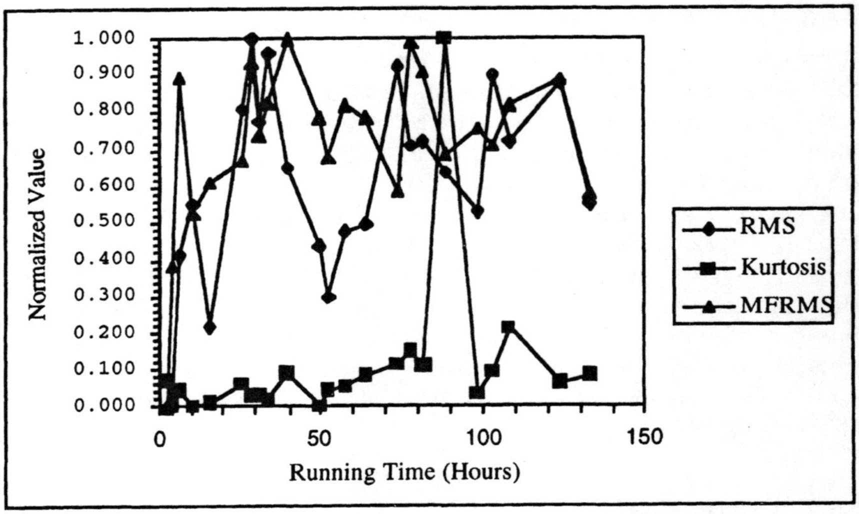

Fig. 5.9 Trending indices plot of the inner race defect bearing.

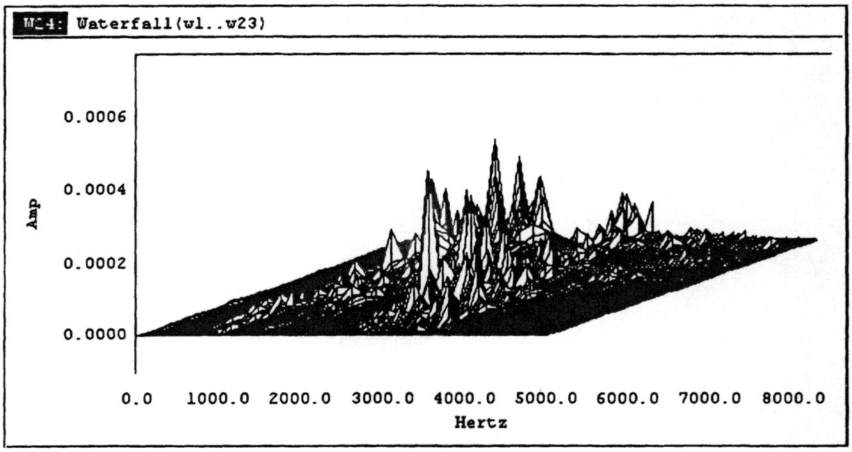

Fig. 5.10 PSD waterfall plot of the inner race defect bearing.

into power spectrum density (PSD) format. They were then plotted as a 'PSD waterfall' graph as shown in Fig. 5.10.

Figure 5.10 is a full-bandwidth PSD waterfall plot covering frequency bandwidth from 0 to 5000 Hz. From the bottom left corner of the plot to the top right corner it displays the gradual changes of the power spectrum of the deteriorating bearing. By drawing a line at 2500 Hz on the waterfall plot two major clusters of frequency peaks are clearly displayed. In order to see the details of the bearing deterioration spectra the original time domain data were high pass filtered or low pass filtered to extract high frequency components or lower frequency components from the raw signals, respectively. For both parts of signals, their PSD waterfall plots were again derived. The low pass PSD waterfall plot is shown in Fig. 5.11, and Fig. 5.12 displays the high pass PSD waterfall plot.

Figure 5.11 illustrates the spectral changes of the low frequency components in the vibration signals. Based on the bearing geometry and

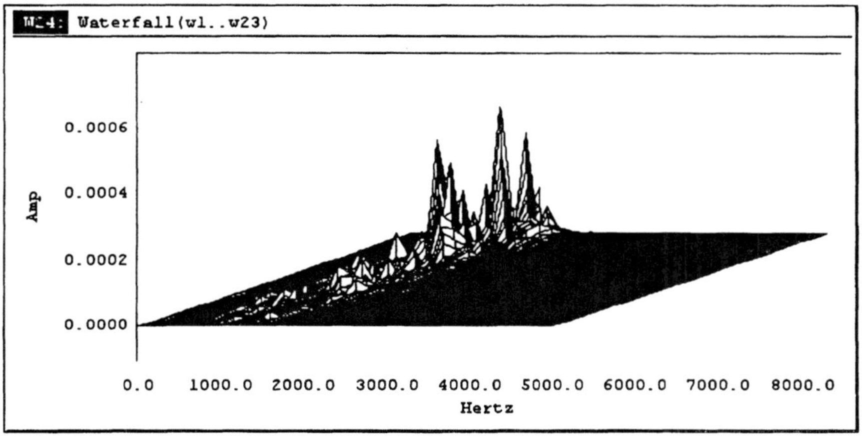

Fig. 5.11 Low pass PSD waterfall plot of the inner race defect bearing.

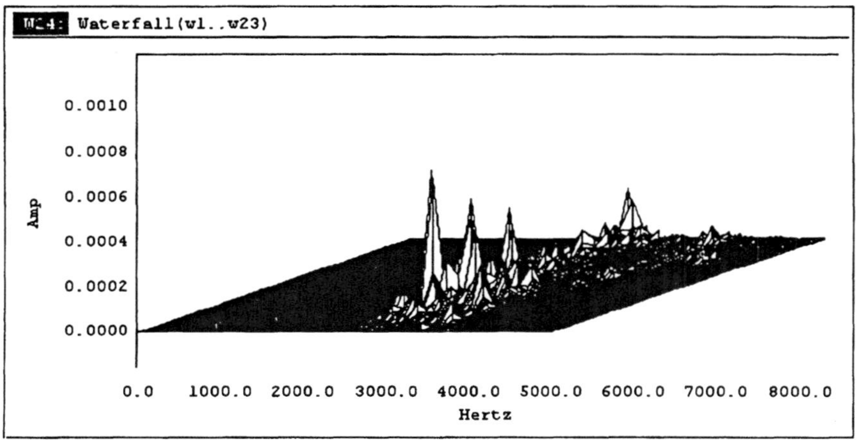

Fig. 5.12 High pass PSD waterfall plot of the inner race defect bearing.

the fundamental shaft speed one can easily estimate a possible ball pass frequency of the defective inner race (BPFI). Thus, a narrow bandwidth spike at the estimated BPFI can be easily identified from Fig. 5.11. As the size of the defect increased, the amplitude of the spike increased. According to Taylor (1980), this was the first stage of the spectral change for a growing inner race defect. As the size of the defect increased to a specific point, because of the modulation effect, the bandwidth of the peak area in the spectrum increased. The increasing bandwidth indicated the second stage of a growing inner race defect. Figure 5.11 mostly displays the first stage of the inner race defect growth and roughly the beginning of the second stage phenomenon.

The extracted high frequency components in the PSD waterfall format, as shown in Fig. 5.12, basically came from the effect of impulse resonance: the shorter an impulse is, the more high-frequency components must be used to represent in spectrum. This means that the spectral spread should occur in high-frequency areas for a very short duration signal (an impulse). The relationship between spectral spread and impulse duration can be used to explain the phenomenon shown in Fig. 5.12. At the beginning of the experiment the spectrum consisted mostly of the high-frequency components, since at this stage the size of the inner race defect was very small and the signals generated from the defect then were mostly short impulses. The high-frequency components gradually faded away as the size of the inner race defect increased. Up to this time the property of impulse no longer existed in the spectrum. The entire spectrum was then dominated by the low-frequency components (see Figs 5.10 and 5.11).

In order to obtain a 'good' bearing deterioration trend, the concept of selective frequency band trending has been employed in this research. The original time domain signals collected from the test stand were processed through a filter to extract the most representative components for inner race defect deterioration. Such preprocessed signals should be better able to reveal the deterioration process of the defective bearing. Therefore, a narrow bandwidth at the bearing ball pass frequency showing in Fig. 5.11 satisfied this purpose. A band pass filter turning at the frequency ranging from 1300 to 2000 Hz was applied to process the original signals. The filtered time signals were then used for CMAC-PEM analysis in the next section. Figure 5.13 shows the band pass PSD waterfall plot of the test bearing. One can see a very clear trend of increasing spikes in Fig. 5.13.

5.4.4 CMAC-PEM performance analysis of bearing deterioration data

The trending indices used in the experiment include RMS, Kurtosis, and MFRMS. Figure 5.14 displays the plot of trending indices from the band pass filtered signals. By comparing Fig. 5.14 to Fig. 5.9 the trending indices from the extracted signals show a better increasing pattern than

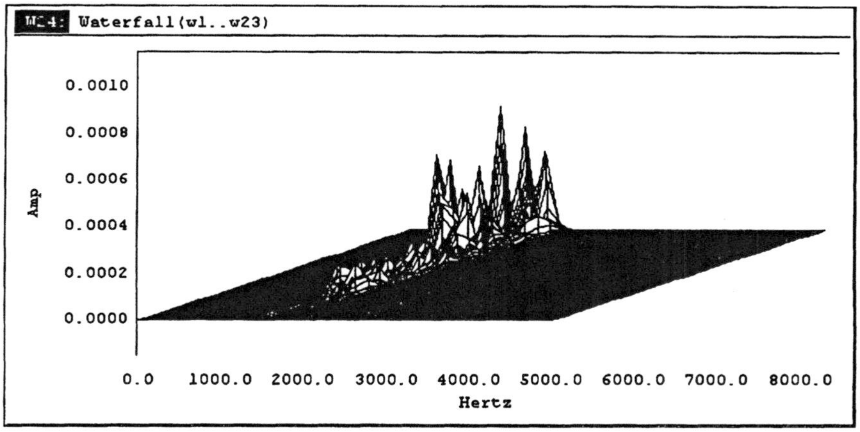

Fig. 5.13 Band pass PSD waterfall plot of the inner race defect bearing.

that of the original signals. This is because the most critical information related to the inner race deterioration had already been extracted from the raw data through the band pass filter. Thus, the vibration trending indices (RMS, Kurtosis, and MFRMS) were able to reflect the growth of the inner race defect better. In addition, when the increasing pattern of each index in Fig. 5.14 is compared, the performances of RMS and MFRMS indices appeared to be better than that of Kurtosis. In other words, RMS and MFRMS were more consistent with the physical change inside the bearing and were able to perform as more dominant factors in testing and training a CMAC network.

Based on the procedure of CMAC-PEM, the data shown in Fig. 5.14 were used to train and test a CMAC network. First, we defined the data sets collected at the beginning of the experiment as the normal condition of the test stand. They were then used to train a CMAC network. The remaining data sets were used as test vectors for the trained CMAC. The test results are shown in Fig. 5.15, where the Y-axis represents the estimated test stand performance level and X-axis is the running time. As expected, Fig. 5.15 depicts a decreasing trend since the test stand performance level was declining due to the growing inner race defect.

According to the assumptions of the CMAC-PEM methodology, Fig. 5.15 actually shows the estimated CMAC reliability measures, CMAC-$R(t)$, of the test stand. Using these measures, the underlying failure model, the Weibull model, and its model parameters β, θ, and c could be obtained. Table 5.2 lists the computed parameters of the model.

Finally, using the reliability functions in Equation 5.4, the estimated reliability value of the Weibull model, $R(t)$, the value of failure distribution function, $F_T(t)$, the value of probability density function, $f_T(t)$, and

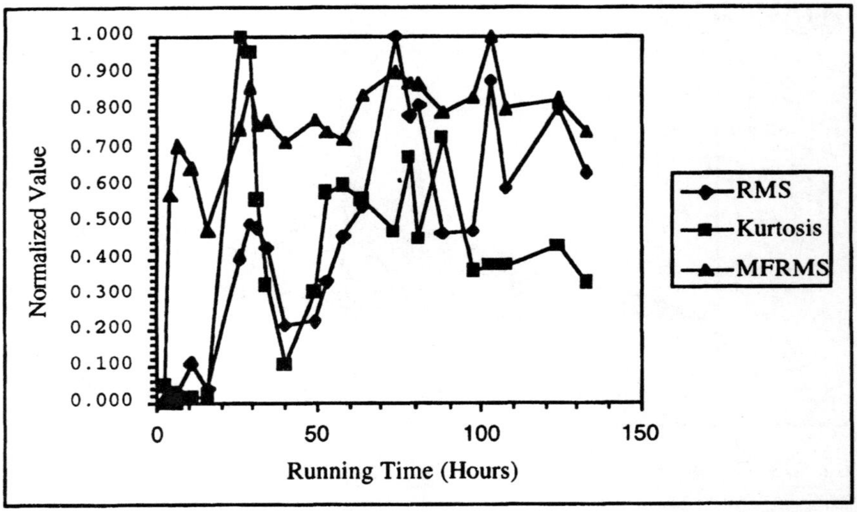

Fig. 5.14 Vibration trending indices plot of the test bearing band pass filtered signals.

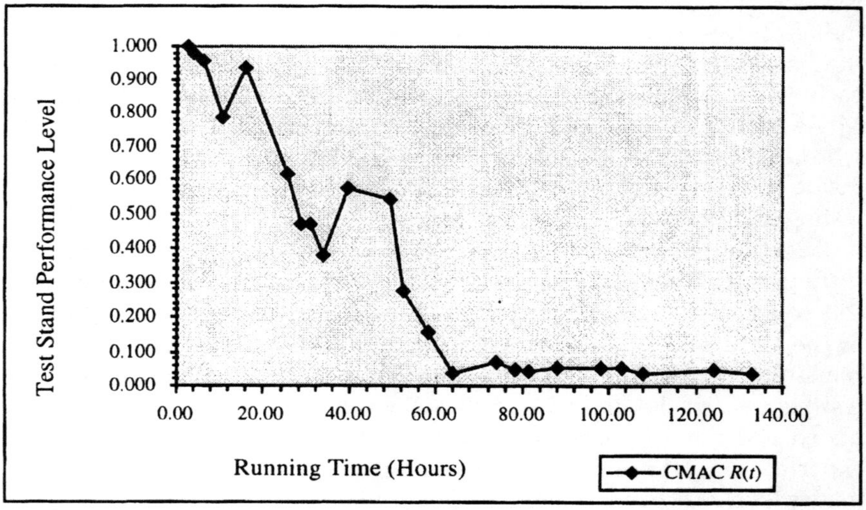

Fig. 5.15 CMAC reliability plot of the inner race defect bearing.

the hazard rate, $h(t)$, were calculated with respect to each specified running time. The final results are displayed in Fig. 5.16. These results provide useful statistical information related to the current condition of the test stand. The model estimated reliability curve indicates the predicted probability of the machine surviving over 150 hours is

Table 5.2 Weibull parameter estimates of the bearing data

Parameter	*Estimate*
β	2.1764
θ	44.1968
c	0

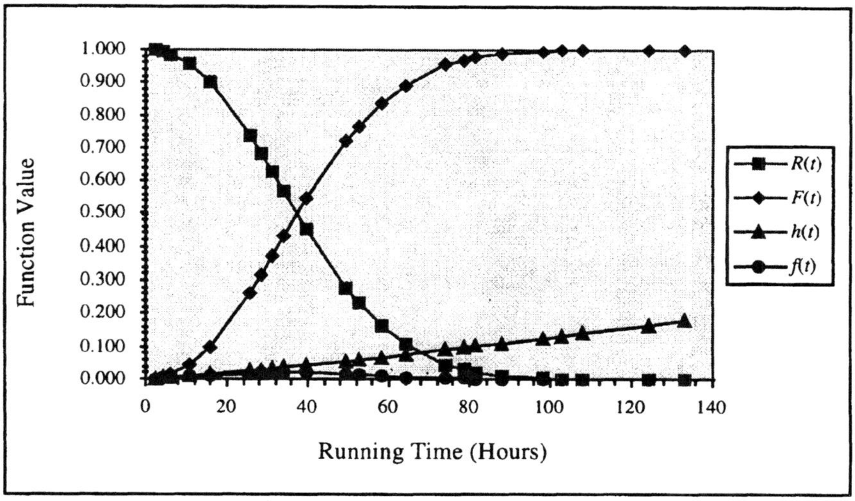

Fig. 5.16 Reliability functions plot of the inner race defect bearing.

approximately zero. In fact, during the experiment the test stand stopped running due to a broken shaft and the experiment was terminated at about 140 running hours.

Finally, the comparison plot of the CMAC reliability curve and the Weibull model calculated reliability curve is given in Fig. 5.17, where a great similarity between them is displayed. The model calculated reliability traces the CMAC estimated reliability very well. In order to see the goodness of fit for the estimated model, a Chi-Square test was performed. The test validated a significant model fit for 95% confidence level.

5.5 PROPORTIONAL HAZARDS MODEL: A VALIDATION TOOL

5.5.1 Introduction

In this section the proportional hazards model (PHM) is proposed as a validation tool to verify the results from CMAC-PEM. The concept of

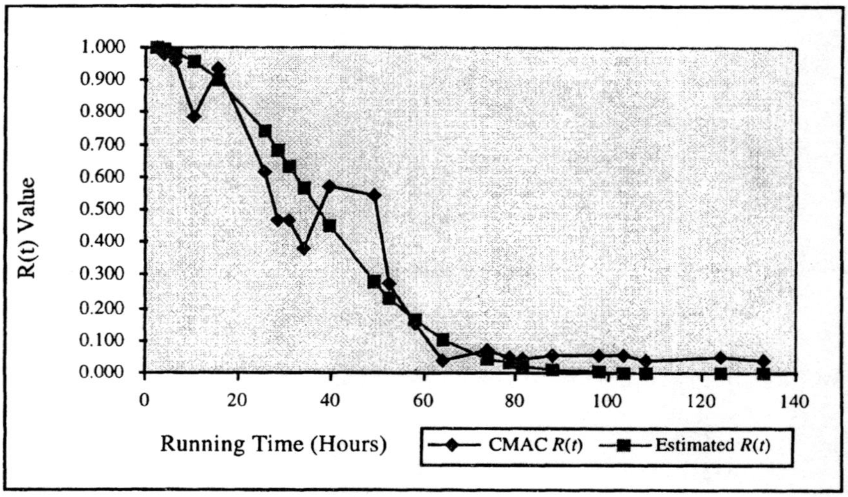

Fig. 5.17 Comparison of CMAC reliability curve and model estimated reliability curve.

hazard modeling is well known in maintenance literature as a method of modeling the process of age-dependent failures. However, a simple (purely time-dependent) hazard model is not always sufficient. Sometimes failures of a system not only correlate with its running time, but also with its diagnostic variables. Diagnostic variables are defined to be measurements reflecting conditions critical to system performance, such as the level of machine vibration and the level of metal particles in engine oil. These diagnostic variables, or so-called concomitant variables, should then be considered when making maintenance decisions. Modeling the effects of these concomitant variables has received considerable attention in research literature, particularly in medical journals.

Proportional hazards modeling, developed by Cox, extends the traditional hazard model to include correlational information supplied by concomitant variables (Cox 1972). It has been widely used in medical diagnostics and first applied to engineering reliability problems by Jardine and Anderson (1985, Jardine *et al.* 1987).

5.5.2 Proportional hazards model

The assumption of the proportional hazards model is the hazard rate of equipment is the product of a baseline hazard rate function $h_0(t)$, depending on equipment age, and a positive function Φ, depending only on the values of a set of concomitant variables. Thus if $\mathbf{z}$ is the vector of concomitant variables, the hazard (or failure) rate function is

$$h(t,\mathbf{z}) = h_0(t)\Phi(\mathbf{z}) \tag{5.7}$$

The most common form for $\Phi(\mathbf{z})$ is

$$\Phi(\mathbf{z}) = exp(y \cdot \mathbf{z}) \tag{5.8}$$

This is roughly the equivalent of the linear model in regression analysis. Alternative models are possible, but Equation (5.8) is the simplest model to constrain $\Phi(\mathbf{z})$ to be positive. Note that $\gamma \cdot \mathbf{z}$ is the dot product of the vectors γ and $\mathbf{z}$. In Equation (5.8) γ represents the coefficients vector of the separate concomitant variables used in the model. Cox suggests estimating the parameters of $\Phi(\mathbf{z})$, and γ, by maximizing an expression he calls conditional likelihood or partial likelihood (Cox 1972). This is the product of the conditional probability of failure of a system which has actually failed at time t, over all such failure times for this system. That is, if R_i is the set of indices to those values of concomitant variables which have not failed or been suspended (censored) just before time t_i, then

$$Li = h(t_i,\mathbf{z}_i) \Big/ \sum_{j \in R_i} h(t_i,\mathbf{z}_j) \tag{5.9}$$

is the required conditional probability for failure time t_i. Note that $\mathbf{z}_i$ is the set of concomitant variables when the system fails at time t_i, and $\mathbf{z}_j$ is the jth set of concomitant variables when the system survives at time t_i. The partial likelihood, L, is the product of these conditional probabilities taken over all failure times:

$$L = \prod_{i \in F} L_i \tag{5.10}$$

where F is the set of indices of all failure times. For the proportional hazard model, the conditional probability for time t_i as given as follows:

$$\begin{aligned} Li &= h_0(t_i) exp(\gamma \cdot \mathbf{z}_i) \Big/ \sum_{j \in R_i} h_0(t_i) exp(\gamma \cdot \mathbf{z}_j) \\ &= exp(\gamma \cdot \mathbf{z}_i) \Big/ \sum_{j \in R_i} exp(\gamma \cdot \mathbf{z}_j) \end{aligned} \tag{5.11}$$

Thus, the partial likelihood for the proportional hazards model is

$$L = \prod_{i \in F} \left[exp(\gamma \cdot \mathbf{z}_i) \Big/ \sum_{j \in R_i} exp(\gamma \cdot \mathbf{z}_j) \right] \tag{5.12}$$

By maximizing L in Equation (5.12) with respect to the parameters in γ, it is possible to estimate the parameters and investigate the effect of the concomitant variables in the model without making any assumption about the form of $h_0(t)$. The $h_0(t)$ is, by definition, the baseline hazard rate with a concomitant variable value of zero. In addition, to derive good results for γ, the variables in the vector $\mathbf{z}$ should be scaled so the average value of each variable is zero (Love and Guo 1991).

In the approach taken by Cox the baseline hazard rate (failure rate) is estimated directly from empirical data (Cox 1972). In the work of Jardine and Anderson the underlying baseline hazard rate is assumed to be Weibull distribution, and the empirical data is used to estimate the shape and scale parameters of the Weibull distribution alone with the coefficients of the concomitant variables (Jardine *et al.* 1987). In this research, the Weibull assumption is applied, and a Weibull-based failure rate is utilized as a baseline hazard rate.

5.5.3 Weibull proportional hazards model (WPHM)

The Weibull distribution has shown itself to be a robust model of time-to-failure for complex systems, and is flexible enough to be fitted to a diverse range of distributional shapes. The hazard rate of the Weibull distribution is given by

$$h_0(t) = \frac{\beta}{\theta}\left(\frac{t}{\theta}\right)^{\beta-1} \tag{5.13}$$

where β is the shape parameter and θ is the scale parameter. The Weibull proportional hazards model is

$$h(t,\mathbf{z}) = \frac{\beta}{\theta}\left(\frac{t}{\theta}\right)^{\beta-1} exp(\gamma\cdot\mathbf{z}) \tag{5.14}$$

Thus, the partial likelihood function for WPHM is specified as Equation (5.15) (Jardine *et al.* 1987).

$$L(\beta,\theta,\gamma) = \left[\prod_i \left(\frac{\beta}{\theta}\left(\frac{t_i}{\theta}\right)^{\beta-1} exp(\gamma\cdot\mathbf{z}_i)\right)\right]\cdot\left[\prod_j exp\left(-\left(\frac{t_j}{\theta}\right)^{\beta-1} exp(\gamma\cdot\mathbf{z}_j)\right)\right] \tag{5.15}$$

Then the parameters β, θ, and γ, maximizing Equation (5.15), are found. It is numerically more tractable to utilize the log of Equation (5.15) for determining the optimal parameters. Therefore the log-likelihood, as shown in Equation (5.16), is the quantity actually maximized.

$$l(\beta,\theta,\gamma) = N\,ln\left(\frac{\beta}{\theta}\right) + \sum_i ln\left(\frac{t_i}{\theta}\right)^{\beta-1} + \sum_i \gamma\cdot\mathbf{z}_i - \sum_j\left(\left(\frac{t_j}{\theta}\right)^{\beta-1} exp(\gamma\cdot\mathbf{z}_j)\right) \tag{5.16}$$

where N is the total number of failure times available. The maximum likelihood estimate is the set of parameter values which maximizes the log-likelihood function.

Once calculated, one can substitute the values of parameters β, θ, and γ into Equation (5.14) to find the proportional hazard rate. The cumulative hazard rate $H(t)$ may then be calculated as follows:

$$H(t) = \int_0^t h(t)dt \tag{5.17}$$

From the definition of Equation (5.2), the reliability $R(t)$ may be calculated as:

$$R(t) = exp(-H(t)) \tag{5.18}$$

WPHM provides another way to calculate the Weibull parameters, which may allow for comparison of its estimates to the results from CMAC-PEM. Therefore, it is used to verify and prove the capability and robustness of the methodology.

5.5.4 WPHM performance analysis of bearing deterioration data

To verify the results of CMAC-PEM in the previous section, the same data set was used again. The Weibull proportional hazards model was formed to fit the data set. The coefficients of the model (β, θ, and γ) may be found via maximum likelihood estimation. The model fitting can be implemented using the Newton-Raphson algorithm (Jardine *et al.* 1987). These routines are readily available in commercial FORTRAN and C libraries. However, the statistical package SAS was chosen. SAS provides a routine called LIFEREG, which can perform proportional hazards model fitting (SAS/STAT User's Guide 1989, p. 997). In addition, the routine performs a number of useful statistical tests on the parameter estimates, calculating standard errors and Chi-Square statistics, and if desired, covariance matrices.

The step-by-step implementation procedure is listed as follows:

1. Identify a set of machine condition measurements which is critical to the machine health. Here vibration trending indices, RMS, Kurtosis, and MFRMS are chosen as the WPHM concomitant variables in order to compare them with the performance of the CMAC-PEM method.
2. Trending indices vectors are formed for all available vibration data. Associated with each sample vector is the length of the test stand running time.
3. The Weibull proportional hazards model is fitted to the data and the model parameters are calculated through the LIFEREG routine in SAS.
4. The model may be refitted in order to improve the parameter estimates by dropping the non-statistical-significant concomitant variables.
5. Choose the estimated parameters from the best fitted model. Calculate the necessary reliability functions and compare results with estimates from the CMAC-PEM method.

Since we used three concomitant variables (RMS, Kurtosis, and MFRMS), the first model derived was:

$$h(t,\mathbf{z}) = \frac{\beta}{\theta}\left(\frac{t}{\theta}\right)^{\beta-1} exp(\gamma \cdot \mathbf{z})$$

$$= exp[\gamma_1(\text{RMS}) + \gamma_2\,(\text{Kurtosis}) + \gamma_3\,(\text{MFRMS})]\,\frac{\beta}{\theta}\left(\frac{t}{\theta}\right)^{\beta-1} \qquad (5.19)$$

where γ_1, γ_2, and γ_3 are concomitant parameters (coefficients) for the three trending indices and t is the running time variable. The analysis results of Model 1 are shown in Table 5.3. The Chi-Square test, provided by the LIFEREG output, is a goodness-of-fit test. The null hypothesis assumed by the LIFEREG routine is that the parameter has zero value. Table 5.3 shows the Chi-Square statistic for parameter γ_2 is not significantly larger than its approximate standard error. Therefore, the corresponding concomitant variable of γ_2 was removed from the model. The second model which removed the Kurtosis variable was then refitted to the data.

The second model is:

$$h(t,\mathbf{z}) = exp[\gamma_1(\text{RMS}) + \gamma_3\,(\text{MFRMS})]\,\frac{\beta}{\theta}\left(\frac{t}{\theta}\right)^{\beta-1} \qquad (5.20)$$

The analysis results of the second model in Table 5.4 shows the Chi-Square statistics are now significant for all parameters indicating both RMS and MFRMS concomitant variables are significant in fitting the model.

In addition, the third model, which contained only the RMS concomitant variable, was tested. Equation (5.21) shows the refitted Weibull proportional hazards model. Its analysis results are given in Table 5.5.

Table 5.3 Weibell proportional hazards model parameter estimates, Model 1

Parameter	*Estimate*	*Std error*	χ^2	$Pr > \chi^2$
γ_1	0.9090	0.3130	8.4341	0.0037
γ_2	−0.0985	0.2696	0.1336	0.7148
γ_3	1.3501	0.4731	8.1428	0.0043
β	2.0961	0.0796	–	–
θ	50.1005	0.1051	1387.49	0.0001

Table 5.4 Weibell proportional hazards model parameter estimates, Model 2

Parameter	*Estimate*	*Std error*	χ^2	$Pr > \chi^2$
γ_1	0.8860	0.3039	8.5004	0.0036
γ_3	1.2951	0.4560	8.0672	0.0045
β	2.0820	0.0797	–	–
θ	50.0987	0.1058	1367.98	0.0001

Table 5.5 Weibell proportional hazards model parameter estimates, Model 3

Parameter	*Estimate*	*Std error*	χ^2	$Pr > \chi^2$
γ_1	1.3260	0.2758	23.1147	0.0001
β	1.8874	0.0886	–	–
θ	51.7406	0.1166	1145.02	0.0001

Table 5.6 Summary of Weibull proportional hazards modeling for the bearing data

	Concomitant parameters			*Weibull parameters*		
	RMS (γ_1)	*Kurtosis* (γ_2)	*MFRMS* (γ_3)	β	θ	*L. L.**
Model 1	0.9090	–0.0985	1.3501	2.0961	50.1005	–19.201
Model 2	0.8860	–	1.2951	2.0820	50.0987	–19.265
Model 3	1.3260	–	–	1.8874	51.7405	–21.695

* Log likelihood

$$h(t,\mathbf{z}) = exp[\gamma_1(\text{RMS})]\,\frac{\beta}{\theta}\left(\frac{t}{\theta}\right)^{\beta-1} \tag{5.21}$$

Table 5.6 summarizes the Weibull proportional hazards modeling results for the test bearing data. From the table one can see that although the Kurtosis variable has been removed from the first model to form the second model, the estimated parameters are not significantly different between Model 1 and Model 2. Also, the values of log likelihood (L. L.) are shown to be slightly reduced, from –19.201 to –19.265, in fact, indicating that Kurtosis is a redundant variable in the model. In other words, RMS and MFRMS are the dominant variables. Thus, there is no need to include the Kurtosis index in the model. This conclusion is consistent with the findings shown in Fig. 5.14, where the RMS and MFRMS plots show better increasing trends to reflect the physical changes inside the bearing than that of Kurtosis. When both Kurtosis and MFRMS indices are not in the model the log likelihood value drops 2.43 points from –19.265 in Model 2 to –21.695 in Model 3, indicating the parameter estimation does not improve by dropping MFRMS. Also, from Model 2 to Model 3 the values of parameters change significantly (see γ_1 in Table 5.6), showing the important role of MFRMS in the model. Therefore, the best model to fit the test stand data would be Model 2.

There are two purposes in implementing WPHM to the bearing data.

The first one is to investigate the individual influences of each concomitant variable in the model. The second is to verify the proposed CMAC-PEM methodology for machine reliability estimation. Table 5.7 lists the Weibull parameters estimated from Model 2 and CMAC-PEM. It shows the calculated parameter values are close. For the slope parameter, β, the two estimates are only slightly different. For the scale parameter, θ, the estimate from CMAC-PEM is 44.1968, which is less than the 50.0987 from WPHM. To visualize the differences the reliability and hazard rate curves from both methods are given in Fig. 5.18 and Fig. 5.19, respectively. Note that in the Weibull proportional hazards model RMS and MFRMS are variables – not constants. In order to plot the curves, the RMS and MFRMS values are assumed to be zero. That is, Equation (5.13) (the baseline Weibull hazard rate) was used instead of (5.14) in calculat-

Table 5.7 Weibull parameter estimates for CMAC-PEM and WPHM

Parameter	*CMAC-PEM estimate*	*WPHM estimate*
β	2.1765	2.0820
θ	44.1968	50.0987
c	0	0

* From Model 2

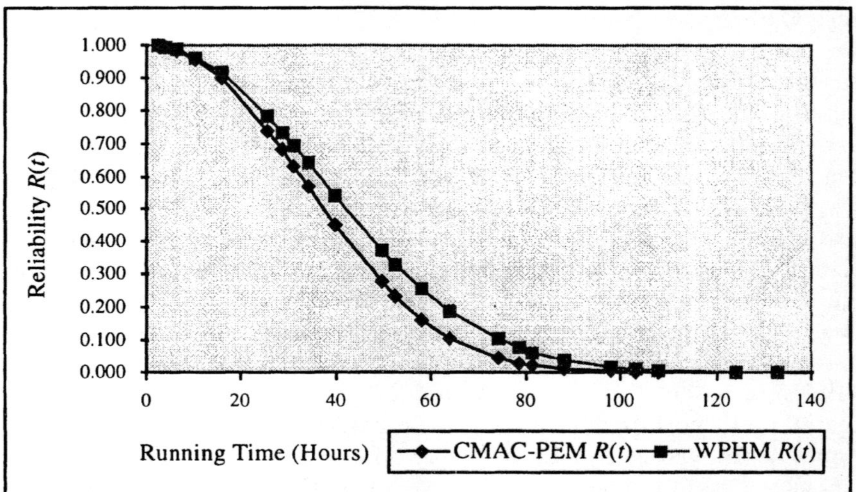

Fig. 5.18 Reliability curves from the CMAC-PEM and the WPHM methods.

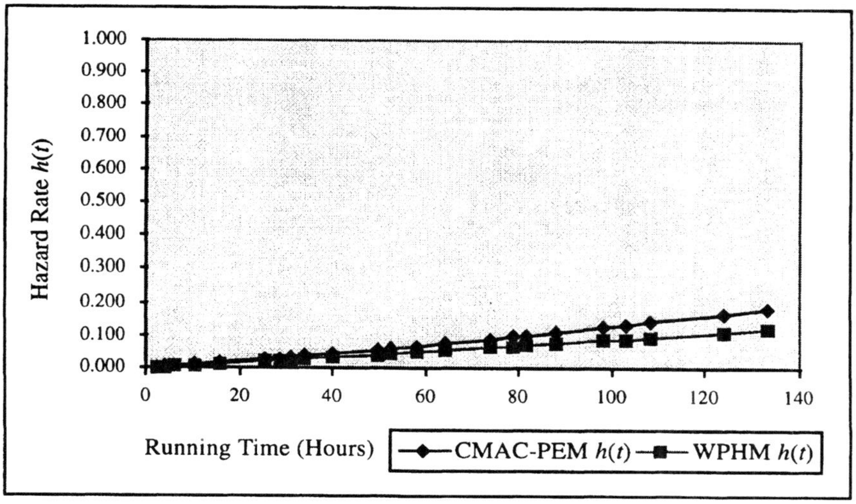

Fig. 5.19 Hazard rate curves from the CMAC-PEM and the WPHM methods.

ing the WPHM reliability and hazard rate curves in Figs 5.18 and 5.19. In fact, these two figures compare the underlying Weibull failure models estimated from the CMAC-PEM method and the WPHM method. The effects of concomitant variables have been removed in these two comparison plots.

Figure 5.20 displays the original CMAC estimated reliability measures and two boundary reliability curves when extreme concomitant variable values were specified. Here, Equation 5.14 was used to calculate the WPHM reliability curves. The U-WPHM-$R(t)$ curve was obtained by assuming the RMS and MFRMS values were from the initial state of the test stand. In contrast, the L-WPHM-$R(t)$ curve was calculated by using the last state RMS and MFRMS readings, which were the values taken before the shaft broke. These two curves represent the worst- and best-case reliability curves for the test stand used in the WPHM method. Note that most CMAC estimated reliability values are located inside the area covered by the two boundary WPHM reliability curves.

By examining Fig. 5.20 one can see the CMAC estimated reliability measures are inside the reliability area calculated from the Weibull proportional hazards model. These results not only prove the useability, but also verify the robustness of the CMAC-PEM methodology in estimating machine reliability. The sensitivity analysis of the model parameters and the concomitant variables is discussed in the next section.

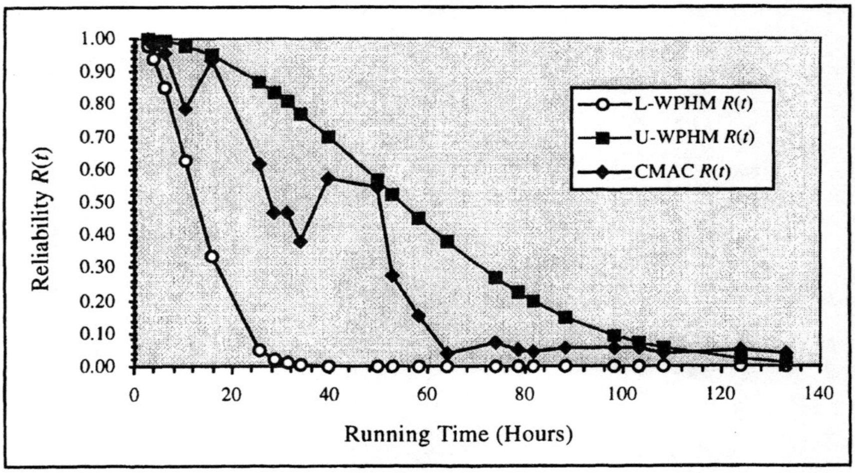

Fig. 5.20 CMAC reliability curve and two boundary WPHM reliability curves.

5.5.5 Sensitivity analysis of WPHM parameters

In this section the influence of each parameter and concomitant variable on reliability curve and hazard rate for WPHM have been examined. Several analysis schemes have been performed. Figure 5.21 shows the effects on the reliability curve of changing β (slope parameter) and θ (scale parameter). The first graph of Fig. 5.21 shows that the slope parameter controls the shape of the reliability curve. When β is increased and θ is kept constant, the shape of $R(t)$ changes from Curve 1 to Curve 5. In the figure Curve 3 and Curve 4 represent the β values estimated from CMAC-PEM and WPHM, respectively. They are close to each other. One special feature of the Weibull distribution should be noted. The intersections of Curve 1 to Curve 5 is at time 44.20, which is the exact constant value of θ.

The second graph of Fig. 5.21 shows keeping β as a constant while increasing θ shifts $R(t)$ to the right. This shifting increases the overall reliability of the system. In other words, the higher the scale parameter is, the longer the life of the system. Thus, θ is also named as the characteristic life. In addition, this graph also explains the $R(t)$ curve deviation between the CMAC-PEM method and the WPHM method (see Fig. 5.18). The characteristic life calculated from CMAC-PEM is shorter than that of the WPHM estimation, which makes the CMAC-PEM $R(t)$ curve shift to the left-hand side of the WPHM $R(t)$ curve.

Figure 5.22 illustrates the same analysis for hazard rate. The first graph of Fig. 5.22 shows that when one increases β and keeps θ as a constant, the shape of $h(t)$ changes from Curve 1 to Curve 5. In fact, this

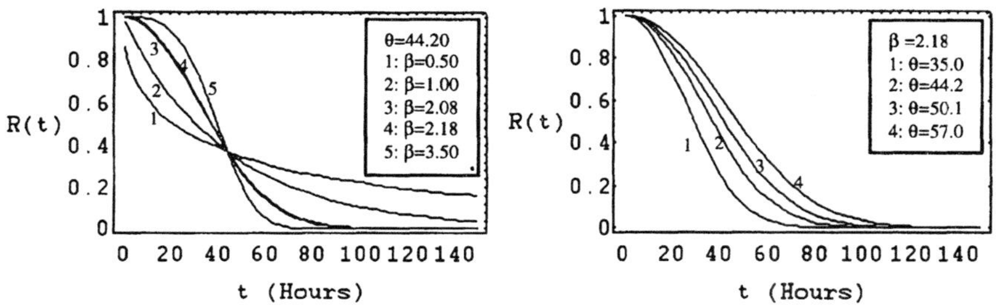

Fig. 5.21 Parameter sensitivity analysis for reliability, $R(t)$.

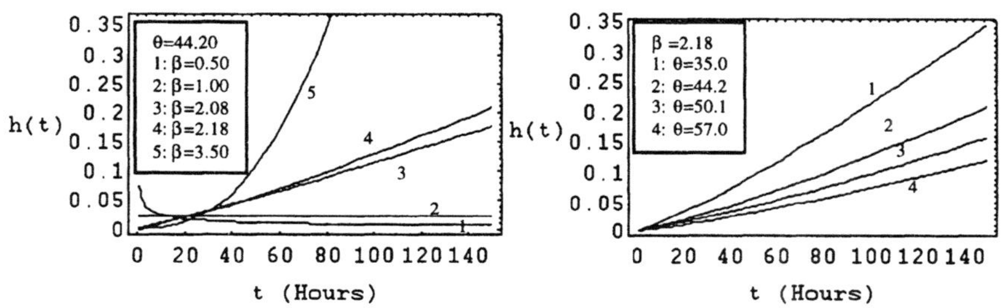

Fig. 5.22 Parameter sensitivity analysis for hazard rate, $h(t)$.

graph illustrates the basic components of the 'bath-tub' curve. When β is less than 1, represented by Curve 1, the system has a decreasing hazard rate. When β equals 1, represented by Curve 2, the system has a constant hazard rate. When β is larger than 1, represented by Curve 3, 4, and 5, the system has an increasing hazard rate.

The second graph of Fig. 5.22 shows that keeping a constant β while increasing θ decreases the slope of the $h(t)$ curve. This explains the result shown in Fig. 5.19, where the deviation between the two $h(t)$ curves basically comes from the difference of θ values.

The last sensitivity analysis results are shown in Fig. 5.23, in which the effects of concomitant variables in the WPHM method are investigated. Both graphs in Fig. 5.23 show that when one variable is constant and the other one is increasing the $R(t)$ curve shifts to the left. Increasing the concomitant variable value, in this case, corresponds to increasing the test stand vibration. This means decreasing the overall reliability of the system. For instance, when the RMS reading increases from 0.038 to 0.071 the chances of the test stand surviving over 60 running hours decreases from approximately 45% to 10%. For the $R(t)$ curve of WPHM in Fig. 5.18, the RMS and MFRMS values are assumed to be zero which is just one special case. The RMS and MFRMS readings in Weibull proportional

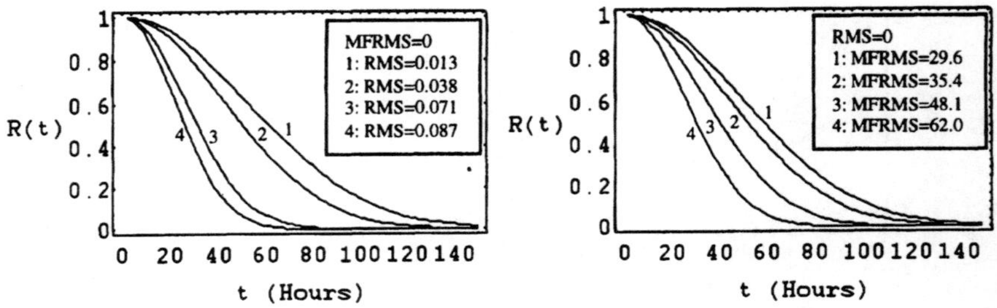

Fig. 5.23 Concomitant variable sensitivity analysis for reliability curve.

hazards model are variables – not constants. Therefore, a better way to illustrate the reliability values for the proportional hazards model is to actually use an area instead of a curve. The example given in Fig. 5.20 shows two boundary $R(t)$ curves tend to wrap the possible $R(t)$ area.

5.6 SUMMARY

The major content presented in this chapter is the proposed CMAC-PEM methodology for machine reliability modeling. Our objective is to combine the traditional reliability modeling method and new vibration-based machine condition monitoring techniques to estimate machine reliability measures. In this chapter a CMAC neural network-based machine performance estimation model is introduced. CMAC-PEM is used to fuse sensory data and to predict machine performance. Then, by using the outputs from CMAC-PEM as overall reliability measures, the underlying Weibull reliability model of the machine can be established. The model provided several reliability statistics, which may be used as basic information for predictive maintenance. A machine vibration imbalance example is provided in the chapter to illustrate the CMAC-PEM methodology. Furthermore, a bearing deterioration experiment was conducted to generate a real-world data set to test the robustness of CMAC-PEM. Finally, a validation tool, the Weibull proportional hazards model (WPHM), has been used to verify the CMAC-PEM results. In-depth analyses of WPHM on the bearing data have been given at the end of the chapter, including sensibility analyses of the WPHM parameters and concomitant variables on the reliability and hazard rate curves.

From this chapter, the author concludes the CMAC-PEM methodology is a reliable and robust system for on-line machine reliability analysis. As described in the chapter, several advantages of CMAC-PEM may outperform other methods and overcome the shortcomings of current methodologies (e.g. complex computation of WPHM). Most importantly, by combining CMAC-PEM and CMDS, one may realize the concept of condition-based predictive maintenance, which will surely play an important role in the fast-changing modern manufacturing environment.

REFERENCES

Albus, J.S. (1975a) A new approach to manipulator control: the cerebellar model articulation controller (CMAC). *Transactions of the ASME Journal of Dynamic Systems, Measurement, and Control*, **97**, September, 220–7.

Albus, J.S. (1975b) Data storage in the cerebellar model articulation controller (CMAC). *Transactions of the ASME Journal of Dynamic Systems, Measurement, and Control*, **97**, September, 228–33.

Albus, J.S. (1981) *Brain, Behavior and Robotics*, BYTE Books, Peterborough, NH.

Archambault, R., and Fillion, R. (1991) Formal tools for improving decisions on the optimal use of machinery. *Proceedings of the 3rd International Machinery Monitoring and Diagnostics Conference*, Las Vegas, Nevada, December, 366–74.

Bloch, H.P., and Geitner, F.K. (1983) *Machinery Failure Analysis and Trouble Shooting*, Gulf Publishing Co.

Braun, S. (1986) *Mechanical Signature Analysis: Theory and Applications*, Academic Press Inc., London.

Cox, D.R. (1972) Regression models and life tables (with discussion). *Journal of the Royal Statistical Society*, Series B, **26**, 187–220.

Cox, D.R., and Lewis, P.A.W. (1966) *The Statistical Analysis of Series of Events*, John Wiley and Sons, New York.

Dyer, D., and Stewart, R.M. (1978) Detection of rolling element bearing damage by statistical vibration analysis. *Transactions of the ASME Journal of Mechanical Design*, **100**, April, 229–35.

Jardine, A.K.S., and Anderson, P.M. (1985) Use of concomitant variables for reliability estimation. *Maintenance Management International*, **5**, 135–40.

Jardine, A.K.S., Anderson, P.M., and Mann, D.S. (1987) Application of the Weibull proportional hazard model to aircraft and marine engine failure data. *Quality and Reliability Engineering International*, **3**, 77–82.

Kelly, A. (1989) *Maintenance and Its Management*, Conference Communication, England.

Lee, J., and Kramer, B.M. (1993) Analysis of machine degradation using a neural network based pattern discrimination model. *Journal of Manufacturing Systems*, **12**(2), 379–87.

Love, C.E., and Guo, R. (1991) Using proportional hazard modeling in plant maintenance. *Quality and Reliability Engineering International*, **7**, 7–17.

Mathew, J. (1989) Monitoring the vibrations of rotating machine elements – an overview. *ASME*, **DE Vol. 18–5**, 15–22.

Mathew, J., and Alfredson, R.J. (1984) The condition monitoring of rolling element bearings using vibration analysis. *Journal of Vibration, Acoustics, Stress, and Reliability in Design*, **106**, July, 447–53.

Mechefske, C.K., and Mathew, J. (1991) Trending the gradual deterioration of low speed rolling element bearings using parametric spectra. *Proceedings of the 3rd International Machinery Monitoring and Diagnostics Conference*, Las Vegas, Nevada, December, 232–320.

Monk, R. (1972) Vibration measurement gives early warning of mechanical faults. *Process Engineering*, November, 135–7.

National Research Council (1992) *The Competitive Edge*, National Academy Press, Washington, DC.

SAS Institute (1989) *SAS/STAT* User's Guide, Version 6, 4th edn, vol. 2, SAS Institute Inc., Cary, NC.

Wong, Y., and Sideris, A. (1992), Learning convergence in the cerebellar model articulation controller. *IEEE Transactions on Neural Networks*, **3**(1), January, 115–120.

CHAPTER 6

Design methodology for self-maintenance machines

Yasushi Umeda, Tetsuo Tomiyama, Tomohiko Sakao and Yoshiki Shimomura

This chapter describes a design methodology for self-maintenance machines. The self-maintenance machine is a machine that can maintain its functions for a while, even though faults happen. In contrast, a traditional machine may not work at all in such a case. The self-maintenance feature increases the availability of the machine, improves the fault-tolerance, and reduces the maintenance costs. For instance it may reduce sudden, unscheduled maintenance. We illustrate the concepts, reasoning methods, and practical examples of self-maintenance machines. First, we discuss the concept of functional maintenance that includes two types of self-maintenance strategies, i.e. control type and function redundant type. Second, a model-based approach that employs qualitative physics is proposed, as the reasoning technique. Third, we illustrate two examples of self-maintenance machines that we have developed; namely, a commercial photocopier with the control-type self-maintenance capability and a prototype of the function-redundant self-maintenance photocopier.

6.1 INTRODUCTION

In this age of automation and computerization, it is remarkable to notice that maintenance tasks are hardly automated and highly laborious. In addition, it is becoming more crucial that machines do not break for safety reasons and possible economic and social damage and, even if they do, they can be repaired quickly. Despite design methodologies for higher reliability and fault-tolerance that aim at fewer faults, once a fault happens machines may cease to operate suddenly. To avoid this unacceptable situation of 'hard-failure,' we should develop soft-fail machines (as opposed to hard-fail) that can still perform their functions even if faults happen. In this article, we propose new design methodologies for self-maintenance machines that exhibit soft-fail features. The concept of soft-failure includes robustness, but it is beyond robustness in that soft-fail machines can operate even with some faults. In this sense, soft-fail machines are fault-tolerant, but not just fault-tolerant; they are functionally fault-tolerant as discussed next.

Here, we introduce the concept of functional maintenance as opposed to physical maintenance. Self-maintenance machines we describe in this chapter exhibit soft-fail features based on functional maintenance.

Functional maintenance aims at recovering the required functions of a machine through trading off functions, whereas traditional repair (physical maintenance) aims at recovering an initial physical state by replacing faulty components, cleaning, etc. It is based on the fact that some physical states besides the initial state of the machine can perform its required functions, though with a low performance.

Therefore, functional maintenance is a more flexible repair strategy than physical maintenance. Functional maintenance is an effective strategy for an industry in which a service engineer will regularly visit customers on-site and maintain machines in a steady state. A typical example of such an industry is the photocopier industry. Suppose a photocopier malfunctions on Friday evening. The service engineer will not (or cannot) show up until Monday morning, but some important documents must be copied immediately. A self-maintenance photocopier will maintain its function 'to make copies' by some means, but it might work slower or the copy quality might be slightly worse until service takes place. The self-maintenance machine, therefore, can reduce sudden, unscheduled visits of service engineers, which will greatly reduce the maintenance costs, while minimizing damage to the customers' satisfaction.

Our approach toward the self-maintenance machine is to add intelligence to the machine, clever enough for functional maintenance. In other words, we append self-maintainability to existing machines as an additional embedded reasoning system.

In this chapter, we discuss concepts of the self-maintenance machines, their reasoning technology, and practical examples. First, section 6.2.1 proposes two types of repair strategies; namely, control type and function-redundant type. Second, for automating fault diagnosing and repair planning, a model-based reasoning technique that employs qualitative physics is illustrated in sections 6.2.2 and 6.2.3. Third, section 6.3 verifies the feasibility and advantages of the self-maintenance machine with two examples; namely, a commercial photocopier that has the control-type self-maintainability and a prototype of the function redundance. Finally, section 6.4 concludes this chapter.

6.2 METHODOLOGY

6.2.1 Fundamentals of the self-maintenance machine

The self-maintenance machine should monitor and diagnose itself. And based on the obtained diagnostic information it must reason about a plant for repair and execute the repair by itself. Therefore, the self-maintenance machine should be equipped with monitoring capability, fault judging capability, diagnosing capability, repair planning capability, and repair executing capability, besides its required main functions.

In order to achieve these five capabilities, we developed an architecture for the self-maintenance machine depicted in Fig. 6.1. In this architecture, the monitoring and repair executing capabilities are given by sensors and actuators, respectively. The embedded reasoning system gives the fault judging, diagnosing, and repair planning capabilities. It should be noted that an existing mechatronics machine such as a photocopier is already equipped with sensors, actuators, and a CPU for controlling purposes. This suggests that we can build a self-maintenance machine just with an additional reasoning system for functional maintenance.

However, this strategy of building self-maintenance machines with additional reasoning capabilities has natural limitations and cannot be perfect. First, the repair operations are limited to those which do not require spare components. This also restricts the kinds of faults that can be repaired. Second, repairs performed by the self-maintenance machine are more or less temporal measures. The self-maintenance machine performs self-maintainability *for a while*, and after a while, it requires repairs by engineers.

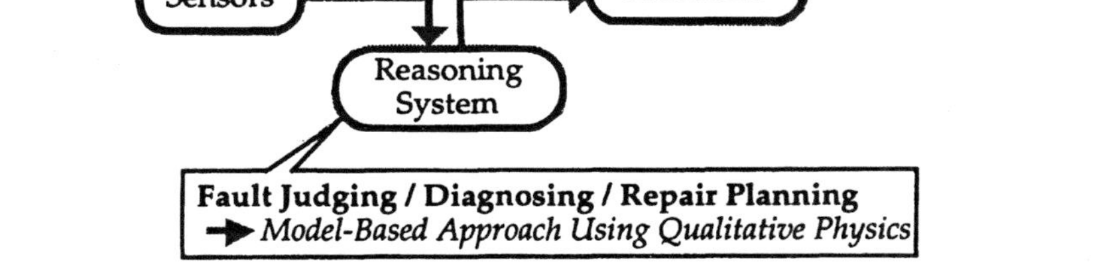

Fig. 6.1 Fundamental architecture of the self-maintenance machine.

Considering these limitations, we propose the following two types of repair strategies for functional maintenance to build a machine that is still functionally operational, although physically it is worn out.

Control type

Some repairs can be accomplished by adjusting a machine's state without changing or reorganizing the structure. This type of repair is achieved by controlling actuators to recover a part or whole of the requested functions, when, for example, deterioration of a component happens.

Function redundancy type

If a machine can reconfigure its behavior to maintain its functions, it would be more tolerant against faults. Traditionally, this has been achieved by component redundancy, which adds redundant components that will replace faulty ones. However, component redundancy often results in undesirable cost, weight, and complexity of the machine [1]. We propose a new repair strategy, called function redundancy, which uses potential functions of components in a slightly different way from the original design to back up and maintain the main functions [2].

Function redundancy can be best explained by an example of an automobile with a manual transmission. Even when its engine does not work, in an emergency the car might operate for a while with its starting motor and battery. While the starting motor has an original function of starting the engine, it has a redundant function of operating the car for a while. This is reconfiguration of behavior, rather than reconfiguration of a structure or adjustment of states. In order for a machine to reconfigure its behavior to maintain the most important functions while sacrificing less important functions during a fault, we need a new design methodology. For this purpose, we have developed a function modeling system called the Function Redundancy Designer [2, 3].[1]

6.2.2 Basic reasoning techniques

In this section, we discuss the methods for judging faults, diagnosing, and planning repair (see [4, 5] for more details). Figures 6.2 and 6.3 show the entire reasoning algorithm and the architecture of the reasoning system, respectively, which are the model-based approach.

[1] We offer the FBS Modeler and the Function Redundancy Designer described in this section as free software. If you are interested, please contact umeda@zzz.pe.u-tokyo.ac.jp by e-mail.

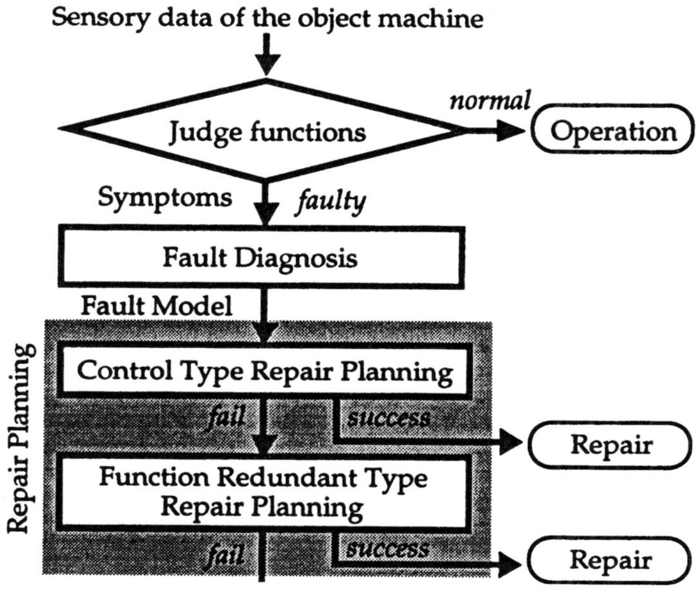

Fig. 6.2 Reasoning algorithm for the self-maintenance machine.

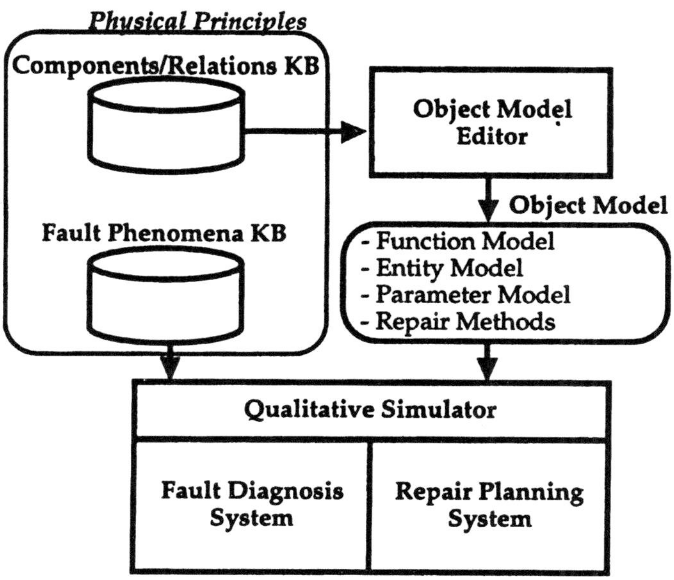

Fig. 6.3 Architecture of the reasoning system.

Fault diagnostic systems employing the model-based approach [6] have the following advantages against conventional heuristics-based expert systems.

- *Reusability of knowledge* Because knowledge dependent on the object machine is localized and isolated in a model, the system can deal with different object machines by changing object models.
- *Robustness of the reasoning system* Applying generic physical principles to the model allows the system to deal with unknown cases that are not described explicitly.
- *Knowledge acquisition* Since models can be prepared from design information and the knowledge about physical principles can be acquired systematically, it is easier to acquire knowledge for the model-based reasoning system than knowledge for the heuristics-based reasoning one.

Qualitative physics [7] plays an important role in the model-based approach as a framework for modeling and reasoning. In this section, we describe our model-based approach that not only diagnoses faults but also generates repair plans.

Knowledge representation

In order to judge and diagnose faults, and to plan repair, the reasoning system requires several kinds of knowledge including functions of the machine, structure, and physical characteristics of the object machine, repair methods that can be executed by the machine, and fault phenomena that might occur on the machine.

We classify these kinds of knowledge into two categories. One is a model of the object machine (*object model*) (see 1, 2, and 3 below) and the other is knowledge about physical principles that govern behaviors of the machine (4). This classification dictates the advantages of the model-based approach mentioned above. The following explains the knowledge employed by our reasoning system in greater detail.

Object model (see Figure 6.4)

1. Function model. The function model represents the functional hierarchy of the object machine. Each function in the function hierarchy has its parametric condition used for judging whether or not the function is performed.
2. Entity model. The entity model represents the topological structure of components of the object machine. This formalization is based on device-oriented qualitative physics [8].
3. Parameter model. The knowledge about components and relations described below (5) is applied to the entity model (2) to generate a

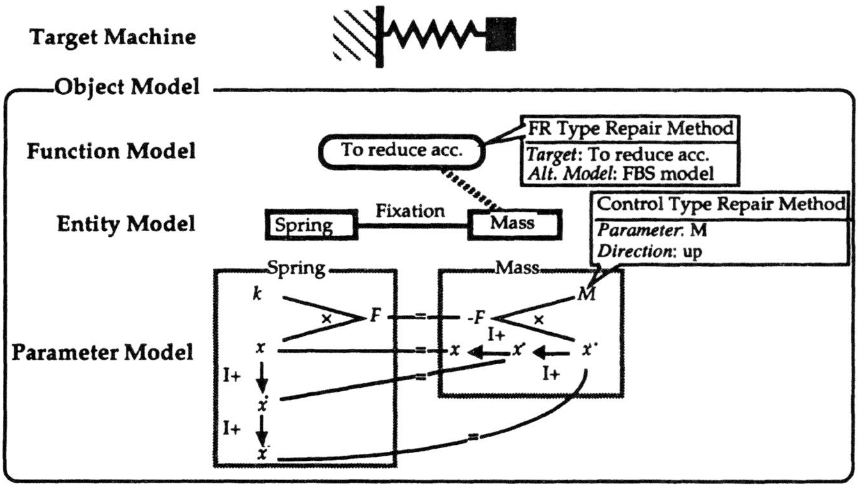

Fig. 6.4 Example of the object model.

network of parameters that represents physical characteristics of the object machine, called a *parameter model*. Since a parameter model describes mathematical relationships among parameters, the reasoning system can be constructed independently of domains.

4. Repair methods. Repair methods represent executable repair methods in the machine. While a control-type repair method means adjustment of an actuator, a function redundancy-type repair method activates a redundant function mode represented by an *alternative model* in which the behavior of the machine is different from the normal model.

Knowledge about physical principles

5. Knowledge about components and relations. This knowledge describes characteristics of components and relations among them in the form of qualitative differential equations. Since this knowledge is generally reusable, the user can easily construct a object model of the machine. The Object Model Editor in Fig. 6.3 supports this process.
6. Knowledge about physical phenomena. We distinguish fault phenomena associated with structural changes of the object machine and ones related to only parametric changes of the differential equations that describe the machine's behavior. An example of the former is a damage of a gear pair, and the latter is exemplified by an abnormal rise of boiler temperature. While the latter does not require special treatment, the former cannot be described in the knowledge about components and relations (5), because the differential equations them-

selves discretely change. We describe both kinds of fault phenomena based on the idea of process in Qualitative Process Theory [9]. For example, a gear breaking phenomenon is described as follows:

Phenomenon
Name: Gear Breaking
Condition: Exist(gear1, Gear) & gear1.torque > Max
Effect: Remove(gear1)

The reasoning method for diagnosis

We define a *fault* as a change of structure or of an attribute of the object machine, a *symptom* as a function that is not performed due to the fault, and a *fault cause* as a physical phenomenon that causes the fault. We diagnose faults by reasoning out physical phenomena that cause the symptoms. The reasoning method for diagnosis consists of the following four steps and is explained by a parameter model depicted in Fig. 6.5 as an example. This method can diagnose multiple faults that are causally related and, therefore, can deal with more complex situations than the traditional single fault assumption [10] can. We call this assumption *single fault process assumption.*

1. Identifying fault symptoms. The reasoning system identifies symptoms by comparing sensory data with the function model. Suppose for Fig. 6.5 increase of parameter A, which causes failure of a function, was detected as a symptom by a sensor.
2. Searching for fault cause candidates. First, the reasoning system searches all changes of parameters and equations that are causally related to the symptom. For example, for increase of A, the following changes can be listed:

   increase of B, C, D, or decrease of E; or
   changes in the equations of $A = B + C$ or $B = D - E$.

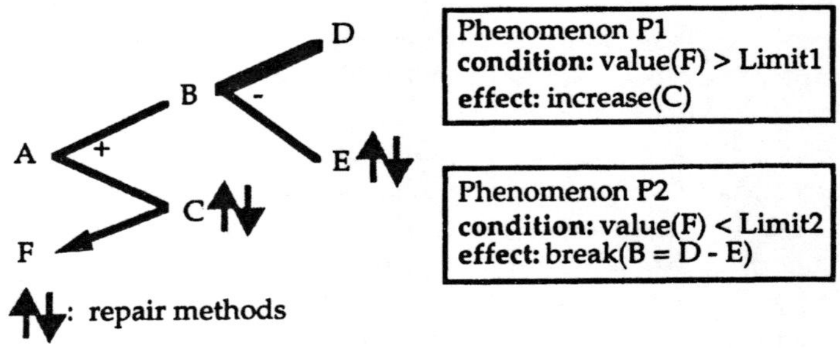

Fig. 6.5 Example of the parameter model.

Second, the system finds all phenomena of which effects include the above changes by searching through the knowledge about physical phenomena. In the example of Fig. 6.5, phenomena *P1* and *P2* are selected. In this way, the reasoning system generates all fault cause candidates.

3. Fault simulation. Next, for each fault cause candidate the system simulates behavior of the machine and builds up its faulty state (*fault model*). This is done by a qualitative simulator based on the extension of the QSIM algorithm [11] that can deal with structural changes caused by phenomena. The fault model provides the repair planning procedure with the information about the structure and the state of the faulty object machine.
4. Identifying the fault cause candidate. The system identifies the most plausible fault cause by selecting a fault model that does not contradict and matches the most with the sensory data.

The reasoning method for repair planning

Control-type repair planning
The reasoning system first tries to recover functions by applying the control-type repair methods. If they fail, the function redundancy repair methods are tried. The control-type repair planning procedure consists of the following five steps.

1. Deciding the repair objective. After the user determines the fault symptom to be removed through functional trade-off, the reasoning system decides the direction of change of the function parameter. For instance, in Fig. 6.5, if function parameter *A* is currently increasing, the objective of repair is to decrease the value of parameter *A*.
2. Generating repair candidates. The system picks up possible repair methods of which parameters and directions can achieve the repair objective by searching through the fault model. In the example against increase of *A*, decrease of *C* and increase of *E* are selected as repair candidates.
3. Repair simulation. The qualitative simulator simulates behavior of the object machine for each repair candidate using fault models with increased or decreased parameters of the repair candidates. As a result, the reasoning system can check whether or not each repair candidate can recover the objective function. If impossible, the repair candidate is excluded from consideration. Then, the reasoning system also checks whether or not the selected repair candidate will cause side effects.
4. Selecting a repair method. From the repair candidates, the reasoning system selects a repair method that can repair or improve the objective function with the least number of side effects.

5. Repair planning for side effects. If side effects cannot be avoided, the system generates secondary repair plans for each side effect by executing repair planning recursively. As a result, we obtain a repair plan as a sequence of one or more repair methods.

Function redundant-type repair planning
When the control-type repair fails to execute, the function redundant-type repair method is used.

1. Selecting a repair method. First, the system selects a function redundancy-type repair method of which the target function is the function to be repaired and of which the alternative model does not use the faulty component by comparing the alternative model with the fault model.
2. Reconstructing the object model. The system reconstructs a new object model by using the alternative model described in the selected repair method, because this type causes structural changes of the object.
3. Executing control-type repair planning. The system conducts the control-type repair planning with this reconstructed object model for parametric tuning.

6.2.3 Reasoning techniques for commercial version of the self-maintenance photocopier

As illustrated in section 6.3.1 below, we developed a commercial product of a self-maintenance photocopier that employs the control-type repair strategy. However, in the course of its development, the following problems were discovered when applying the techniques described in the previous section to the commercial machine as embedded software.

1. Both the size and the speed of the prototype reasoning system were unacceptable as embedded software.
2. The prototype reasoning system used a qualitative space mapping for converting numerical sensory data into qualitative values (see Fig. 6.6). However, in the real world, it is not easy to obtain the mapping statically because of inaccuracy and instability of the sensors and the constantly changing environment around the machine. Incorrect mapping results in mistakes in fault diagnosis.
3. The qualitative representation of states of the machine is too rough to judge whether a repair operation is correctly executed or not.

To solve these problems, we developed the following new techniques: namely, virtual case based reasoning for problem 1, *fuzzification of qualitative values* and the *imitational fault method* for problem 2, and *division of the quantity space* and *job elements* for problem 3 (see [12] for more details).

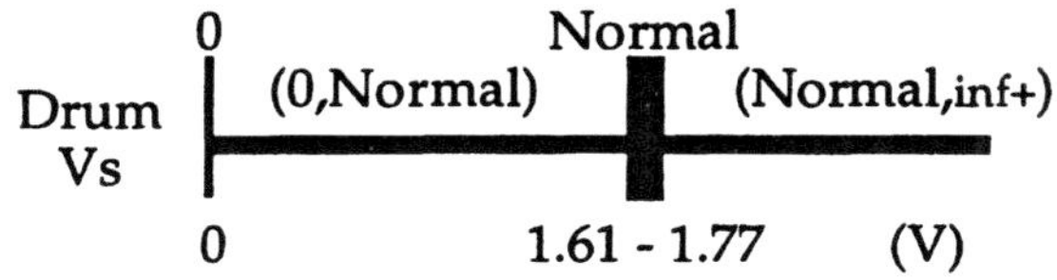

Fig. 6.6 Example of qualitative space mapping in the prototype.

Virtual case

To improve the reasoning speed and to reduce the system size, we took the knowledge compilation approach and adopted the single fault assumption. We developed a tool called a virtual case compiler. A virtual case represents a typical state of the object machine as a tuple of sensory parameters, their value, and a corresponding fault that causes this state. The compiler generates all possible cases through qualitative simulation about all faults and all operations of the actuators. Since this virtual case based reasoning replaces the qualitative simulator and the reasoning boils down to looking up a table, we can reduce the size of embedded software and increase the reasoning speed.

Fuzzification of qualitative values

In order to solve the problem of instability of the qualitative space mapping, we developed two techniques; namely, the fuzzification of qualitative values and the imitational fault method. The fuzzificiation of qualitative values is a method to relate numerical sensory data to qualitative values by means of fuzzy theory [13]. Fuzzy qualitative sensory data makes smooth transition against numerical sensory data. This also makes the reasoning system more robust against inconsistency among sensory data.

Imitational fault method

The imitational fault method dynamically calibrates the fuzzy membership functions of sensors, which are used for translating numerical sensory data into fuzzy qualitative values, through imitationally caused faults every time a repair operation is executed.

Division of the quantity space

Since virtual cases are ordered as the result of diagnosis, we can divide the quantity space of the target machine into some areas according to the order of cases. Here, the quantity space is a vector space of all sensory parameters in the machine. Fig. 6.7 depicts an example of this divided

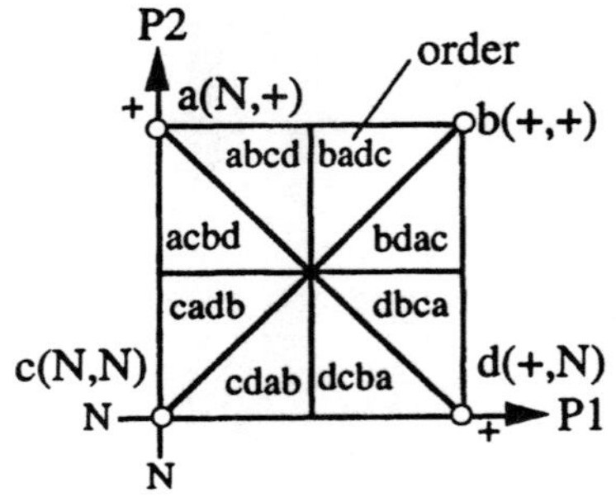

Fig. 6.7 Example of the divided qualitative space.

space that consists of two sensory parameters *P1* and *P2* and, therefore, that has four cases, namely *a*(*N*, +), *b*(+, +), *c*(*N*, *N*), and *d*(+, *N*), if only the positive regions of the parameters are considered. These four cases can be ordered according to the machine's state. For instance, with a certain set of the sensor values, case *b* is the most plausible, case *d* is the second-most, and so on. This order, *bdac*, corresponds to an area in this quantity space. As a result, because the machine's state is represented as an order of the cases rather than just as a case, the machine's state is identified more precisely than the prototype system can be. This feature makes repair operations in the commercial machine more accurate.

Job elements

Each repair operation that controls an actuator means a transition of areas in the divided quantity space. However, there are possible and impossible transitions from one area to neighboring areas according to the physical characteristics of the object machine. We call these transition rules *job elements*. The job elements simplify the reasoning algorithm for repair planning, because the planning is accomplished just by searching for appropriate job elements and combining them. Therefore, we can reduce the size of embedded software and increase the reasoning speed.

Figure 6.8 shows the reasoning algorithm of the commercial self-maintenance photocopier. First, the parameter model of the photocopier is constructed (during the design stage) and virtual cases are compiled from the parameter model with the virtual case compiler. The job elements and the fuzzy qualitative space maps are also prepared in the design stage. Before operation, first the machine collects sensory data and translates these numerical data into fuzzy qualitative values by using the fuzzy qualitative space maps. During fault diagnosis, the system orders virtual cases according to fuzzy qualitative sensory

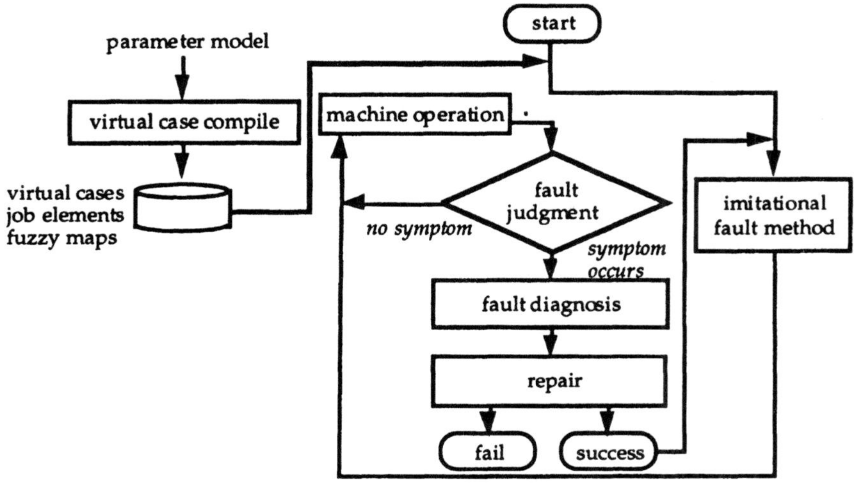

Fig. 6.8 Reasoning algorithm of the commercial self-maintenance photocopier.

values. The resulting order of the cases corresponds to an area in the divided quantity space.

After fault diagnosis, the system executes repair planning and operation with job elements. First, the system searches a path from the current machine's state to a goal state in the divided qualitative space by combining job elements. Then, the system manipulates the actuators to change the state of the machine along the path. After executing repair, the system calibrates the fuzzy qualitative space map for each sensor with the imitational fault method.

6.3 EXAMPLES OF SELF-MAINTENANCE MACHINES

This section illustrates two examples of the self-maintenance machines that implement our concepts.

6.3.1 The commercial self-maintenance photocopier

We developed a commercial photocopier (Fig. 6.9) that executes control type repairs. Figure 6.10 shows the structure of the photocopier, which includes three sensors and three actuators, and Fig. 6.11 its parameter model that describes the physical characteristics of the photocopier.

The reasoning system, developed in the C and assembler languages, takes at most 10 minutes for a diagnosis and a repair, including revision of the fuzzy qualitative space mapping. The system size is about 280

Fig. 6.9 Exterior of the self-maintenance photocopier.

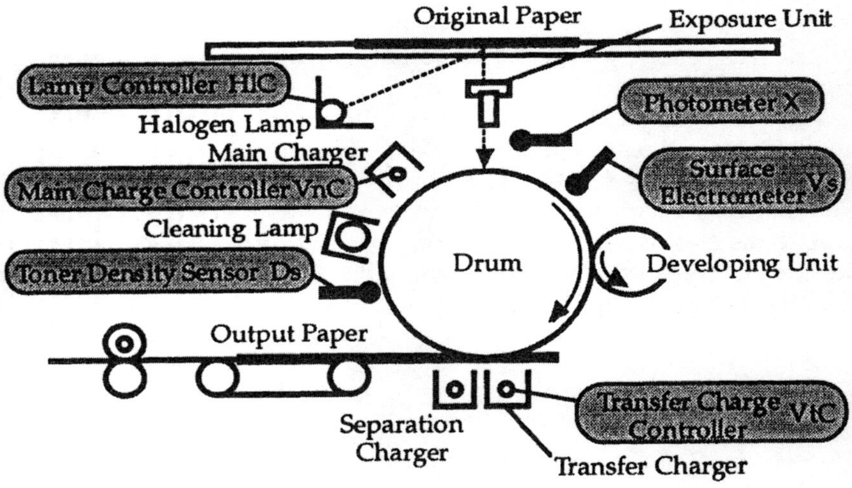

Fig. 6.10 Structure of the self-maintenance photocopier.

kbytes. The reasoning system includes 16 virtual cases concerning two symptoms. The results are acceptable as an embedded system.

Figure 6.12 compares a faulty output image and a repaired one. In this example, the reasoning system correctly identified the symptom 'increase of parameter *Os*,' which tells that the density of the output image is too high, and the fault cause 'deterioration of the halogen lamp' (see Fig. 6.11). As the repair procedure, the system first tried to increase

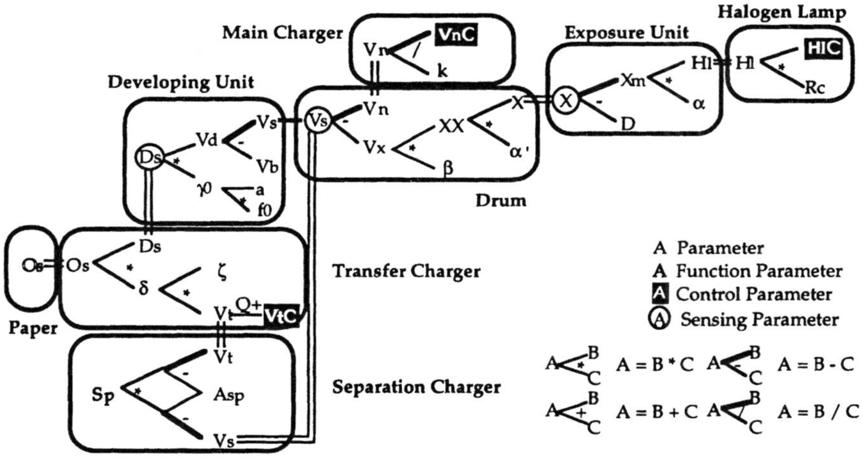

Fig. 6.11 Parameter model of the self-maintenance photocopier.

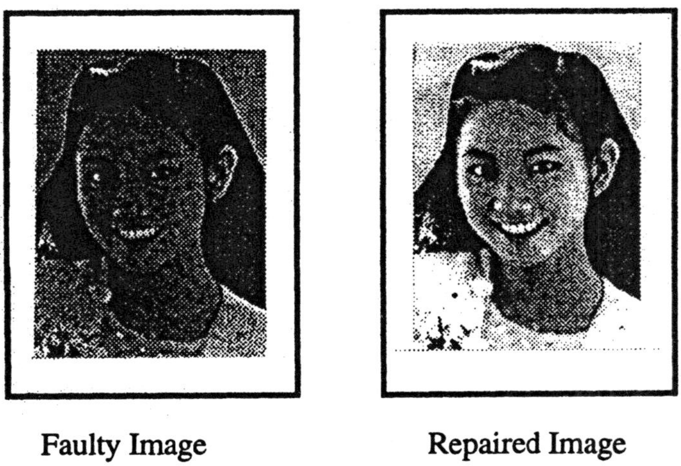

Fig. 6.12 The results of repair operation of the self-maintenance photocopier.

the voltage of the halogen lamp controller. However, the controller saturated without recovering the correct density of the paper because of deterioration of the lamp. Then, the system executed another repair operation in which the main charger controller decreased the main charge voltage and it succeeded in recovering the required function.

This commercial self-maintenance photocopier has been available to the market worldwide since April 1994. Although we have not evaluated payoff of the product, we predict decreases of down-time and maintenance costs that will pay off the cost of the development. At least, service engineers, particularly in the US, prefer this photocopier to traditional ones, because it reduces their unscheduled visits while maintaining the level of service. Some discussion can be made as follows:

1. This machine can recover from faults related to the quality of image, but it cannot repair itself in the situation where some crucial components do not work. This is not due to the limitations of the reasoning system but due to the limitation of repair execution. The functional redundancy type can deal with such a fault.
2. The reasoning system succeeded in diagnosing and generating repair plans flexibly, owing to the model-based approach. It could choose a repair plan that was not explicitly related to the observed fault.
3. Because the reasoned repair method is qualitative and the reasoning system selects only one repair method at one time, the repair execution may fail when more than two parameters must be simultaneously adjusted. To avoid this, the repair planning reasoning system should reason out multiple repair methods at one time.

The main difference between traditional control techniques and the control-type self-maintenance machine is their objectives; that is, the former aims at keeping a given state of a machine, while the latter aims at keeping functions of the machine. The following are advantages of the control-type self-maintenance machine:

1. When faults change the characteristics or structure of the object machine, the transfer function of the traditional control techniques should also be changed. Our model-based approach can do so, whereas this is difficult for the traditional control techniques.
2. The control-type of self-maintenance machine switches its repair plans flexibly, even when an actuator saturates.
3. The functional trade-off dynamically changes functions to be recovered (i.e. control objects) so that repair becomes possible.

6.3.2 The function redundancy-type self-maintenance machine

We also developed a prototype of a functionally redundant self-maintenance photocopier that has a redundant function for the target function 'to charge the drum'. This function redundancy, which is found by the Function Redundancy Designer mentioned in section 6.2.1, denotes that the target function can also be performed by the transfer charger of which the original function is 'to transfer toner from the drum to the output paper,' by rotating the drum twice for a copy process (see Fig. 6.10). This is possible, because the working principle of the transfer charger is the same as that of the main charger. To add this redundancy, we only needed to replace the transformer for the transfer charger (because the transformer should control the output voltage of the charger) and to install an additional control sequence program. If we wanted to achieve the same performance by the component redundancy, we should have needed to add an additional main charger, a transformer for the additional charger, and so on. This means that function redundancy can achieve more fault-tolerance with a much smaller number of additional components.

Figure 6.13 illustrates an example operation. In this example, because the main charger is broken, the machine could not produce a copy. By applying the function redundancy of the transfer charger, it succeeded in repairing its function 'to take a copy,' but copying speed was sacrificed to half as a result of functional trade-off, because the drum should rotate twice. This fault cannot be repaired with the control type described in section 6.3.1.

The functionally redundant photocopier extends the capability of repair execution of the control type and this can be achieved by adding a smaller number of additional software-based functions than hardware-based component redundancy. However, since modifying the control sequence programs is required to achieve behavioral changes of this prototype for exhibiting function redundancy. Therefore, developing control programs would become a burden, if a machine uses n function redundancies that require 2^n control sequences. In order to solve this problem, we have developed a methodology for automatically generating control sequence programs [14]. This enables us to build a function redundancy-type self-maintenance machine with increased repair capabilities that can dy-namically generate function redundancies during operation and to execute them according to its faulty state. We demonstrated the effectiveness of this approach by developing a prototype of such a photocopier [14].

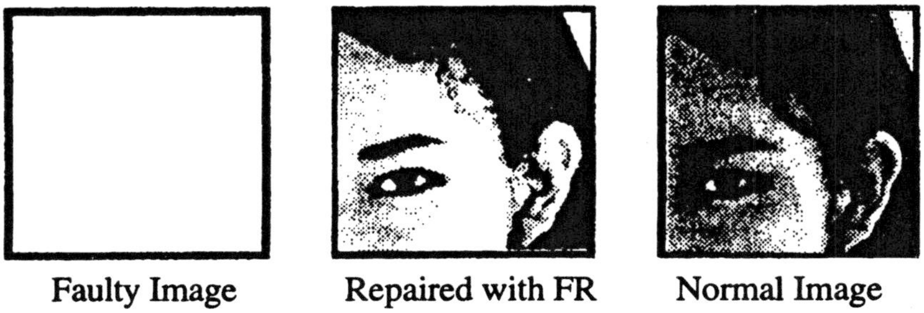

Fig. 6.13 The results of the function redundancy type repair operation.

6.4 CONCLUSIONS

In this chapter, we proposed self-maintenance machines as a design strategy for machines that are highly reliable and robust against faults and discussed their concepts, reasoning techniques, and practical examples. The primary goal of the self-maintenance machine was to cut down maintenance costs, so that the customer is troubled as little as possible while the manufacturer can benefit from the reduction of unscheduled maintenance services. To summarize, two key concepts can be listed:

- Functional maintenance, as opposed to traditional maintenance, which leads to the repair strategies such as control type and function redundancy type.
- Embedded intelligence to operate a machine more cleverly.

These features enabled us to build a self-maintenance machine just by adding a reasoning system to an existing mechatronics machine, such as a photocopier. We demonstrated the advantages and verified feasibility of the idea of the self-maintenance machine with two examples. The commercial self-maintenance photocopier demonstrated the feasibility in an industrial context. The function redundancy type clarified that function redundancy adds more redundancies to a machine and, therefore, more fault-tolerance by adding a smaller number of components than the component redundancy.

A potential candidate of future self-maintenance machines might be the modular type that has characteristics including autonomy, reconfigurability, self-organization, and high fault-tolerance. We have also developed a prototype of the modular type self-maintenance machine and verified its high fault-tolerance [15].

ACKNOWLEDGEMENTS

We would like to thank members of our laboratory at the University of Tokyo for their contribution to the self-maintenance project. Particularly, many thanks go to Hiroyuki Yoshikawa (formerly President of the University of Tokyo) for his careful suggestions and comments. Also, we would like to thank Takashi Sato, Yuichi Koike, Yasuo Motegi, Hiroshi Nanto, and Shinsuke Kondoh for their contributions to the development of the software and machines. The prototypes of the self-maintenance photocopiers were developed with support from Mita Industrial Co., Ltd. The research on the modular machine was supported by IMS (Intelligent Manufacturing Systems) project/GNOSIS consortium.

REFERENCES

[1] Lockner, A., and Hancock, P.D. (1990) Redundancy in fault tolerant systems. *CIME Mechanical Engineering*, May, 76–83.

[2] Umeda, Y., Tomiyama, T., and Yoshikawa, H. (1992) A design methodology for self-maintenance machines based on functional redundancy, in *Proceedings of Design Theory and Methodology (DTM'92)*, Taylor, D.L., and Stauffer, L.A., eds, **DE-Vol. 42**, ASME, 317–24.

[3] Tomiyama, T. and Umeda, Y. (1993) A CAD for functional design. *Annals of CIRP*, **43**(1), 143–6.

[4] Umeda, Y., Tomiyama, T., Yoshikawa, H. and Shimomura, Y. (1994) Using functional maintenance to improve fault tolerance. *IEEE Expert, Intelligent Systems & Their Applications*, **2**(3), 25–31.

[5] Umeda, Y., Tomiyama, T., and Yoshikawa, H. (1995) A design methodology for self-maintenance machines. *Journal of Mechanical Design*, ASME, **117**(3), 355–62.

[6] Winston, H.A., and Clark, R.T. (1994) AGETS MBR: an application of model-based reasoning to gas turbine diagnostics. *Proceedings of the Seventh Innovative Applications of Artificial Intelligence Conference*, AAAI, 181–9.

[7] Weld, D., and de Kleer, J. (1989) *Readings in Qualitative Reasoning about Physical Systems*, Morgan-Kaufmann, San Mateo, CA.

[8] de Kleer, J., and Brown, J.S. (1984) A qualitative physics based on confluences. *Artificial Intelligence*, **24**, 7–83.

[9] Forbus, K. (1984) Qualitative process theory. *Artificial Intelligence*, **24**, 85–168.

[10] Genesereth, M.R. (1984) The use of design descriptions in automated diagnosis. *Artificial Intelligence*, Vol. **24**(1–3), 411–36.

[11] Kuipers, B. (1986) Qualitative simulation. *Artificial Intelligence*, **29**(3), 289–338.

[12] Shimomura, Y., Tanigawa, S., Umeda, Y., and Tomiyama, T. (1995) Development of self-maintenance photocopiers. *AI Magazine*, **16**(4) 41–53.

[13] Zadeh, L.A. (1965) Fuzzy sets. *Information Control*, **8**, 338–53.

[14] Sakao, T., Umeda, Y., and Tomiyama, T. (1996) Model-based automatic generation of sequence control programs from design information. *Proceedings of The Tenth International Workshop on Qualitative Reasoning*, Farquhar, A., and Iwasaki, Y., eds, Technical Report WS-96-01, AAAI Press, Menlo Park, CA, 206–15.

[15] Sakao, T., Kandoh, S., Umeda, Y., and Tomiyama, T. (1996) The development of a cellular automatic warehouse. *Proceedings of IEEE/RSJ International Conference on Robotics and Systems (IROS'96)*, IEEE, 324–31.

CHAPTER 7

Integrated prognostics, maintenance and life-extending control of continuous-time production processes

Asok Ray and Shashi Phoha

Global economic competition has resulted in increased emphasis on the maintenance and life extension of existing plant equipment to reduce major capital costs and increase productivity. The economic and technological challenges for reliable and cost-effective production in continuous-time and discrete production processes are met through a combination of failure prognostics, information-based predictive maintenance, and life-extending control. These three interrelated issues are addressed in this chapter. The technical approach is based on augmentation of system-theoretic techniques for synthesis of optimal decision and control policies with governing equations and inequality constraints that model the mechanical behavior of structural materials for the purpose of damage representation. The major challenge here is to characterize the dynamic properties of the materials and then utilize this information in a mathematical form such that it is readily applicable to synthesis of robust decision and control systems for integrated prognostics, maintenance, and life extension.

7.1 INTRODUCTION

With the pressure of global economic competition, increased emphasis is being put on maintenance and life extension of existing plant equipment for continuous-time and discrete production processes in an effort to reduce major capital costs and increase productivity. The economic and technological challenges for reliable and cost-effective production are met through a combination of failure prognostics, information-based predictive maintenance, and life-extending control. These three interrelated concepts can be realized in an integrated manner by taking advantage of the synergism of the disciplines of *operations research and controls* and *structures and materials* sciences. Cross-disciplinary research in systems and materials sciences allows formulation of decision and control policies based on validated sensor signals and phenomenological models of material degradation such as those due to thermo-mechanical fatigue and creep.

A major goal of operation and maintenance (O&M) decision systems for complex processes is to achieve high productivity with increased reliability, availability, component durability, and maintainability. The specific requirements are:

- achieving the target productivity within the scheduled time;
- extension of the service life of the controlled process;
- increase of the mean time between major maintenance actions.

Therefore, O&M decision systems need to be synthesized by taking mission objectives, (e.g. productivity and performance), service life, and the overall production costs into consideration. The current state-of-the-art of decision systems synthesis focuses on improving productivity and diagnostic capabilities under constraints that often do not adequately represent material properties of critical plant components. The reason is that traditional design is based upon the assumption of conventional materials with invariant characteristics.

As the science and technology of materials continue to evolve, the design methodologies for life-extending systems must have the capability of easily incorporating an appropriate representation of material properties. This requires augmentation of the system-theoretic techniques for synthesis of decision and control laws with governing equations and inequality constraints that would model the mechanical properties of the materials for the purpose of damage representation and failure prognosis. The major challenge in this type of research is to characterize the damage generation process, and then utilize this information in a mathematical form for synthesizing algorithms of robust decision and control, diagnostics, and risk assessment in complex mechanical systems.

The discussions above evince the need for interdisciplinary research in prediction and mitigation of structural damage in mechanical components as an augmentation to decision systems for predictive maintenance. From this perspective, the research work reported and cited in this chapter is oriented toward the development of a methodology to achieve optimized trade-off between system productivity and structural durability of the plant under control. The task is to synthesize a robust decision system by maintaining the damage of critical component(s) within prescribed limit(s) while allowing for external disturbances and uncertainties in modeling the dynamics of process and material damage.

The research work presented in this chapter focuses on continuous-time production processes such as chemical and processing plants, pulp and paper mills, and large rolling mills. This novel concept can be extended to discrete production processes such as manufacturing of glass containers and light bulbs, semi-conductors, ceramic-enamel coatings on appliances, and food products. In these applications, both product quality and cost effectiveness can be improved by simultaneously controlling the manufacturing process and the manufacturing equipment.

The technical approach is based on augmentation of system-theoretic techniques for synthesis of optimal decision and control policies with governing equations and inequality constraints that model the structural properties of the materials for the purpose of damage representation. The major challenge here is to characterize the dynamic properties of structural materials and then utilize this information in a mathematical form such that it is readily applicable to robust decision and control systems for prognostics, maintenance, and life extension. The technology of prognostics, maintenance, and life extension is also directly applicable to non-manufacturing continuous-time processes such as automative, aircraft, and nuclear and fossil-fueled power plants. From an application perspective, major benefits of this technology are as follows:

- prediction of impending failures of critical plant components in real time resulting in enhanced safety, operational reliability, availability, and maintainability;
- reduction of the life cycle cost by optimization of the plant operation and maintenance schedule;
- optimized trade-off between the system performance and structural durability of the plant under control to achieve the mission objectives;
- facilitation of design revisions as new technologies of active control and structural materials evolve.

This chapter is organized in four sections including the introduction. Section 7.2 presents a methodology for formulation of operation and maintenance (O&M) decision strategies of continuous-time and discrete production processes. The basic concepts of continuous-time damage modeling are derived in the stochastic setting. It is also shown how analytical models of plant and damage dynamics can be utilized for failure prognosis, risk analysis, and decision-making for information-based predictive maintenance. Finally, section 7.2 presents an architecture of the integrated decision and control system based on both continuously varying and discrete-event concepts. Section 7.3 presents examples of verification of the damage model by comparison with the statistical data of fatigue tests in a laboratory environment. Section 7.4 summarizes the chapter and concludes the major benefits and applications of this advanced decision and control technology.

7.2 METHODOLOGY

Although a significant amount of research has been conducted in the individual fields of operations research and controls, structural mechanics, and analysis and prediction of materials damage, integration of these three disciplines did not receive much attention in the past. In view of the requirements for structural durability and flight control of advanced aircraft, Noll *et al.* (1991) have suggested the need for interdisciplinary research in the fields of active control technology and structural durability, specifically aeroelasticity and fatigue life analysis. Lorenzo and Merrill (1991) have delineated the role of life-extending control or reliable operation of reusable rocket engines. Very recently, optimal decision and control systems have been formulated for damage mitigation and life extension of rocket engines based on systems theory and thermomechanical fatigue analysis (Ray *et al.*, 1994a, 1994b; Ray and Wu, 1994; and Dai and Ray, 1994a, 1994b, 1995). These evolving technologies focus on three major areas, namely, failure prognostics, information-based predictive maintenance, and life extending control, which are briefly described below.

Failure prognostics

Methods for failure diagnostics in diverse applications have been reported by many researchers including Willsky (1976), Himmelblau (1978), Iserman (1984), Basseville (1988), Patton *et al.* (1989), and Ray and Luck (1991). Failure diagnostics procedures have been formulated based on several different concepts, detection filters such as (Massoumnia, 1986), analytical redundancy (Ray *et al.*, 1983), and management of sensor and functional redundancy (Ray and Desai, 1986). Although sensor-based diagnostics play an essential role for safe operations, they alone may not provide sufficiently advanced warnings of impending failures under certain circumstances. For example, if the cause of failure in a critical plant component such as the blades of a gas turbine is high cycle fatigue, then detection of a failure signature derived from the vibration monitoring sensors could be too late to avoid an emergency plant shutdown. Impending failures can be predicted based on an analytical model of the material properties even though the available sensors may not be able to detect any change in the plant behavior. Specifically, prognoses of impending failures enhance plant and personnel safety, improve availability and maintainability, and reduce operational cost. Therefore, in this context, failure diagnosis is implied to be included within failure prognosis. To facilitate planning of operation and maintenance in aging plants, failure prognostics must have the capability of easily incorporating an appropriate representation of material properties as these materials monotonically degrade with time under both normal and off-normal operating conditions.

Information-based predictive maintenance

The maintenance of complex continuous processes such as chemical plants and large rolling mills is currently performed at preassigned intervals following three methods: (1) calendar based; (2) operating time based; and (3) failure based. Calendar-based maintenance is primarily aimed at natural degradation of materials where the system components are periodically inspected and repaired or replaced. Examples are aging and wear of rubber gaskets and corrosion of equipment components in an aggressive environment. Operating-time-based maintenance is scheduled in response to the usage time of the equipment where inspection is conducted in a sequential order at varying time intervals, and parts are replaced independent of the actual operating conditions and the environment. This approach is often very conservative because the specific operating conditions which stress a mechanical system are not taken into consideration in scheduling the maintenance actions. For example, a power plant experiences higher stress as it follows load changes than when operated at a fixed base load for the same time period. A major disadvantage of the operating-time-based maintenance approach is the enforced loss of availability for being overly conservative. Therefore, in a time-critical environment, the plant maintenance engineer may defer or skip the regular inspections for immediate availability of critical machinery and thereby increase the risk of an unscheduled shutdown. Failure-based maintenance, as its name indicates, relies on identified failure of specific components or detected faults in plant operations. A major disadvantage of failure-based maintenance is that it may lead to a catastrophic accident before the fault is detected and corrective measures are taken.

Information-based predictive maintenance (IPM) takes into account the operating environment and the measurement history of critical plant variables in scheduling maintenance actions. In the current state of the art, applications of IPM are rather uncommon. An example that can be related to the IPM concept involves the usage of an acceleration-counter, popularly known as the g-counter, installed on tactical aircraft. The g-counter records the number of times a specified level of airframe acceleration is exceeded. The maintenance engineer uses the flight recorder data to guess the remaining service life of the stressed components.

Implementation of IPM must be made far more rigorous than this example in order that the full benefits are realized. For example, since operating stress is the most important factor in degradation of materials, making maintenance decisions based on the history of operating stresses incurred will reduce unnecessary inspections and part replacements while increasing the availability and decreasing the risk of unexpected failures.

The decision policy of IPM, if appropriately implemented, offers significant benefits compared to other existing forms of maintenance. The earlier example of the g-counter is illustrative. Since the g-counter merely records the number of excursions past a specified level of acceleration, it does not provide the time history of the airframe acceleration or its derivatives (i.e. jerky movements). This information is crucial for damage prediction. However, there are unsolved problems in implementing the IPM concept that include: (1) analytical development of life prediction models in the stochastic setting; and (2) risk analysis and formulation of maintenance decisions based on the predicted value of the remaining service life.

Life-extending control

In additions to failure prognosis and information-based predictive maintenance, on-line damage prediction can be utilized for service life extension and operation planning of the plant. In general, performance, mission objectives, maintenance cost and service life must be taken into consideration for operation and control of continuous-time production processes such as chemical and processing plants. However, the current state-of-the-art in synthesizing decision and control policies focuses on improving plant performance without adequately representing the material properties of the critical plant components. The reason is that process planning and control is traditionally based upon the assumption of invariant characteristics of conventional materials. Given the requirements for high performance, longer expected life, and availability of improved materials which may have significantly different damage characteristics relative to conventional materials, the lack of appropriate knowledge about the material properties would lead to either of the following:

- less than achievable performance due to overly conservative operations; or
- over-straining of the structure leading to unexpected failures and drastic reduction of the service life.

In view of the above, the objective of life-extending control, also referred to as damage-mitigating control (Ray *et al.*, 1994a, 1994b, 1994c), is to operate the plant in an optimal sense such that a trade-off between the service life and the performance is achieved. The major benefit is extension of the functional life of critical plant components accompanied by enhanced safety, operational reliability, and availability.

The challenge here is to synthesize a life-extending control policy along with failure prognostics and information-based predictive maintenance to enhance plant and personnel safety, improve availability and operability, and reduce operational cost (Ray and Weng, 1995). This decision and control system may have a hierarchical structure and is required to be robust, i.e., tolerant of a variety of errors resulting from parametric and non-parametric uncertainties in modeling of plant and damage dynamics, and continuously varying or discrete-event disturbances. This goal is achieved via formulation of decision and control policies based on validated sensor signals and phenomenological models of material degradation such as those due to thermomechanical fatigue and creep.

7.2.1 Modeling of damage dynamics in the continuous-time setting

The damage prediction system as depicted in Fig. 7.1 is an essential ingredient of the decision support system for plant operation and maintenance (Ray *et al.*, 1994a, 1994b, 1994c). It consists of a plant model, structural model(s) of the critical component(s), and the appropriate damage model(s). The plant model is a finite-dimensional state-space representation of the continuous dynamic process and is continuously calibrated by one-line sensor data from the plant instrumentation system. The plant states are inputs to the component structural model which generates the necessary information for the damage model. The output of the component structural model is the *structural load vector* which, for example, consists of time-dependent stress, strain, and temperature at critical point(s) of the structure. Typically, the structural dynamic equations represent the mechanical and thermal load conditions (Dai and Ray, 1995; Ray and Dai, 1995). The damage model is a continuous-time representation of the estimated damage and predicted life so that this model can be integrated with the plant and structural dynamic models in the state-variable setting. The objective here is to include the effects of time-dependent damage rate and damage accumulation at the critical points of plant components which are subjected to

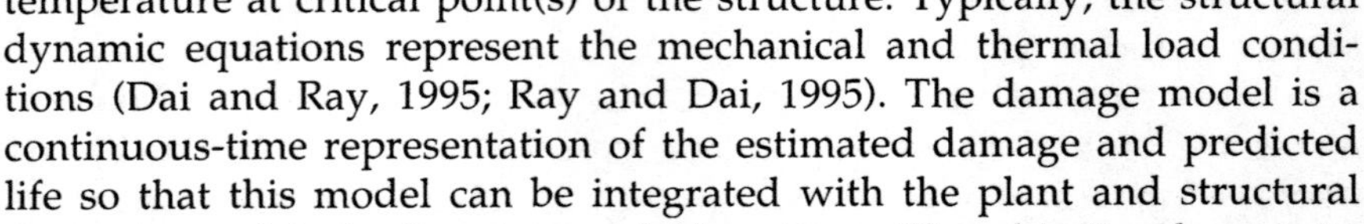

Fig. 7.1 The damage prediction system.

varying-amplitude and random load. The damage state vector $v(t)$ indicates the accumulated damage in terms of micro-cracking, macroscopic cracks, inelastic strain, etc., due to fatigue and creep-plasticity at the critical points. The time derivative of damage, $\dot{v}(t)$, indicates how the instantaneous load is affecting the critical structures of the plant (Tangirala *et al.*, 1995).

The component structural model and the damage prediction model are governed by a set of nonlinear differential equations after spatial discretization of the nonlinear partial differential equations and the associated boundary conditions. These differential equations must satisfy the local Lipschitz condition (Vidyasagar, 1992) within the domain of the plant operating range. The combination of the component structural model and the damage prediction model in Fig. 7.1 is referred to as the structural damage model in the sequel. Note that the plant dynamic model is unaffected by accumulation of structural damage and hence there is no inherent feedback from the damage model to the plant model as seen in Fig. 7.1. The rational is that the physical phenomenon of material damage in a plant component does not alter its macroscopic mechanical behavior (e.g. stiffness constant) within the span of its service life. An example is fatigue-induced microcracks in a turbine blade that do not alter the natural frequency until the end of its useful life. That is, in a vast majority of cases, when a change in the frequency becomes detectable, the blade life is almost expended.

A general structure of the plant and damage dynamics and the constraints on damage rate and accumulation are represented in the deterministic setting as:

Task period: Starting time t_o to final time t_f

$$\text{Plant dynamics: } \frac{dx}{dt} = f(x(t), u(t)) \ \forall t \geq t_o; \text{ given } x(t_o) = x_o \tag{7.1}$$

Damage dynamics:

$$\frac{dv}{dt} = h(v(t), q(x(t), u(t))) \text{ such that } h \geq 0 \ \forall t \geq t_o; \text{ given } v(t_o) = v_o \tag{7.2}$$

$$\text{Damage rate tolerance: } 0 \leq h(v(t), q(x(t), u(t))) \leq \beta(t) \quad \forall t \in [t_0, t_f] \tag{7.3}$$

$$\text{Accumulated damage tolerance: } [v(t_f) - v(t_o)] < \Gamma \tag{7.4}$$

where: $x \in \mathbf{R}^n$ is the plant state vector;
$u \in \mathbf{R}^m$ is the control input vector;
$v \in \mathbf{R}^r$ is the damage state vector;
$q \in \mathbf{R}^p$ is the structural load vector; and
$\beta(t) \in \mathbf{R}^r$ and $\Gamma \in \mathbf{R}^r$ are specified tolerances for the damage rate and accumulated damage, respectively.

Modeling of fatigue damage has been a topic of intensive research for several decades. Many researchers cited in Suresh (1991) have proposed empirical and semi-empirical models based on observed experimental data and attempted to provide a physical interpretation to these models. Modeling of fatigue damage dynamics via nonlinear stochastic differential equations is a relatively new area of research, and an extensive list of the literature representing the state of the art is cited by Sobczyk and Spencer (1992). One approach is to randomize an established deterministic model of fatigue damage dynamics to generate the necessary stochastic information.

The deterministic fatigue crack growth model is based on the concept of the short crack growth model (Newman, 1981; Newman, Swain and Phillips, 1986). While the Paris model (Paris and Erdogan, 1963) is valid in the macro-crack range, the Newman model represents the fatigue crack growth process down to microcracks in the order of 10 μm. The Newman model for crack growth is of the form:

$$\left.\begin{aligned} &\frac{d\mu}{dN} = C_1 \times (\Delta K_{\text{eff}}{}^{C_2}) \times \left(\frac{1 - \left(\dfrac{\Delta K_0}{\Delta K_{\text{eff}}}\right)^2}{1 - \left(\dfrac{K_{max}}{C_5}\right)^2} \right) ; \mu(N_o) = \mu_o > 0 \\ &\Delta K_0 = C_3 \left(1 - C_4 \frac{S_o}{S_{max}} \right) ; K_{max} = S_{max}\sqrt{\pi\mu F} ; \Delta K_{eff} = (S_{max} - S_o \sqrt{\pi\mu F}) \end{aligned}\right\} \quad (7.5)$$

where μ is the mean crack length; N is the number of cycles; the crack growth rate $\frac{d\mu}{dN}$ is the so-called derivative of μ with respect of N as commonly used in the fracture mechanics literature (Bannantine *et al.*, 1990); S_0 is the crack opening stress; F is the correction factor for finite width of the specimen; S_{max} is the maximum applied remote stress; and C_i, $i = 1, ..., 5$ are material parameters. As an alternative to the functional form of Equation 7.5, a table-lookup form can determine $\frac{d\mu}{dN}$ in terms of ΔK_{eff}. In this setting, the crack growth rate is expressed as:

$$\frac{d\mu}{dN} = exp(ml\,n(\Delta K_{eff})+b);\ \mu(N) = \mu_o > 0 \quad (7.6)$$

where m is the slope and b is the intercept of the linear interpolation of the (log scale) ΔK_{eff}-dc/dN look-up table. Details of this method are reported by Newman, Swain and Phillips (1986). It should be noted, however, that any other deterministic model of fatigue crack growth can be used with the only criterion being that the expected value of the crack growth profile is accurately represented. For example, several researchers

(Spencer and Tang, 1988, 1989; Enneking, 1991) have used a cubic polynomial fit in $\ln(\Delta K_{eff})$ to determine crack growth rate. Following Ray *et al.* (1994a), the crack growth equation is expressed in the continuous-time setting as:

$$\frac{d\mu}{dt} = \frac{\left(\partial\Phi/\partial S\right)\left(dS/dt\right)}{1 - \partial\Phi/\partial\mu}\ ; \mu(t_0) = \mu_0 > 0 \qquad (7.7)$$

where Equation (7.5) or Equation (7.6) is represented $\frac{d\mu}{dN} = \Phi(\mu, S, S_0)$. A common practice is to define the time-dependent fatigue crack damage as the ratio of the mean crack length and the critical value of the crack length which is a function of the material, component geometry, and the operating condition (Bolotin, 1989).

It has been shown by several researchers, including Bolotin (1989) and Spencer and Tang (1988, 1989), that the process of fatigue crack growth can be modeled by nonlinear stochastic differential equations satisfying the Itô conditions [Oksendal, 1992]. The information generated from these models can be used to formulate algorithms for damage prediction and risk analysis. Specifically, Kolmogorov forward and backward diffusion equations, which require solutions of nonlinear partial differential equations, have been proposed by several researchers, cited in Sobcyzk and Spencer (1992), to generate the statistical information required for risk analysis of mechanical structures. These nonlinear partial differential equations can only be solved numerically; however, the numerical procedures are computationally intensive as they rely on a fine-mesh model using finite-element or combined finite-difference and finite-element methods (Enneking, 1991). Therefore, although this approach might be useful for making off-line maintenance decisions, it is not sufficiently fast for on-line damage monitoring, failure prognosis, and remaining service life prediction. To enhance the numerical efficiency, we propose to adopt the underlying principle of extended Kalman filtering (Jazwinski, 1970) in which the first two moments of the stochastic damage state are computed on-line by constructing the stochastic differential equations in the Wiener form as opposed to the Itô form. This can be achieved following the two-state model of Spencer and Tang (1988) provided that the shaping filter is constructed with additive white Gaussian noise. The underlying principle of extended Kalman filtering can be used with or without any damage sensor(s). The absence of any useful sensor data is equivalent to having the inverse of the intensity of measurement noise covariance tend to zero; in that case, the filter gain approaches zero. Consequently, the conditional density function generated by the filter becomes identical to the prior density function whose evolution is governed by the Kolmogorov forward equation (Jazwinski, 1970).

The stochastic differential equation for the crack growth process $c(\omega,t)$ is expressed in terms of the deterministic damage dynamics as:

$$\frac{dc(\omega,t)}{dt} = \exp\left(z(\omega,t) - \frac{\sigma_z^2}{2}\right)\frac{d\mu}{dt} \quad \forall t \geq t_o;\ \text{given } E[c(\omega, t_o)] = \mu_o \text{ and } cov[c(\omega,t_o)] = P_o \quad (7.8)$$

where ω and t represent the sample point and time of the stochastic process, respectively. The auxiliary process $z(\omega,t)$ is assumed to be stationary Gauss–Markov of variance σ_z^2 implying that the crack growth rate is a lognormal-distributed Markov process. The rationale for this assumption is as follows: (1) fatigue crack growth rates in many metallic materials are observed from experimental data to be approximately lognormal-distributed and highly correlated over a short period of time (Sobczyk and Spencer, 1992); and (2) the crack propagation phenomenon is modeled as a Markov process by including history-dependent crack opening stresses or reference stresses (Ray *et al.*, 1994a) as an additional input. Yang and Manning (1994) recommended the lognormal distribution model of fatigue crack propagation statistics in aluminum alloys for scheduling of aircraft maintenance because: (1) this approach is moderately conservative; (2) it is mathematically tractable and can be easily implemented; (3) a small number of replicate fatigue tests are adequate to calibrate the model; and (4) reasonable crack growth prediction for structural details have been obtained previously from this model.

It follows from Equation (7.8) that $E\left[{}^{dc(\omega,t)}/_{dt}\right] = {}^{d\mu}/_{dt}$ due to the Gaussian distribution of $z(\omega,t)$. This is in agreement with the governing equation of mean crack growth in Eq. (7). The correlated stationary Gauss–Markov process $z(\omega,t)$ is modeled by using a first-order linear shaping filter which is driven by the additive zero-mean white Gaussian noise $z(\omega,t)$ of intensity Q_o as:

$$\frac{dz(\omega,t)}{dt} = \xi\, z(\omega,t) + w(\omega,t) \ \forall t \geq t_o;\ \text{given } E[z(\omega, t_o)] = z_o \text{ and } cov[z(\omega,t_o)] = \frac{Q_o}{2\xi} \quad (7.9)$$

where the filter parameter ξ is a measure of 'coloredness' of the auxiliary random process $z(\omega,t)$. As ξ is increased, $z(\omega,t)$ becomes more uncorrelated. Specifically, if ξ is set equal to zero, then $z(\omega,t)$ becomes a Brownian motion process which is perfectly correlated. In the extreme case, as * tends to infinity, $z(\omega,t)$ approaches zero implying no uncertainties in the fatigue crack growth process. Experimental data show that $z(\omega,t)$ remains highly correlated over a long period and eventually may become uncorrelated (Lin and Yang, 1983). This phenomenon is realized if $0 < \xi << 1$.

The constant parameter z_o in Equation (7.9) is determined from the environmental condition under which the fatigue crack growth takes place. Specifically, z_o is set to zero at the nominal environmental condition (e.g. laboratory air at room temperature) under which experiments are conducted to determine the fatigue damage parameters in Equation (7.3). However, in a more hostile environment (e.g. humid and salty air at the ocean surface), the increased damage rate can be realized by setting z_0 to a positive value. Similarly, in a more benign environment (e.g dry air), z_0 should be set to a negative value to realize the relatively slower growth of fatigue cracks.

The stochastic differential equations (7.8) and (7.9) are now combined to yield the augmented stochastic vector $\chi = [c\ z]^T$ as a vector diffusion Markov process with additive noise, which can be interpreted in the Wiener sense instead of the Itô sense. Then, linearization of the augmented state equation yields the local state transition matrix which would lead to computation of the first two moments of $\chi(\omega,t)$ conditioned on the history of the stress and the mean and variance of $\chi(\omega,t_0)$. This provides full statistical characterization of the crack length at each instant of time based on the assumption of the lognormal distribution which is described by the first two moments.

The measurement, generated from any available sensor(s) (e.g. strain gauges and acoustic emission transducers) at the instant t, is modeled as an algebraic function of the damage state vector χ and additive sensor noise υ:

$$y(\omega,t) = g(\chi(\omega,t)) + \upsilon(\omega,t) \tag{7.10}$$

where the deterministic function $g(\bullet)$ represents the sensor model; and $\upsilon(\omega,t)$ is zero-mean white noise of known intensity representing measurement noise. Since $\upsilon(\omega,t)$ is a consequence of electronic and acoustic noise, it is assumed to be independent of the material uncertainties $w(\omega,t)$, i.e. $E[w(\omega,t)\ \upsilon(\omega,t)^T] = 0\ \forall t, \tau \geq t_0$. Conditional mean and variance of the crack length $c(\omega,t)$ can be obtained in real time via Equations (7.8), (7.9), and (7.10).

Yang and Manning (1994) recommended the lognormal distribution to model the complete statistics of fatigue cracks from the first two moments for scheduling of airframe maintenance in terms of risk because: (1) it is conservative; (2) it is mathematically simple and can be easily implemented on a computer; (3) a small number of replicate fatigue tests are adequate to calibrate the model since it does not require a correlation distance parameter; and (4) reasonable crack growth predictions for structural details have been obtained previously using this model. Further details of the stochastic model of fatigue crack dynamics are reported by Ray and Tangirala (1997).

7.2.2 Failure prognosis, risk analysis, and decision-making

Failure prognosis and risk analysis are envisioned to be an augmentation of failure diagnosis to predict impending failures of critical plant components. To facilitate operation and maintenance (O&M) planning of production processes, the failure prognostics system must have the capability of easily incorporating an appropriate representation of material properties as these materials monotonically degrade with time under both normal and off-normal operating conditions. Prediction of both the damage rate and hence the accumulated damage are affected by plant disturbances and sensor noise in addition to the errors incurred in modeling of damage dynamics. Furthermore, the failure prognostics system must be cognizant of the plant operational requirements, mission objectives, and the limitations and inaccuracies of the predicted damage information.

The damage calculations are based on the current plant state, damage state, and any additional available information, such as history of measurements of the local strain and temperature, as discussed in the beginning of section 7.2. Therefore, robustness of the prognostic decisions is dependent on uncertainties in each of the plant, component structural and damage models in Fig. 7.1. Traditionally, the risk index and residual service life (Bolotin, 1989) of mechanical structures and machine components are calculated off-line on the basis of statistical models of the material properties, operating conditions and any major disruptions in the plant. It follows from Equation 7.2 that the service life is finite as the damage acumulation is monotone with time. Furthermore, scheduling of plant operation and maintenance is based on the fact that damage accumulation is likely to be accelerated in the event of disruptions. Therefore, on-line monitoring of damage allows refinements of the risk index and residual service life prediction (Bogdonoff and Kozin, 1985; Bolotin, 1989) as time progresses. Specifically, damage prediction assists the plant engineer in making dynamic decisions regarding the safety procedures, mission accomplishments, and the time interval between major maintenance actions. In the absence of any accident or major disruption after the current time, the service life can be obtained from the stochastic differential equations of plant and damage dynamics. However, in the presence of anticipated accidents, computation of remaining service life becomes more complex because the damage accumulation depends not only on the transient states of the plant and damage dynamics as a result of the disruption, but also on the instant of its occurrence. The approach proposed here is to identify the type of distribution (e.g. lognormal) from experimental data, and then generate the requisite statistical information from the time series of the first two moments of the plant and damage states. Although this information does not provide exact statistics, it serves to generate early warnings for impending failures (Ray, 1990) as well as for making maintenance decisions (Yang and Manning, 1994). In the context of the crack growth model presented in the latter part of section 7.2.1, the concept of service life prediction is explained below.

We define M hypotheses as disjoint regions of normalized crack length (i.e. the fatigue crack damage defined as $v(t) = c(t)/c^*$) where c^* is the critical crack length beyond which the crack growth rate becomes very large rapidly leading to complete rupture. Therefore if the initial crack length is c_o, then the first (M–1) hypotheses can be represented by the following disjoint intervals in the space of the crack length:

$$c_o < c(t) \le c_o + \frac{(c^*-c_o)}{M-1}; \; c_o + \frac{(c^*-c_o)}{M-1} < c(t) \le c_o + 2\frac{(c^*-c_o)}{M-1}; \cdots ; c_o + (M-2)\frac{(c^*-c_o)}{M-1} < c(t) \le c^* \quad (7.11)$$

The last (i.e. the Mth) hypothesis is chosen as the region $c(t) > c^*$ which is popularly known as the unstable crack region in the fracture mechanics literature (Bannantine *et al.*, 1990). Each of these M hypotheses represents a distinct range in the entire space of normalized crack lengths from an initial value of c_o till the rupture occurs, and together they form an exhaustive set of mutually exclusive regions in the damage state space. The probability that the jth hypothesis, H_j, holds at any given instant of time, t, is computed via the instantaneous probability distribution function $F_c(*;t)$ of the crack length $c(t)$.

$$P_j(t) = F_c\left(c_o + j\frac{(c^*-c_o)}{M-1}\right); t - F_c\left(c_o + (j-1)\frac{(c^*-c_o)}{M-1}t,\right); \; j=1,2,..,M-1; \text{ and } P_M(t) = 1 - F_c(c^*;t) \quad (7.12)$$

If the crack length is assumed to be lognormal-distributed (Sobczyk and Spencer, 1992), then the instantaneous probability distribution function F_c of the stochastic process $c(t)$ can be determined from the first two moments of crack length that are generated as continuous functions of time by the extended Kalman filter. Therefore, the probability that a given hypothesis is valid can be computed on-line as a function of time. Using the distribution function F_c, we can compute the remaining service life $T(t;\, Y_d(t);\, \varepsilon)$ at any specified time instant, t, based on an anticipated plant output profile $Y_d(t)=\{y(\tau)\!: \tau^3 t\}$ and a confidence level ε. This implies that if the plant operation is scheduled to yield the anticipated output, then $T(t;\, Y_d(t);\, \varepsilon)$ is the maximum time of operation such that the probability that the crack length $c(t + T)$ does not exceed c^* is greater than ε. This is mathematically represented as:

$$T(t;\, Y_d(t);\varepsilon) = sup\{\tau \in [0,\infty)\!: P[c(t) + \tau) \le c^*] > \varepsilon\} \quad (7.13)$$

The above task of remaining life prediction is accomplished by using the same extended Kalman filter formulation described earlier. Given the expected value and variance of crack length at time t, the extended Kalman filter can be simulated using these values (as initial conditions) and the anticipated load profile to find the maximum remaining time of operation T such that $P[c(t + T) \le c^*] > \varepsilon$.

The above procedure can be used to provide valuable information to the higher level of decision-making or to a human operator about the confidence with which the plant operations can be planned for a specified amount of time. This information is also of considerable importance to maintenance scheduling or in making decisions on how to reschedule the plant operations to avoid an untimely shutdown. The major implications of the proposed service life prediction and risk analysis are:

(1) generation of early warnings resulting in diminished risk of loss of human life and plant equipment; (2) enhancement of plant availability by avoiding an emergency shutdown; and (3) reduction of the plant operational cost by dynamic rescheduling of the maintenance plan via on-line assessment of damage in the critical components.

The above information may also serve as inputs to a hybrid (i.e. combined continuous-time and discrete-event) decision system for information-based predictive maintenance as explained next in section 7.2.3. This decision system may incorporate nonlinear stochastic models of the plant and damage dynamics in a continuous setting coupled with a discrete-event model representing accidents and disruptions as uncontrollable events and maintenance actions as controllable events.

7.2.3 Integrated decision and control system architecture

The integrated decision and control system for prognostics, information-based predictive maintenance and life extension is built upon the principle of hybrid supervisory control (Ramadge and Wonham, 1987; Garcia, 1993). This concept of supervisory control offers a systematic and rigorous approach to hierarchical control of hybrid processes (Ray and Weng, 1995) where the continuous-time process, e.g. a chemical or processing plant, is controlled at a lower level of hierarchy, and the decisions for operation and maintenance are made at upper levels. The rationale for this approach is that the inputs to the upper levels are a combination of continuous signals and discrete events and that the resulting decisions are implemented as discrete events. Figure 7.2 shows the general concept

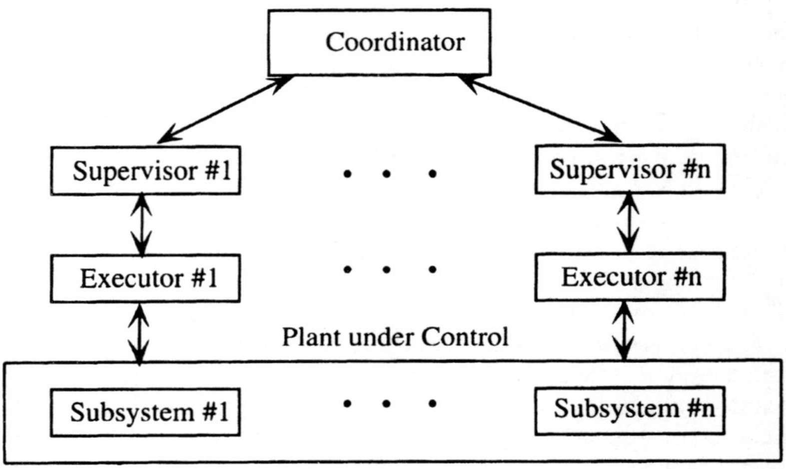

Fig. 7.2 Concept of the integrated decision and control system.

of the integrated decision and control system which is hierarchically structured in three levels of operation, namely, coordination, supervisory, and execution (Garcia *et al.*, 1994; Garcia and Ray, 1996). As their names suggest, these three levels perform different tasks and are partially ordered according to the principle of increasing intelligence with decreasing precision (Saridis, 1989). While the lowest level (i.e. the execution level) is generally of high precision and low intelligence, the higher levels are of increasing intelligence and diminishing precision. There is exactly one coordinator for control of the entire process but there may be one or more supervisor(s) over which the coordinator exercises control. Each subsystem, representing a major plant component, e.g. the distillation column in a chemical plant or the furnace-boiler in a paper mill, whose performance and service life are to be controlled belongs to the supervisory level. Each supervisor, in turn, controls its own executor which communicates with the continuously varying process of the plant subsystem. The operating principle of the supervisory control system is briefly described below.

A sequence of discrete commands, called strategy patterns, which are controllable events and provide the information regarding the desired plant performance and service life, are issued to each individual supervisor by the coordinator. Strategy patterns represent the coordination level decision for plant operation and maintenance actions to mitigate detrimental effects of actual or anticipated failures and changes in the operation strategy. Similarly, a sequence of discrete commands, called control patterns, which are controllable events and represent the supervisory level actions in response to the strategy patterns from the coordinator and the events of subsystem operation generated by the discrete-event interface, are issued to the execution level by each supervisor. The control patterns are interpreted at the execution level to generate continuous signals to modify the control signals in the inner loop.

The architecture of the proposed integrated supervisory control system is presented in Fig. 7.3 in which only one supervisor and one executor are shown for clarity. It consists of two loops, namely, the outer loop and the inner loop. The decisions on plant operation and maintenance are made at the coordination level in the outer loop. This information is communicated up to the human users and down to the supervisor(s). Life-extending control of a given subsystem via damage mitigation is the sole function of a supervisory level controller which has its own executor for communicating with the continuously varying control system within the inner loop. The role of the executor is to modify some of the control signals within the inner loop so that the detrimental effects of the uncontrollable environmental events, e.g. failures, accidents, or operational disruptions, acting upon the plant are mitigated. It is also responsible for selecting a feedforward and feedback control policy based on the control strategy received from the supervisor.

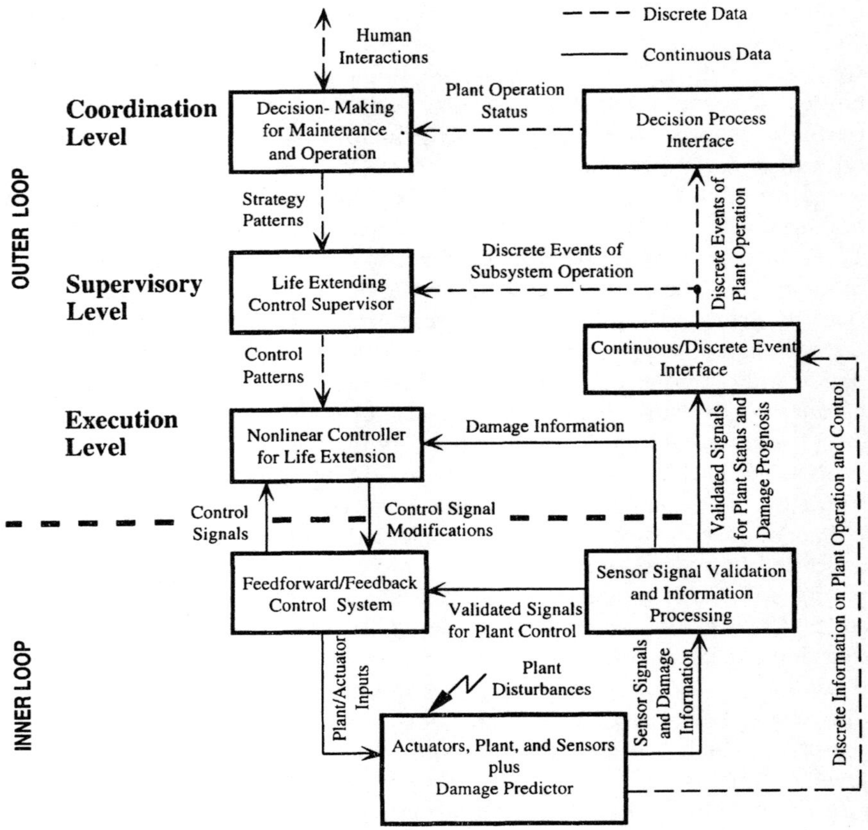

Fig. 7.3 Architecture of the integrated decision and control support.

The functions of the coordination, supervisory, and execution levels in the outer loop are realized as minimization of maintenance cost and the probability of subsystem failures that may result in loss of plant availability. To accomplish these tasks, the coordinator exercises control over the supervisors, and each supervisor on its executor by enabling or disabling events through discrete choices of strategy patterns and control patterns, respectively. Discrete-event phenomena at these three levels are modeled as interacting *finite-state automata* following the recent work in this field (Ramadge and Wonham, 1987; Cislak *et al.*, 1988; Ozveren and Willsky, 1989; Garcia, 1993). In this setting, for example, the states of an automation reflect status, such as the material degradation with respect to the need for maintenance, repair, or replacement, of a plant subsystem; and the events interconnecting the states correspond to discrete and abrupt changes in the equipment state, or gradual degradation of the material condition as it crosses a predetermined threshold of the predicted remaining life. The automation for each of the interacting process at three levels is represented by a quintuple, $G = (Q, \Sigma, \delta, q_0, Q_m)$, where Q is the finite set of states, Σ is the finite set of events (i.e. the alphabet of the automation), δ is the state transition function, q_0 is the initial state, and Q_m is the set of final states. The set of events is further partitioned into the set of *controllable* events and the set of *uncontrollable* events. Controllable events are events that may be prevented by some control action from the higher level. An example is taking a component off-line (control action) in order to prevent catastrophic failure (controllable event). An uncontrollable event is one that is not directly controllable from the higher level, for example exogenous inputs to the system and unpredicted failures.

The inner loop consists of two major functional modules: sensor signal validation and information processing; and feedforward/feedback control. The sensor signal validation and information processing module feeds validated sensor signals to the control system module within the inner loop, and plant and damage information to the interface and life-extension controller modules in the outer loop as seen in Fig. 7.3. The discrete-event interface at the supervisory level receives both continuously varying signals of validated sensor data and predicted damage, and discrete information of plant operation status such as past incidents of equipment failures and unscheduled plant outage. The interface generates discrete events which are, in turn, inputs to both the supervisory system and the decision process interface at the coordination level.

7.2.4 The control module within the inner loop

The control module has two functions, namely, feedforward control (FFC) and feedback control (FBC). The FFC provides an open loop control policy, e.g. the sequence of control commands during start-up, shutdown and load maneuvering of a plant, which can be preprogrammed in the control system. Since FFC is based on the nominal plant model with no feedback, its ability to overcome any disturbances and modeling uncertainties is limited. Therefore, the FBC which closes the control loop is necessary to compensate for these disturbances and uncertainties. However, for wide range operations, FBC alone suffers from loss of performance as the control system becomes excessively conservative to guarantee stability over the full range. This problem is alleviated in the proposed control system in the inner loop which unifies the concepts of FFC and FBC to compensate each other's disadvantages. This FF/FBC architecture is intended for robust performance and damage mitigation when the plant is operated over a wide range. For narrow range operations, however, the FBC alone is sufficient.

Feedforward control (FFC) policy

Given the initial states of the plant and damage, the FFC policy is obtained via nonlinear programming (Gill *et al.*, 1981) by minimizing a specified cost functional under the prescribed constraints of damage rate and damage accumulation. The objective is to minimize the cost functional J without violating the prescribed upper bounds of the damage rate and the accumulated damage to identify an optimal control sequence $\{u_k: k=0,1,2,\ldots,N-1\}$. The cost functional J is to be chosen in an appropriate form representing a trade-off between the system performance and the damage. The optimization problem is then formulated as follows:

$$\text{Minimize:} \qquad J = \sum_{k=0}^{N-1} J_k\,(\tilde{x}_{k+1}, \dot{v}_k, u_k) \tag{7.14}$$

Subject to:

$$(v_N - v_0) < \Gamma;\ 0 \le h(v_k, q(x_k), k) < \beta(k);\ \text{and}\ u^{\min} \le u_k \le u^{max}\ \text{for}\ k=1,2,3,\ldots,N \tag{7.15}$$

where $\tilde{x}_k = x_k - x_k^d$ is the deviation of the plant state vector x_k from the respective point x_k^d of the desired state trajectory; and $\tilde{u}_k = u_k+1 - u_k$ is the incremental change in the control effort representing the rate of actuator movements with u_N set equal to the final steady state value u_{ss} of the control input vector u_k: $u^{\min}$ and $u^{\max}$ are the lower and upper limits of u_k; and $\beta(k) \in \mathbf{R}^r$ and $\Gamma \in \mathbf{R}^r$ are specified tolerances for the damage rate and accumulated damage, respectively, in Equations (7.3) and (7.4). Optimization in Equations (7.14) and (7.15) must be carried out off-line for several maneuvers, and Fig. 7.4 shows the feedforward block for a given maneuver. The feedforward control policy is selected based on the control patterns received by the executor from its supervisor.

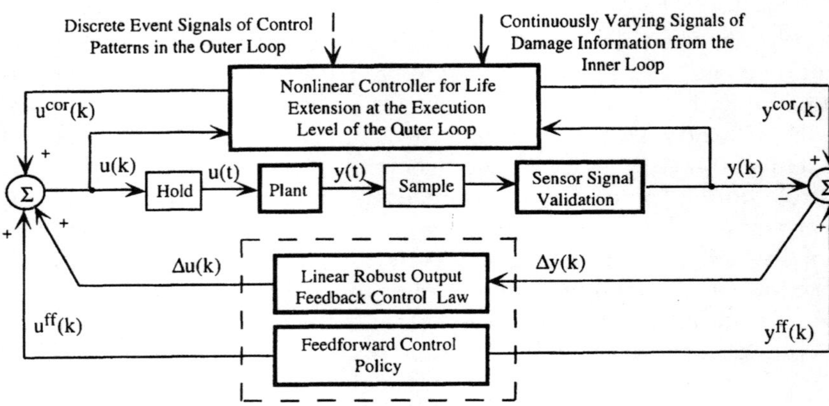

Fig. 7.4 Structure of the life-extending control system.

Feedback control (FBC) policy

The structure of the life extending control system is shown in Fig. 7.4 in which the feedforward and linear feedback control (FF/FBC) are exercised in the inner loop to ensure robust performance; the nonlinear controller for life extension in the outer loop is identical to the execution level controller for life extension in Fig. 7.3. In FF/FBC, u^{ff} and y^{ff} represent the sequences of the plant input and plant output, respectively, generated as an optimal solution of the open-loop control problem via nonlinear programming. As mentioned above, a feedback controller is necessary to compensate for the plant modeling uncertainties and disturbances such that the trajectory of the actual plant output, y, should be close to that of the desired plant output y^{ff} which serves as the reference trajectory. The resulting error in the plant output is an input to the feedback controller and the signal Δu modifies the plant/actuator input in an attempt to compensate for errors resulting from plant disturbance and noise.

The linear robust feedback control (FBC) policy is formulated by using one or more techniques of robust multi-input multi-output (MIMO) control synthesis, e.g. H_2-based LQG/LTR (Stein and Athans, 1987), H_2/H_∞ optimization (Doyle *et al.*, 1989), *l*-synthesis (Packard and Doyle, 1993; Stein and Doyle, 1991; Skogestad *et al.*, 1988), or l_1-synthesis (Diaz-Bobillo and Dahleh, 1993), which rely on approximation of the plant dynamics by a linear time-invariant model. In this case, the FBC is synthesized based on a plant model which is linearized at a desired point on the plant trajectory, and the (slowly) time-varying plant parameters, if any, are averaged over the time horizon of interest. Selection of the frequency-dependent weighting parameters in the FBC synthesis procedure, in turn, depends on the region over which the accuracy of the linearized model is acceptable. Since the damage rate in the closed-loop control system may violate the specified constraints due to the additional compensation generated by the feedback control effort Δu, the nonlinear controller is incorporated into the system to reduce the damage rate and accumulation. This is accomplished by control signal modifications y^{cor} in the reference trajectory and u^{cor} in the control effort. The rationale for modifying the reference trajectory $\{y^{ff}\}$ is that, due to plant disturbances and sensor noise, it may not be possible to follow this trajectory without violating the damage constraints. The additional control effort u^{cor} is intended to provide a fast corrective action to the control input, u, whenever necessary.

The control signal modifications, y^{cor} and u^{cor}, from the outer loop can be considered as exogenous inputs to the linear robust control system in the inner loop. Since the plant is nonlinear and the feedback control system is synthesized based on a model linearized at an operating point, y^{cor} and u^{cor} must be norm-bounded such that the postulated bounds of plant uncertainty are not exceeded. This assures the stability and performance robustness of the control system. A possible approach to synthesis of the nonlinear control system in Fig. 7.4 is to postulate a control structure and then optimize the control parameters based on a cost functional which depends on the control patterns from supervisory level, the control signals from FFC/FBC, and damage information from the sensor signal validation and information processing module.

7.3 EXAMPLE

We present, in this section, an example of how to tune and verify the stochastic damage model with statistical data on fatigue tests. We used the data of Virkler *et al.* (1979) in which the tests were conducted in an environment of dry air under a constant load amplitude of 4200 lbs with a peak load of 5250 lbs. Each test specimen was made of center-cracked 2024-T3 aluminum alloy panels 22 in long, 6 in wide and 0.10 in thick as shown in Fig. 7.5. A stress raiser of half length 0.05 in was machined in the center of the specimen. The tests were conducted on 68 specimens under very tight quality control to ensure that each test was identical including environmental conditions. Fig. 7.6 shows that, even under these tight conditions and constant amplitude loading, the fatigue life of the specimens had a large variance. However, the data in Fig. 7.6 correspond to macrocracks. Statistical data of short cracks (Newman, Swain and Phillips, 1986) need to be collected for complete validation of the crack growth model.

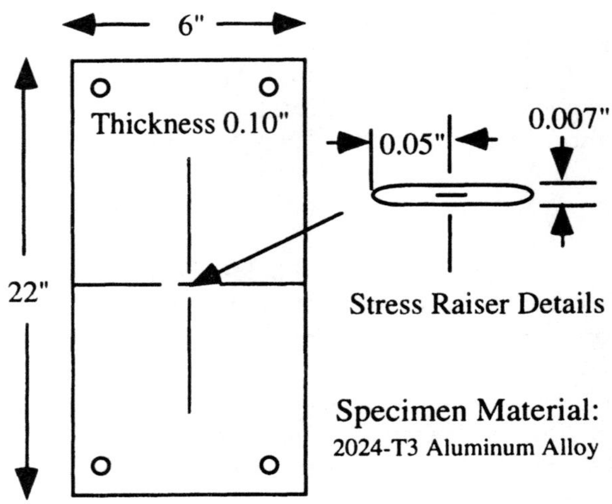

Fig. 7.5 Test specimen for the Virkler experiments.

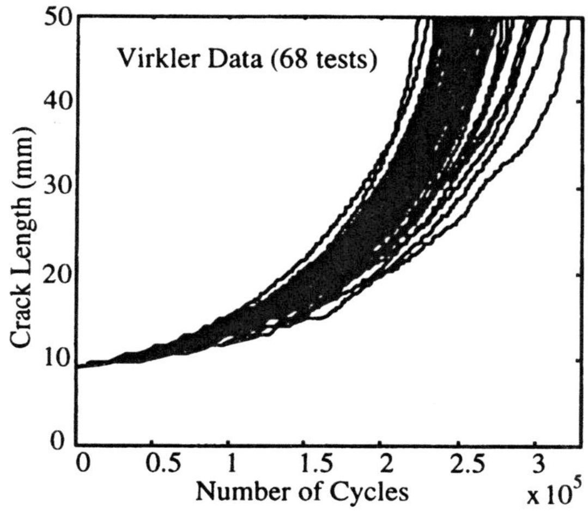

Fig. 7.6 Statistical data of fatigue crack growth.

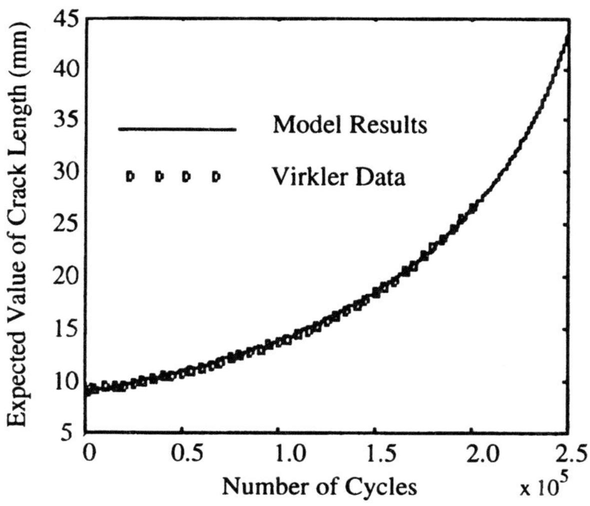

Fig. 7.7 Expected value of crack length.

Figure 7.7 compares the profile of the mean crack length generated from the test data with that of the expected value derived from the extended Kalman filter. The deterministic fatigue crack growth model used in the extended Kalman filter formulation uses a table lookup to compute the crack growth rate as a function of ΔK_{eff} following the procedure outlined by Newman, Swain and Phillips (1986). Figure 7.8

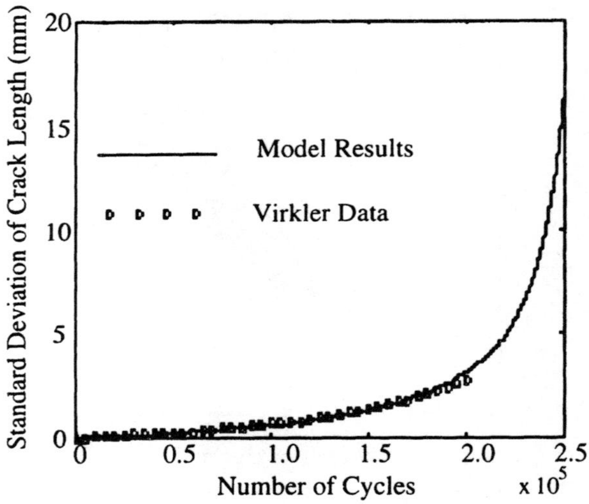

Fig. 7.8 Standard deviation of crack length.

compares the profile of the standard deviation of crack length generated from the experimental data with that derived from the model. The model results are in very close agreement with the experimental data for the first two moments of the crack length as seen in Figs. 7.7 and 7.8. The rapid rise in the standard deviation after 11 500 seconds (i.e. 230 000 cycles) in Fig. 7.8 is due to the fact that unstable crack growth begins here and the specimen is about to break. These results of damage model validation have been generated with no sensor data. That is, the inverse of the intensity of the measurement noise, υ, in Equation (7.10) has been taken to be zero.

The results in Figs. 7.7 and 7.8 are comparable with those reported by Spencer and Tang (1988), Spencer *et al.* (1989), and Enneking (1991) who numerically solved the Kolmogorov diffusion equations to obtain the statistics of crack length. The solution of the partial differential equations of Kolmogorov is computationally two orders of magnitude more expensive than that of the ordinary differential equations encountered in the extended Kalman filter (EKF) which required less than two seconds of CPU time on a Pentium-90 computer for the entire period of 12 500 sec, i.e. 250 000 cycles. It should be noted that the Kolmogorov equations yield the conditional density as a function of time while the extended Kalman filter formulation yields a time series of only the first two moments of the crack length. However, the CPU time requirement is virtually unchanged if the two parameters of the lognormal distribution are also computed based on the first two moments provided by the Kalman filter. Since Kalman filtering provides a recursive solution, the memory requirements are also significantly lower. This fact coupled with the accuracy of the results make the proposed method ideally suited to real-time sensing of fatigue damage.

As mentioned in the previous section, the instantaneous PDF of the crack length $c(t)$ is determined from the first two moments of crack length assuming that the distribution of crack length is lognormal. Based on these statistics of $c(t)$, the conditional probability of any one of the (M) mutually exclusive and exhaustive hypotheses is computed as a function of time. To elucidate the concept of hypothesis testing for remaining life prediction, the damage state space is partitioned into M segments starting from the initial crack length c_0. Probabilities of the M hypotheses having the random crack length $c(t)$ at a given time t located in one and only one of these segments is computed on-line. Given that c_0 = 9.0 mm with probability 1 and the critical crack length c^* = 49.8 mm, the 11 hypotheses (M=11) are depicted in Table 7.1. The time evolution of probability of the hypotheses for the simulation of the Virkler experiment described above is shown in Fig. 7.9 where each successive hypothesis

Table 7.1 Damage prediction in the probabilistic setting

Description	*Range*
Hypothesis H_1	9.00 mm ^{2}c(t) < 13.08 mm
Hypothesis H_2	13.08 mm ^{2}c(t) < 17.16 mm
•	•
•	•
•	•
Hypothesis H_{10}	45.72 mm ^{2}c(t) < 49.80 mm
Hypothesis H_{11}	49.80 mm ^{2}c(t) (Unstable crack growth)

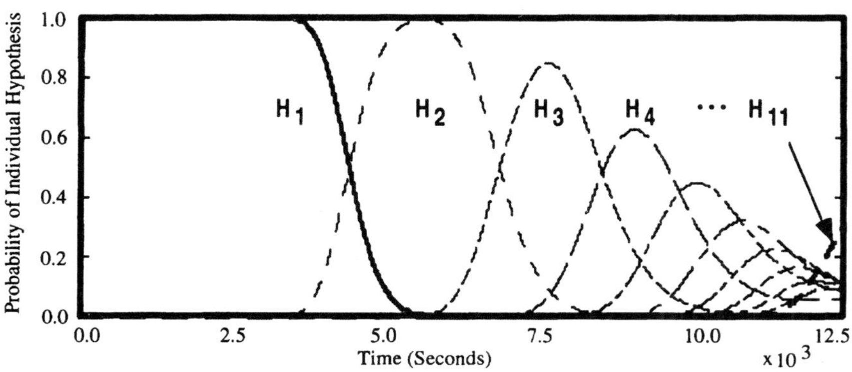

Fig. 7.9 Probabilities of the individual hypotheses as functions of time.

starts at time t=0 sec on the left and ends at time 12 500 sec (i.e. 250 000 cycles) on the right. The plot of H_1 begins with a probability close to 1 and later diminishes as the crack grows with time (i.e. number of cycles). The probability of each of the remaining hypotheses H_2 to H_{10} is initially close to zero and then increases to a maximum and subsequently decreases as the crack growth process progresses with time. The probability of the last hypothesis H_{11} (on the extreme right in Fig. 7.9) of unstable crack growth beyond the critical crack length initially remains at zero and increases rapidly only when the specimen is close to rupture. At this stage, the probability of each of the remaining hypotheses, H_2 to H_{10}, diminishes. The plots in Fig. 7.9 also show the effects of the increasing variance of crack length on the probability of a hypothesis. The higher variance of crack length after about 7500 seconds (i.e. 150 000 cycles), as seen in Fig. 7.8, is manifested in the spreading of the probability over several hypotheses. This information can be used in a higher level decision-making module (possibly a discrete-event supervisor [Garcia *et al.*, 1995]) to make decisions of failure prognosis, life-extending control, and information-based predictive maintenance, or generate warnings and alerts.

To elucidate the concept of hypothesis testing for remaining life prediction, we partition the damage state space starting with the initial crack length into M segments. Each of the M hypotheses of having the random crack length $c(t)$ at a given time t being located in one and only one of these segments are mutually exclusive and exhaustive, and can be computed on-line as a function of time. Given that the initial crack length c_o = 9.0 mm with probability 1, the critical crack length c^* = 49.8 mm, and the 11 hypotheses (M=11) are depicted in Table 7.1. The time evolution of probability of the hypotheses for the simulation of Virkler's experiment described above is shown in Fig. 7.9 where each successive hypothesis starts at time t=0 sec on the left and ends at time 12 500 sec (i.e. 250 000 cycles) on the right. The plot of H_1 begins with a probability close to 1 and later diminishes as the crack grows with time (i.e. number of cycles). The plot of each of the remaining hypotheses H_2 to H_{10} is initially close to zero and then increases to a maximum and subsequently decreases as the crack growth process progresses with time. The plot of the last hypothesis H_{11} (on the extreme right in Fig. 7.9) of unstable crack growth beyond the critical crack length increases rapidly when the specimen is close to rupture. At this stage, the probability of each of the remaining hypotheses, H_2 to H_{10}, diminishes. The plots in Fig. 7.9 also show the effects of increasing variance of crack length on the probability of a hypothesis. The higher variance of crack length after about 7500 seconds (i.e. 150 000 cycles) is manifested in the spreading of the probability over several hypotheses. The real-time information is used in the higher level decision module of a discrete-event supervisor for failure prognosis and maintenance.

In its current form, the deterministic damage model is valid only for constant amplitude loads. Additionally, the stochastic damage model is tuned for particular constant amplitude load and due to the lack of experimental data; the parameter ξ for stress amplitudes other than that used in the Virkler experiment could not be determined. These restrictions do not allow the computation of the remaining life for arbitrary load profiles. Since the deterministic damage model is valid for constant amplitude loads, it can be used to predict fatigue crack growth accurately for step changes in the load amplitude, i.e. a change in the load applied to the structure from an amplitude of 5250 lbs to say 5500 lbs from one cycle to another. This does not affect the accuracy of the deterministic damage model since the structure can be assumed to be freshly loaded with the previous value of crack length as an initial condition. Preliminary work reported by Tang and Spencer (1989) has shown that the parameter ξ is dependent on the maximum stress applied to the structure; it was conjectured that a constant value of ξ can be used for small changes in the stress ratio of the loading applied to the structure without causing large errors. Experimental data at different load levels must be available for testing this conjecture.

Figures 7.10 and 7.11 show the profiles of expected value and standard deviation of crack length, respectively, for the original loading of

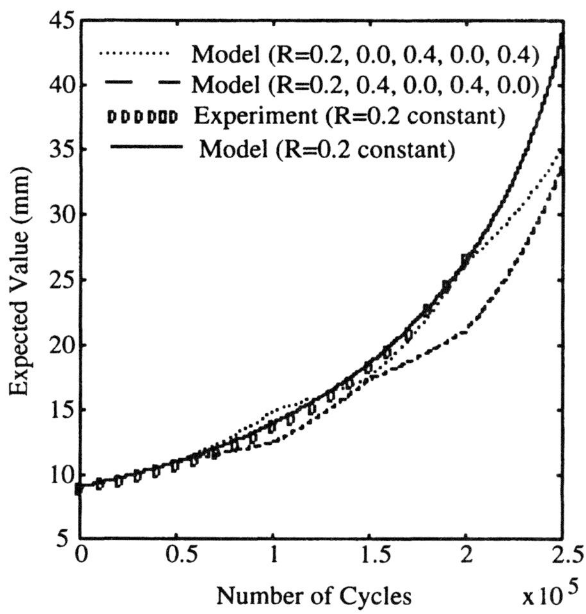

Fig. 7.10 Expected value of crack length under varying amplitude loading.

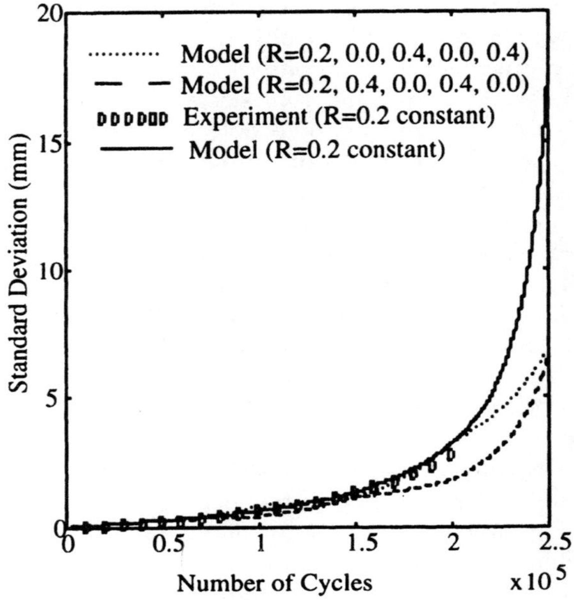

Fig. 7.11 Standard deviation of crack length under varying amplitude loading.

the Virkler experiment along with two additional load profiles which are constructed by changing the stress ratio (i.e. the ratio of minimum applied stress to the maximum applied stress) after 50 000, 100 000, 150 000 and 200 000 load cycles. The effects of changing the stress ratio, *R*, for a given peak stress were investigated. A smaller *R* causes a higher stress amplitude and hence a larger crack growth rate (due to increased ΔK_{eff}) even though the mean stress is decreased. Similarly, a larger *R* results in a smaller crack growth rate due to a reduced stress amplitude. The sequence effect of fatigue crack growth (Suresh, 1991) is clearly evident whereby the fatigue damage accumulation is dependent on the order in which different stress amplitudes are applied to the structure. It is interesting to note that the amount of fatigue damage accumulation during the test period of 250 000 cycles is less in both cases of variable amplitude loading than that for the original constant amplitude case. This also results in a lower standard deviation of the crack length.

Figure 7.12 shows the probability of two hypotheses, H_2 and H_6, out of the 11 hypotheses defined above as functions of time for the Virkler and two additional load profiles. Figure 7.13 shows the effects of changing the stress ratio on the remaining life under three different scenarios. The first scenario, represented by the dashed line, predicts the remaining life

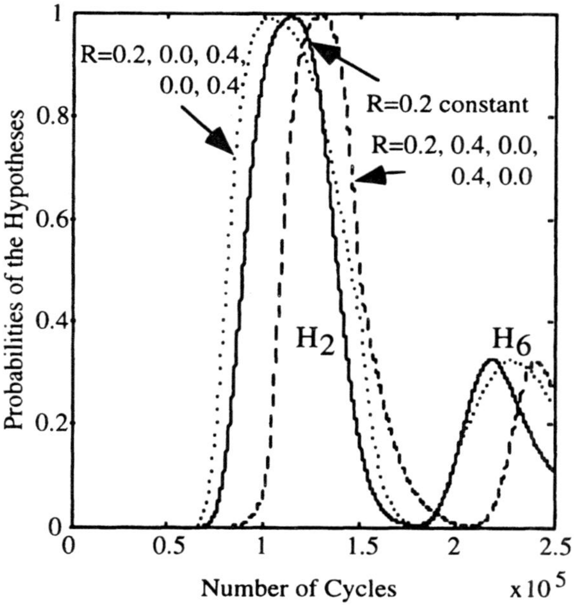

Fig. 7.12 Probabilities of hypotheses H_2 and H_6 under varying amplitude loading.

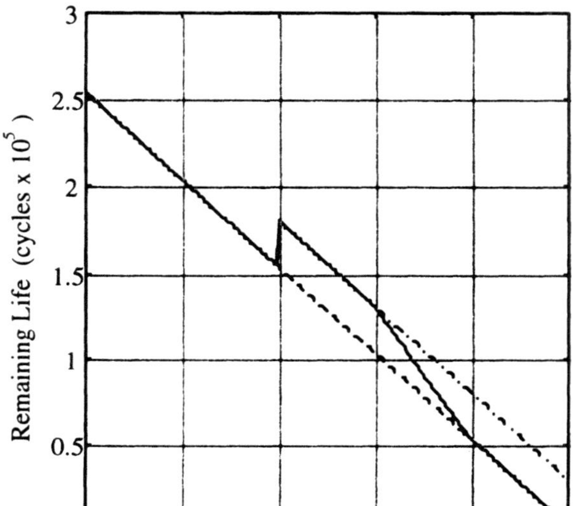

Fig. 7.13 Predicted remaining life of a specimen under varying amplitude loading.

when the stress ratio R in the anticipated load profile is held constant at 0.2 until 200 000 cycles are applied and then changed to 0.5 for the remaining 50 000 cycles up to the final stage at 250 000 cycles. This is a straight line of slope –1 since the remaining life prediction is based on the anticipated load profile which is identical to the actual load applied to the specimen for the entire period of 250 000 cycles. The second scenario represented by the dash-dot line is identical to that for the dashed line up to 100 000 cycles and then differs in the following sense. After 100 000 cycles, the anticipated stress ratio R in the scenario of the dash-dot line is changed to 0.5 at 150 000 load cycles and held constant until 250 000 load cycles have elapsed. This causes the remaining life prediction to show a step increase at 100 000 cycles due to the lower applied stress for part of the entire load sequence. This change is actually implemented at 150 000 cycles and hence the dash-dot line remains at a constant slope after the initial step change at 100 000 cycles. The third scenario, represented by the solid line, is generated by anticipating, at the instant of 100 000 cycles, the same change in value of R to take place at 150 000 cycles but, unlike the second scenario (i.e. the dash-dot line), this anticipated change does not actually happen. Since the fatigue damage rate is higher due to the larger stress amplitude of the applied load, the predicted remaining life starts ramping down at 150 000 cycles and ending at 200 000 cycles when the stress level is known. The solid line merges with the dashed line at 200 000 cycles since, from that moment onward, the applied load and anticipated load are identical for the two scenarios.

7.4 CONCLUSION

This chapter presents an integrated concept of prognostics, maintenance, and life extension to enhance safety, operational reliability, and availability of continuous-time production processes such as chemical and processing plants, pulp and paper mills, and large rolling mills. This concept can be extended to discrete production processes such as the manufacture of glass containers and light bulbs, semi-conductors, ceramic-enamel coatings on appliances, and food products. It is also directly applicable to non-manufacturing continuous-time processes such as automotive, aircraft, and nuclear and fossil power plants. Furthermore, as the science and technology of materials evolve, the proposed approach for systems synthesis will be capable of easily incorporating an appropriate representation of the material properties. Extended life coupled with enhanced safety, high performance, and reduced maintenance cost will have a significant economic impact on diverse industrial applications. This research is an important step toward synthesis of control and diagnostic systems based on sensory and operational information such that the plant can be inexpensively maintained, and safely and efficiently maneuvered under diverse operating conditions. Salient features of the proposed technology of integrated decision and control are summarized below:

- real-time prediction of impending failures of plant components resulting in increased safety and reliability;
- increase of the mean time between major maintenance actions resulting in reduced operational cost;
- substantial extension of service life with no significant loss of plant performance; and
- optimized trade-off between the system performance and structural durability of the plant to achieve the mission objectives.

From an application perspective, major potential benefits of the proposed research relative to technology enhancement of production processes are as follows:

- enhanced safety, operational reliability, availability, and maintainability;
- reduction of the life cycle cost by optimization of the plant operation and maintenance schedule;
- flexibility in scheduling of plant operation and maintenance; and
- facilitation of design revisions as new technologies of active control and structural materials evolve.

Potential applications of integrated prognostics, maintenance, and life-extending control of continuous-time production processes include a wide spectrum of industrial applications where results of the proposed research can enhance safety and productivity accompanied by reduced life cycle cost. For example, in chemical plant and paper mill operations, life extension and reduced maintenance can be achieved by placing a constraint on the level and rate at which the plant can be operated to meet the production demand and, in the event of impending failures, early warnings can be generated to avoid an emergency shutdown. Similarly, for manufacturing of glass products by a continuous melting process, an integrated approach to prognostics, maintenance, and life extension will extend the useful life of a critical plant component, e.g. the tank, and increase overall productivity. As a non-manufacturing example, life-extending control of automotive engine and transmission systems will reduce stresses in the drive-train components while engine performance remains virtually unchanged. The result will be a combination of extended life and smaller drive-train mass with associated emission reduction and savings in the fuel consumption. Another example is life extension and reduced maintenance of fossil power plants which are required to respond to megawatt demand and pollutant emission constraints.

ACKNOWLEDGEMENTS

The authors acknowledge the technical contributions of their graduate students and colleagues in this research work. Some of the results on stochastic modeling of damage dynamics were generated by Dr. Sekhar Tangarila. The authors are grateful to Professor B.M. Hillberry of Purdue University for providing the fatigue test data. The research work reported in this chapter was supported in part by:

- National Science Foundation under Research Grant No. ECS-9216386.
- National Science Foundation under Research Grant No. DMI-9424587.
- National Science Foundation under Research Grant No. CMS-9531835.
- Department of Energy Contract No. DE-FC36–95GO10064.
- Electric Power Research Institute under Contract No. EPRI RP8030–5.
- NASA Lewis Research Center under Grant Nos. NAG 3–1240, NAG3–1673, and NAG3–1799.
- Office of Naval Research under Grant No. N00014–90-J-1513.

BIBLIOGRAPHY

Bannantine, J.A., Comer, J.J. and Handrock, J.L. (1990) *Fundamentals of Metal Fatigue Analysis*, Prentice-Hall.

Basseville, M. (1988) Detecting changes in signals and systems – a survey. *Automatica*, **24**, 309–26.

Bogdonoff, J.L. and Kozin, F. (1985) *Probabilistic Models of Cumulative Damage*, Wiley Interscience, New York.

Bolotin, V.V. (1989) *Prediction of Service Life for Machines and Structures*, ASME Press.

Cieslak, R., Desclaux, C., Fawaz, A.S. and Varaiya, P. (1988) Supervisory control of discrete-event processes with partial observations. *IEEE Transactions Automatic Control*, **33**(3), March, 249–60.

Dai, X. and Ray, A. (1994a) Damage-mitigating control of the space shuttle main engine: Part I – Life prediction of the main thrust chamber wall. *ASME Winter Annual Meeting, Chicago*, and a revised version to appear in *Journal of Dynamic Systems, Measurement and Control*, September 1996.

Dai, X and Ray, A. (1994b) Damage-mitigating control of the space shuttle main engine: Part II – Formulation of an optimal policy. *ASME Winter Annual Meeting, Chicago*, and a revised version to appear in *Journal of Dynamic Systems, Measurement and Control*, September 1996.

Dai, X. and Ray, A. (1995) Life prediction of the thrust chamber wall of a reusable rocket engine. *Journal of Propulsion and Power*, 11(6), November–December, 1279–87.

Diaz-Bobillo, I.J. and Dahleh, M.A. (1993) Minimization of the maximum peak-to-peak gain: the general multiblock problem. *IEEE Transactions on Automatic Control*, **38**(10), October, 1459–82.

Doyle, J.C., Glover, K., Khargonekar, P. and Francis, B. (1989) State-space solu-

tions to standard H2 and H∞ control problems. *IEEE Transactions on Automatic Control*, **34**(8), August, 831–47.
Enneking, T.J. (1991) *On the Stochastic Fatigue Crack Growth Problem*, PhD Dissertation, Department of Civil Engineering, University of Notre Dame, Notre Dame, Indiana.
Garcia, H.E. and Ray, A. (1996) State-space supervisory control of reconfigurable discrete event systems. *International Journal of Control*, **63**(4), 767–97.
Garcia, H.E., Ray, A. and Edwards, R.M. (1994) A reconfigurable hybrid system and its application to power plant control. *IEEE Transactions Control Systems Technology*, **3**(2), 157–70.
Gill, P.E., Murray, W. and Wright, M.H. (1981) *Practical Optimization*, Academic Press, London.
Himmelblau, D.M. (1978) *Fault Detection and Diagnosis in Chemical and Petrochemical Systems*, Elsevier, Amsterdam, The Netherlands.
Iserman, R. (1984) Fault detection based on modeling and estimation methods – a survey. *Automatica*, **20**, 387–404.
Jazwinski, A.H. (1970) *Stochastic Processes and Filtering Theory*, Academic Press, New York.
Lin, Y.K. and Yang, J.N. (1983) On statistical moments of fatigue crack propagation. *Engineering Fracture Mechanics*, **18**(2), 243–46.
Lorenzo, C.F. and Merrill, W.C. (1991) Life extending control: a concept paper. *American Control Conference*, Boston, MA, June, 1080–95.
Massoumnia, M.A. (1986) A geometric approach to synthesis of failure detection. *IEEE Transactions on Automatic Control*, **31**, 839–46.
Newman, J.C. (1981) A crack-closure model for predicting fatigue crack growth under aircraft loading. *Methods and Models for Predicting Fatigue Crack Growth under Random Loading*, ASTM STP 748, 53–84.
Newman, J.C., Swain, M.H. and Phillips, E.P. (1986) An assessment of the small-crack effect for 2024-T3 aluminum alloy, in *Small Fatigue Cracks*, Ritchie, R.O. and Lankford, J. (eds).
Noll, T., Austin, E., Donley, S. Graham, G. Harris, T., Kaynes, I., Lee, B. and Sparrow, J. (1991) Impact of active controls technology on structural integrity. *32nd AIAA/ASME/ASCE/AHS/ASC Structures, Structural Dynamics, and Materials Conference*, Baltimore, MD, April 1991, 1869–78.
Oksendal, B. (1992) *Stochastic Differential Equations: An Introduction with Applications*, 3rd edn., Springer-Verlag, Berlin.
Ozveren, C.M. and Willsky, A.S, (1989) *Aggregation and Multi-Level Control in Discrete Event Dynamic Systems*, Massachusetts Institute of Technology, Technical Report No. LIDS-P-1902.
Packard, A. and Doyle, J. (1993) The complex structured singular value. *Automatica*, **29**(1), 71–109.
Paris, P.C. and Erdogan, F. (1963) A critical analysis of crack propagation laws. *Journal of Basic Engineering*, Trans. ASME, **D85**, 528–534.
Patton, R., Frank, P. and Clark, R. (1989) *Fault Diagnosis in Dynamic Systems*, Prentice-Hall, NJ.
Ramadge, P.J. and Wonham, W.M. (1987) Supervisory control of a class of discrete event processes. *SIAM Journal of Control and Optimization*, **25**(1), January, 206–30.
Ray, A. (1990) Multi-level hypotheses testing for signal validation in continuous-time processes. *ASME Journal of Dynamic Systems, Measurement and Control*, December, 787–90.
Ray, A. and Dai, X. (1995) *Damage-Mitigating Control of a Reusable Rocket Engineer*

for High Performance and Extended Life, NASA Contractor Report 4640, NASA Lewis Research Center, Cleveland, OH.

Ray, A., Dai, X., Wu, M-K., Carpino, M. and Lorenzo, C.F. (1994a) Damage-mitigating control of a reusable rocket engine. *AIAA Journal of Propulsion and Power*, **10**(2), March–April, 225–34.

Ray, A. and Desai, M. (1986) A redundancy management procedure for fault detection and isolation. *ASME Journal of Dynamic Systems, Measurement and Control*, September, 248–54.

Ray, A., Geiger, R., Desai, M. and Deyst, J. (1983) Analytic redundancy for one-line fault diagnosis in a nuclear reactor. *AIAA Journal of Energy*, July–August, 367–73.

Ray, A. and Luck, R. (1991) Signal validation in multiply-redundant systems. *IEEE Control Systems Magazine*, February, 44–49.

Ray, A. and Tangirala, C.S. (1997) A nonlinear stochastic model of fatigue crack dynamics. *Probabilistic Engineering Mechanics*, **12**(1), 33–40.

Ray, A. and Weng, C-K. (1995) Integrated prognostics, maintenance and life extending control of power plants. *ISA POWID/EPRI Symposium*, La Jolla, CA, June, 67–75.

Ray, A. and Wu, M-K. (1994) Fatigue damage control of mechanical systems. *SMART Materials and Structures*, **3**(1), March, 47–58.

Ray, A., Wu, M-K., Carpino, M. and Lorenzo, C.F. (1994b) Damage-mitigating control of mechanical systems: Part I – Conceptual development and model formulation. *ASME Journal of Dynamic Systems, Measurement and Control*, **116**(3), September, 437–47.

Ray, A., Wu, M-K., Carpino, M. and Lorenzo, C.F. (1994c) Damage-mitigating control of mechanical systems: Part II – Formulation of an optimal control policy and simulation. *ASME Journal of Dynamic Systems, Measurement and Control*, **116**(3), September, 448–55.

Saradis, G.N. (1989) Analytic formulation of the principle of increasing precision with decreasing intelligence for intelligent machines. *Automatica*, **25**, 461–67.

Skogestad, S., Morari, M. and Doyle, J.C. (1988) Robust control of ill-conditioned plants: high purity distillation. *IEEE Transactions on Automatic Control*, **33**(12), December, 1092–1105.

Sobczyk, K. and Spencer, B.F., Jr (1992) *Random Fatigue: Data to Theory*, Academic Press, Boston, MA.

Spencer, B.F., Jr. and Tang, J. (1988) A Markov process model for fatigue crack growth. *ASCE Journal of Engineering Mechanics*, **114**(12), 2134–57.

Spencer, B.F., Jr., Tang, J. and Artley, M.E. (1989) A stochastic approach to modeling fatigue crack growth. *The AIAA Journal*, **27**(11), 1628–35.

Stein, G. and Athans, M.A. (1987) The LQG/LTR procedure for multivariable feedback control design. *IEEE Transactions on Automatic Control*, **AC-32**(2), February, 105–14.

Stein, G. and Doyle, J.C. (1991) Beyond singular values and loop shapes. *AIAA Journal of Guidance, Control and Dynamics*, **14**(2), February, 105–14.

Suresh, S. (1991) *Fatigue of Materials*, Cambridge University Press, Cambridge, U.K.

Tangirala, S., Ray, A. and Carpino, M. (1995) Damage mitigating control of mechanical systems: experimental verification of the concept. *SMART Materials and Structures*, **4**(2), June, 139–46.

Vidyasagar, M. (1992) *Nonlinear Systems Analysis*, 2nd edn, Prentice-Hall, Englewood Cliffs, NJ.

Virkler, D.A., Hillberry, B.M. and Goel, P.K. (1979) The statistical nature of fatigue crack propagation. *ASME Journal of Engineering Materials and Technology*, **101**(2), April, 148–53.

Willsky, A.S. (1976) A survey of design methods of failure detection in dynamic systems. *Automatica*, **12**, 601–11.

Yang, J.N. and Manning, S.D. (1994) Aircraft fleet maintenance based on structural reliability analysis. *Journal of Aircraft*, March–April, **31**(2), 419–25.

CHAPTER 8

An integrated automated root cause identification fuzzy neural network reasoning for quality control

Farhad Tadayon and Jay Lee

Early recognition of assembly interface problems can noticeably save time and money by eliminating the red tape and procedural activities among different and often redundant organizations. The Automated Root Cause Identification System (ARCIS) is a quality control expert database system which can suggest and identify the root cause of an assembly variation problem based on the history of previous occurrences and in conjunction with the on-line multivariant statistical information obtained from principal clusters of variations. A fuzzy Cerebellar Model Arithmetic Computer (CMAC) neural network system serves as a 'watch-dog' to perform feature extraction of the output from the ARCIS and to compare its pattern with previous patterns. As a result, the trends of performance degradation can be measured through this integrated methodology.

8.1 INTRODUCTION

In today's manufacturing world, a major global concern for every industry is to reduce defects and improve quality. The major source of defects is variability of the system from what it should be, and not necessarily from ideal conditions. Variability comes from changes in the conditions in which parts are produced. Variability and defects are directly related. When variability increases, the effectiveness or efficiency of the product degrades and the number of defects elevates. Imagine that in the production cycle of an item the raw materials were perfect, the design was flawless, tools were 100 per cent precise, and its components were manufactured with no mistakes. In this case, all the components could be assembled correctly with zero defects.

In practice, however, materials are not ideal, workmanship varies from one location to another, tools do not have the required precision, machines do not produce identical parts, parts do not fit, resulting in the assembly shop having numerous defective parts. Variation is a fact of the real manufacturing world. Not controlling it can paralyze the output of an industry. Small variations accumulate from part to part significantly, and sometimes the excess of this accumulation can destroy the purpose and objective of the product. The impact of the variability is felt especially in the sub-assembly or assembly shops where parts are put together. Part variability might not cause that much of a problem when two or three parts are being assembled, but when the number of components increases, major complications occur. A good example is in the assembly of an airplane. There are about two million parts and according to Kuiper (1990), based on industry data, there could be about 140 000 defects in the parts manufactured for each plane. This type of situation signifies the importance of improving the overall production process which would eventually reduce the defects. Defects result not only from limitations on parts and materials, but also from design.

It is a fact that we are surrounded by variations. As Lefevre (1991) points out, variation can be good or bad: with no variation the world would be dull, but with too much variation the situation would be chaotic. In a production environment, variation is justly considered as an enemy. Person-to-person, part-to-part, time-to-time, and project-to-project variations as well as variability in the processes, equipment, and environment cause problems which cost money and time. In the manufacturing environment, variation of component parts causes major problems in fitting them together, and in combination with variation in assembly processes, more problems occur. These variations can result in schedule delays, rework, repair, scrap, and other non-value added cost and time.

Detecting the source(s) of variations is the key to understanding and resolving the problem. Experts agree that identifying the root causes of variability problems consumes most of the variation control activities. There are few methods and techniques which are often used for identifying the root cause of these problems. In an assembly or production environment, when a problem has been identified and has resolved, occasionally a report is generated and eventually filed away. It is very common that the problem or one similar to it could reoccur. Unfortunately, most of the time, the activity of solving the problem would start from scratch, which would cost the company unnecessary resources. There are also situations where the procedures to identify the real cause are quite abstract and based on the skill of the investigator (person-to-person variation).

When a misfit or interference problem occurs in an assembly process, the production will either slow down or shut down depending on the severity of the problem. In a unitized job shop, only one assembly unit is worked and completed at a time. Therefore, the slowing of the process or closing down of the shop would have an immediate effect on the production of the next assembly shop, and eventually the entire production schedule will be delayed.

In a large job shop assembly as in airplane or aerospace industries, a production defect results in the generation of a rejection note by the quality inspector. Generally, the inspector is not responsible for identifying the organization unit to investigate the defect. The liaison group would look into the problem. Occasionally, this group would come up with a solution, but frequently the rejection note is routed to another organization such as product engineering, planning, manufacturing, tooling, etc. If any of these organizations do not agree that the problem is related to their activity and involvement, then the rejection note would be transferred to another group to take some kind of action. Therefore, it is possible that the paperwork might be floating from one area to another for many days while the defect has caused schedule delays and tremendous cost to the production line. Early identification of the cause of the defect can noticeably improve the situation by knowing what it is that requires to be investigated right away.

An objective of this chapter is to provide methodologies that can be used to locate the main causes of typical assembly variation which prevent an acceptable fit for an assembly. ARCIS (Automated Root Cause Identification System) is an expert assist database system which can suggest and identify the root cause of an assembly variation problem based on the history of the previous occurrences and in conjunction with the on-line multivariant statistical information obtained from the principal clusters of variations. Implementing such a system could archive different ongoing assembly problems and their corresponding corrective actions. Managers and analysts can better investigate the performance and the quality improvement of their shops or industry by accessing this type of data.

Recognition of the root causes of the part and assembly variations are either qualitative, such as tool wear, drawing errors, poor workmanship, etc., or quantitative, in which the root causes arise mainly from dimensional variability. In the latter case, dimensionally inaccurate parts or assembly are the main driver of the problem, and measurements are necessary to identify the sources of the variations.

Learning and refining from trends and patterns of occurrences are essential elements of a quality control process. An integrated fuzzy Cerebellar Model Arithmetic Computer (CMAC) can function as a neural network system to perform feature extraction of the output from the ARCIS and compare its behavior with previous patterns. The multidimensional and often nonlinear data from ARCIS is directed to the fuzzy CMAC input arena. The system assigns weights and when they are selected, a more refined and adjusted root cause suggestion is reported.

8.2 METHODOLOGY

In order to provide a system which would assist the expert or the manufacturing problem investigator to identify the root causes of the assembly variation, the following four categories for the methodology, as shown in Fig. 8.1, are established.

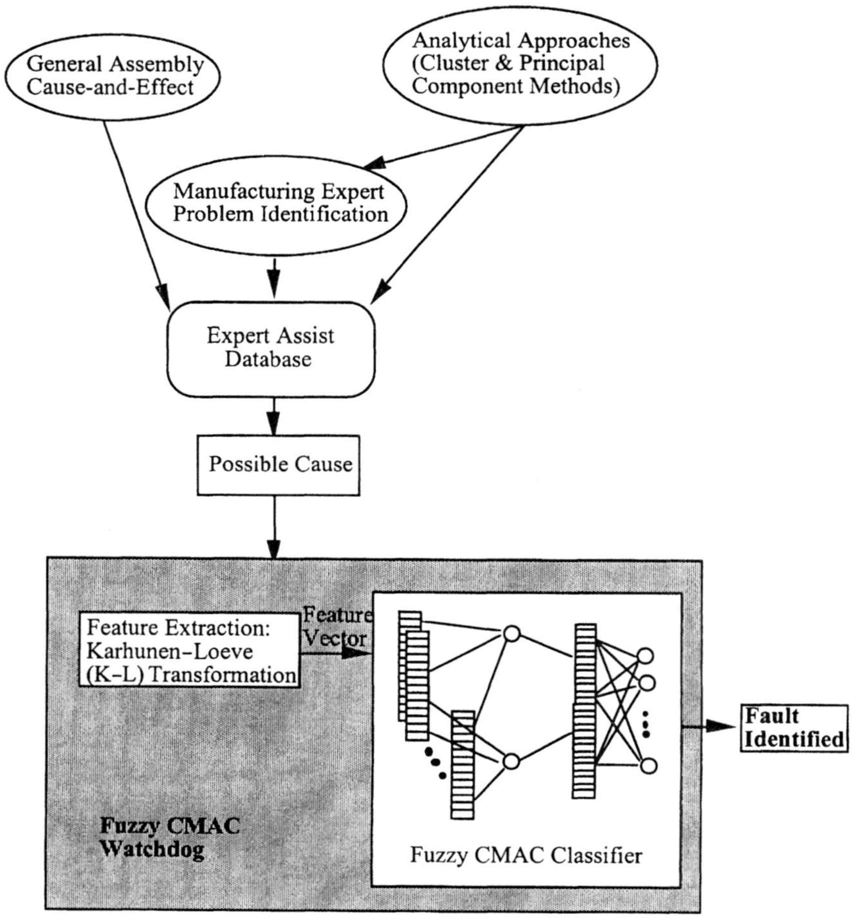

Fig. 8.1 ARCIS-CMAC diagram.

1. General assembly cause-and-effect.
2. Analytical approaches.
3. Manufacturing expert problem identification.
4. Expert assist database.
5. Fuzzy neural networks reasoning for performance measurement.

8.2.1 General assembly cause-and-effect

The methodology used in this approach is considered qualitative because of the lack of dimensional data. Manufacturing defects, mainly assembly products, are investigated based on historical data. This type of information can be extracted from the recorded rejection tags or reports. The data is collected for different types of production models. Pareto analysis can be undertaken to detect the top few defects which contribute to the majority of the problems. After identifying the key defects, main causes of these defects are studied. Cause-and-effect diagrams can be used to demonstrate their relationship.

Again, in addition to experts' input, documented past history can provide information about the specific defects and their relationship with certain causes. In order to make a fair comparison, it is recommended to normalize the frequency of the defects and their causes relative to each other among different models or products. For example, for a certain type defect, model A has more defects than model B, and this is because more model A products are being made in that period of time.

8.2.2 Analytical methods

The primary approach for this section of the study considers two multivariate statistical methods. Cluster analysis is implemented initially to group the key variables of the assembly misfit. Then the principal component analysis is executed on those variables. These two methods, in conjunction, define a procedure for locating the root variables for assemblies that contain many data points at different locations or parts which cause the assembly misfit.

8.2.3 Manufacturing expert problem-cause identification

The results of the analytical approaches would indicate the location and the pattern of the key variables. The structure and the methodology depends mainly on the knowledge of the process experts. The purpose is to capture the manufacturing information of the investigators, which is based on many years of experience, for that specific assembly. It is highly recommended to conduct independent surveys or interviews with the different people involved, such as the operators, shop supervisor, quality inspector, to identify and group the most frequent problems and their relationship with the specific causes for each cluster or zone of variations for the assembly.

8.2.4 Expert assist database

The analysis can identify the location of the parts or components which are the key variables. Then geometric and process characteristics for the identified principal components which could cause the assembly variations are described based on the manufacturing experts' knowledge. If a number of causes could contribute to the problem, then different weighting factors would be assigned to them.

The incoming assembled parts are examined by the coordinated measuring machine (CMM) system to collect numerical variation data (deviation from nominal). The measurement for every variable or location is checked in the corresponding principal component equation(s). The highest principal value identifies the significant cluster, where the variation occurs most. Upon entering or assigning the cluster number, the system shows a possible defect distribution for that cluster. If the user believes that the problem is other than the suggested one, the system can access the general cause-and-effect data and suggest possible causes for the identified defect by the user. Otherwise, the system awaits a selection. If no selection is made, then the system recommends one or a list of causes, based on the most probable defect. If the user is interested in a specific problem, upon the selection, the possible root cause distribution of that defect is presented.

8.2.5 Fuzzy neural networks reasoning

To illustrate the proposed neural network system, the following sections briefly present the fuzzy CMAC neural network architecture.

The Cerebellar Model Arithmetic Computer (CMAC) neural network

The first CMAC was established by J. Albus (1981) based on a model of the human memory and neuromuscular control system. The CMAC is a perceptron-like associative memory that performs nonlinear function mapping over a particular region of a function space. A CMAC network has the capability to rapidly learn an unknown nonlinear mapping by examples, and to reproduce multiple outputs in response to multiple inputs. Because of its 'smart' table look-up mechanism, and its hash-code based mapping structure, CMACs are able to cope with high-dimensional input/output applications with little deterioration of their processing speed.

Lee and Kramer (1993) and Lee (1992, 1995) have proposed an innovative CMAC-based pattern discrimination model for machine degradation monitoring. As illustrated in Figure 8.1, the proposed fuzzy CMAC watchdog to machine degradation monitoring is an integrated learning, monitoring, and recognition technique. The system is based on a novel neural network structure with two major components: feature extraction, and fuzzy CMAC classifier. The inputs to the fuzzy CMAC watchdog consist of the data characterized by the ARCIS configuration. The outputs will be the confidence value that reflects the performance degradation and the fault detection results.

The feature extraction block is responsible for extracting and enhancing features (behavioral patterns) that are embedded in the collected current and historical sensory data. The Krahunen–Loeve (K–L) nonlinear transformation is used to aggregate these data and generate a feature vector with a reduced dimension that contains enriched information on assembly process behavior. The reduced dimension of this feature vector significantly reduces the burden of the neural network classifier in data processing.

The fuzzy CMAC classifier takes the low-dimensional feature vector as its input and produces fault detection signals as the results of pattern classification. This classifier uses the novel fuzzy CMAC architecture which has been successfully used in many applications. Its prominent feature lies in the extremely fast learning rate (at least 10 times faster) in comparison with the commonly used multi-layer perceptron neural network. It is believed that the proposed fuzzy CMAC watchdog architecture is well suited for self-learning in real time.

Figure 8.2 shows a typical CMAC network schematically, where it is designed to approximate a single-valued mapping $f: u \text{ -> } z$ from a multidimensional real-valued input space to a multidimensional real-valued output space. Architecturally, an overall CMAC mapping $f: u \text{ -> } z$ can be split into three separate modules, namely, input encoding mapping $E(.)$: $u \text{ -> } x$, hash-code based mapping $H(.)$: $x \text{ -> } y$, and a weighted summation function W: $x \text{ -> } y$. The real valued output vector of the CMAC is generated by:

$$z = W^T Y = W^T HX = W^T HE(u)$$

A CMAC network needs training for determining appropriate values of its weight matrix elements in order to represent a given nonlinear function f over the region of interest. A supervised learning scheme is used on observed training data pairs $\{u_0, z_0\}$:

$$W_i + 1 = W_i + \beta_y (Z_{di} - Z_i)^T q$$

where q is the total number of weights activated by u_0, and β is the learning rate between 0 and 1.

Fuzzy systems

Fuzzy set theory was initially proposed by L. Zadeh (1975) as a tool to model the imprecision that is inherent in human reasoning, especially

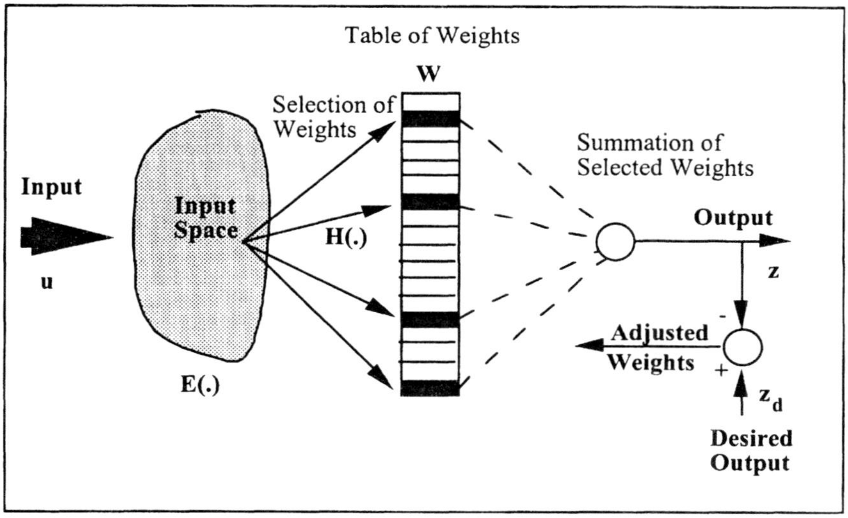

Fig. 8.2 A simplified schematic diagram of a CMAC neural network.

when dealing with complexity. A fuzzy system consists of an input fuzzification module, a set of fuzzy-based rules based on which the approximate reasoning technique is used to make a decision, and an output defuzzification module. The fuzzy approximate reasoning proceeds through the following steps:

1. *Input fuzzification*: in the encoding process, sensor input values are transformed to 'firing strength' of predefined linguistic labels based on the membership function concept.
2. *Knowledge-based inference mechanism*: the fuzzy knowledge base recommends a decision based on its rules' firing strength. A fuzzy system allows designers to incorporate human knowledge into the decision-making process.
3. *Output defuzzification*: because of the partial matching attribute of fuzzy rules and the fact that the preconditions of rules do overlap, more than one fuzzy can fire simultaneously. The consequences of multiple rules can be combined by fuzzy defuzzification procedure.

8.3 EXAMPLES

Tadayon and Liu (1993) and Tadayon (1996) have demonstrated the analytical methodologies and their application in an airplane door assembly shop. They have shown that variation of the critical components or parts at any location can cause problems in obtaining the acceptable and

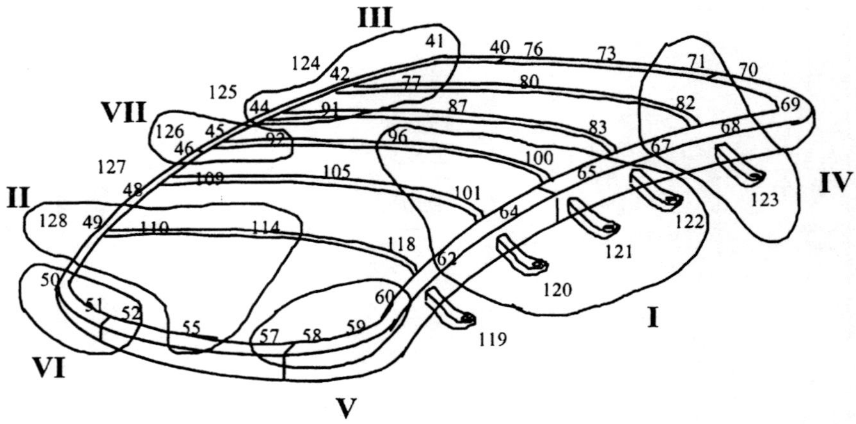

Fig. 8.3 Door assembly cluster formation.

required contour of the final door assembly. For example, by the help of cluster analysis, major clusters of the measured locations, as shown in Figure 8.3, identify significant clusters or zones of the variation caused by different parts or components of the assembly. The principal component analysis for every cluster defines the variation behavior of different parts which would lead to the specific location of the misfit problem of the door assembly.

Based on the location and the nature of the assembly, certain problems could occur at those identified key locations which would contribute to the overall assembly misfit. Possible problems based on their certain characteristics at those key locations can be identified by the manufacturing experts. For example, if the analysis showed that the root cause of the variation is located near the area where two parts are fastened together, then the misfit of the assembly could be due to the misalignment of the parts, mislocation of the fastener holes, engineering mismatch, etc. This set of expert knowledge could be organized and grouped in the form of a knowledge base data set.

The expert assist database incorporates different weight factors for different causes of a problem. Table 8.1 illustrates that the weights, shown as percentages in the table, reflect the probability of causes for a problem which can happen in every cluster.

This database could be self-adaptive, meaning that additional collected data from the other door frames and their component analysis results could be added to the database, which will increase the size and the knowledge capacity of the system for suggesting the root cause of the assembly misfit. This can also be incorporated by interfacing to the fuzzy neural system. The fuzzy CMAC neural network system monitors the

Table 8.1 Principal causes

Cluster	*Cause*	*Weight*
1	Tool wear	60%
1	Wrong tool	40%
4	Vendor	30%
4	Mislocated tool	70%
8	Mislocated part	20%
.	–	–
.	–	–
.	–	–

trend of measured doors and the associated root cause analysis from the ARCIS system to provide a more consistent and accurate diagnostic system for the assembly misfit conditions.

8.4 CONCLUSION

The major objective of this chapter is to introduce and test the concept of an Automated Root Cause Identification System (ARCIS). The central idea behind such a system is the need for a dynamic on-line variation detection system. Consider a scenario in an assembly plant, where an incoming assembly is being checked or inspected at different stages. Different locations on the assembly are being measured by an automated inspection system, for instance a computer-aided coordinate measuring machine or a vision system. The collected data, such as the variation measurements from the defined nominal engineering values for different locations, is routed automatically to the computing process of the system for analysis.

In the dynamic or production mode, specific location measurements or variables are fitted into the principal component of each cluster of the static database. Clusters with the highest principal value would define locations of the potential problem areas. One or a set of problems could occur at this identified location. The problems are then categorized relating to one or more of five categories of machine/tool, material, environment, methods, and/or people related issues. For example, if material is an issue, the quality control database will indicate that the vendor did not deliver good parts during the previous week. Therefore, if the system shows that the assembled parts were selected from the batch received during that time, the vendor is the more probable cause of the problem. It is possible to branch further into more specific causes.

Consider a situation in which the problem deals with the method category, and planning was identified as the primary cause. Additional information from the planning database would indicate that the assembly was routed incorrectly, and operation B (deburring the drilled holes) was omitted, and this is determined to be the root of the problem.

The success, sensitivity, reliability, and detailed root cause identification of the system depends on many factors. The integration and networking of the root case identification system with other design and manufacturing integrated on-line systems is essential. Links with systems such as the engineering design database, computer-aided manufacturing system, computerized process planning, on-line tool and machine monitoring systems, and even on-line management information about the training and the skills of the operators, are important components of a comprehensive system.

Another important element which is very essential for any root cause identification system is a substantial and growing data history. In another words, the system should be self-adapting, meaning that as the data for each unit is collected and fed into the static database, the system could be built to automatically update the clustering and the principal components. Also, if new problems and their associated causes should occur, the system needs to be updated. The fuzzy CMAC neural network of the integrated system works as an inspector/auditor by comparing past and present patterns, and measures its performance degradation. This action can be utilized to flag the need for updating.

REFERENCES

Albus, J.S. (1981) *Brains, Behavior, and Robotics*, BYTE Publications, Peterborough, NH.

Kuiper, K.F. (1990) Defect reduction and process improvement. *Quality Breakthrough, a Boeing Commercial Airplane Group Publication for Boeing Suppliers* (Report No. 10), 1–12.

Lee, J. (1992) *In Process Machine Degradation Monitoring and Fault Detection Using a Neural Networks Approach*, DSc Dissertation, George Washington University.

Lee, J. (1995) Modern computer-aided maintenance of manufacturing equipment and systems: review and perspective. *Journal of Computers and Industrial Engineering*, **28**(4), 793–811.

Lee, J. and Kramer, B.M. (1993) Analysis of machine degradation using a neural networks based pattern discrimination model. *Journal of Manufacturing Systems*, **12**(3), 379–87.

Lefevre, H.L. (1991) Variation in the service sector. *Quality Progress* (Special Issue), 60–2.

Tadayon, F. (1996) Assembly problem identification by using cluster analysis. *International Journal of Production Research*, **34**(3), 879–88.

Tadayon, F. and Liu, M.C. (1993) Principal component analysis: a tool for assembly management. *Computers and Industrial Engineering*, **25**(1–4), 77–80.

Zadeh, L.A. (1975) *Fuzzy Sets and Their Applications to Cognitive and Decision Processes*, Academic Press, New York.

CHAPTER 9

Maintenance using activity-based costing (ABC)

Alice S. Tsai

9.1 INTRODUCTION

Intense global competition and advanced computer technologies bring cost pressures on American businesses. Compared to their American counterparts, educated foreign workers can perform high-technology jobs for far lower wages [1]. Since computers and telecommunications have minimized geographical boundaries and time, American businesses can export even high-technology jobs overseas with lower or minimum costs. Because these businesses must control costs and measure performance, managers can use activity-based costing (ABC) and related activity-based management (ABM) to plan competitive strategies and make critical decisions in complex operations.

ABC complements traditional cost accounting. ABC is an alternative method to measure and assign costs to products produced or services rendered. The major similarity is that both cost methods directly assign costs of direct (raw) material and direct labor to produce or services in the exact same way. The main difference is that traditional cost accounting allocates overhead costs using bases, such as direct labor hours or dollars, whereas ABC assigns overhead costs based on specific activities or processes to produce certain products or render services. ABC can provide a more accurate, detailed assignment of overhead costs to products and services based on corresponding activities or processes.

For maintenance cost, traditional cost accounting treats it as a single cost in overhead without detailed analysis. In contrast, ABC focuses on activities or processes of business operations and can analyze maintenance cost in greater detail. Using ABC, management can minimize non-value-added activities and strive toward continuous improvement and re-engineering of business operations. Thus, ABC has evolved into an activity-based management (ABM) tool to help managers deploy resources, improve their usage, and optimize performance.

With increasing world competition and computerization, the traditional business paradigm is no longer viable. To meet future challenges and global competition, progressive businesses are implementing new management practices such as: activity-based costing (ABC), continuous improvement, re-engineering, 'downsizing,' 'rightsizing,' benchmarking, teaming, empowerment, or strategic alliance. The 'bottom-line' goals of these efforts are to reduce costs and increase profits.

To meet global competition, businesses must juggle conflicting priorities, such as: maintain quality, increase profit, reduce cost, eliminate waste, improve productivity, provide faster response, and offer diversified products or services. For maintenance operations, businesses must improve effectiveness, efficiency, and economy of maintenance functions.

The dynamic business world is forcing changes in the organizational structure of corporations. Compared to yesterday's bureaucratic hierarchy, today's corporations have flatter structures with shorter chains of command and tighter controls. In re-engineered organizations, empowered employees work on cross-functional teams and as a network. In maintenance, these movements have resulted in *total productive maintenance*.

Since many businesses are 'outsourcing' and marketing their products or services across national boundaries, they take advantage of global economies of scale. In becoming 'virtual' enterprises, 'world-class' businesses obtain international value-added services, maintain global supply chains, and collaborate in worldwide joint ventures. For maintenance, these strategies emphasize better use of logistics and new technologies.

Due to revolutionary information technology, global networking and instant electronic data interchange are compressing time and distance for international businesses. Artificial intelligence produces new adaptive controls and decision-support systems. For computer-aided maintenance, businesses must train employees and use advanced technology. The goals of maintenance are to attain zero breakdowns in operation and zero defects in equipment and maintenance service.

With increasing automation, maintenance cost has become a major cost of total costs of goods produced or services rendered. In 'factories of the future,' robots under control of computers can perform nearly all manufacturing operations. Thus, in a 'light-out' factory, maintenance cost is a major overhead cost. Depending on the type of industry, maintenance cost ranges from 15% to 40% of total manufacturing costs [2]. In a typical factory, maintenance costs can exceed $10 million, and the number of maintenance staff can exceed 100 employees. However, to many managers, maintenance remains a necessary evil, instead of a strategic opportunity.

In summary, this chapter will present and apply accounting concepts to maintenance mainly in manufacturing operations. However, because service operations are simpler than manufacturing, we will discuss maintenance in the service industry, where appropriate. We will look at the ABC procedures to measure and assign costs to maintenance activities and then assign these maintenance costs to products produced or services rendered. Finally, we will conclude with an introduction to activity-based management (ABM).

9.2 METHODOLOGY

9.2.1 Issues and challenges: traditional costing and activity-based costing (ABC)

To analyze some basic accounting issues and challenges, we will now discuss some past and current methods in cost accounting and compare traditional costing to activity-based costing (ABC).

Cost accounting measures resources exchanged or consumed, and ultimately reports on income and expenses for products or services. Management uses cost accounting for four main purposes: revenue decisions, profit analysis, inventory valuation, and cost control. For example, revenue decisions and profit analysis include optimizing pricing, product mix, and profit-cost-volume analysis.

Cost accounting accumulates costs by jobs or processes, using two cost methods: job-order costing and process costing. However, some productions fall in between these two methods, such as the production run of thousands of XYZ brand of Pentium computers, each with a speed of 166 MHz.

Job-order costing records costs of a one-of-a-kind, special-order, or customized product or service. Each job is discrete and readily identifiable, such as in manufacturing a specialized machine or maintaining this machine. Each job has a separate identification number and cost record. Original source documents, such as stores requisitions, time sheets, and work orders, provide costs chargeable to each job.

Process costing records costs for homogeneous (like) units of a product or service. Each unit of product or service is identical to another within the same batch (lot). In mass production or continuous processing, homogeneous units of a product have the same amounts of materials, labor, and overhead. For example, the oil, steel, and rubber industries use process costing. To calculate the average unit cost for inventory, we divide total manufacturing costs by number of units produced. However, this calculation of unit cost becomes complicated, when inventory consists of work-in-process, shrinkage, spoilage, or waste.

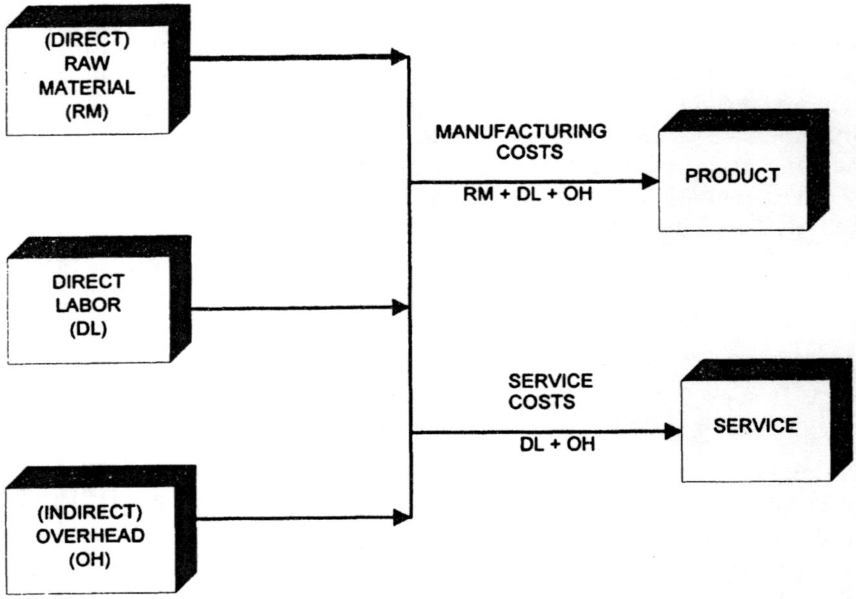

Fig. 9.1 Costs of a product or service.

In the manufacturing industry, a finished product has three major types of costs: direct (raw) materials, direct labor, and overhead. In the service industry, the services rendered include two major types of costs: direct labor and overhead (Fig. 9.1). For direct material cost, material requisitions record raw materials directly used to produce a product. Similarly, for direct labor cost, time cards record labor hours directly used to produce a product or service. Thus, costs of raw materials and direct labor are *direct* costs, *directly traceable* to the products produced or services rendered. On the other hand, overhead are *indirect* costs that are neither direct labor nor direct material; overhead costs are *not directly traceable* to the products produced or services rendered. Examples of overhead are maintenance costs, cutting oil (indirect material), material handling (indirect labor), and utilities.

A decade ago, the manufacturing industry was labor intensive, using low technology and low automation. During this period, direct labor cost exceeded one third of the cost of a manufactured product. Consequently, traditional cost accounting was known as 'direct-labor-hour' accounting. Traditional costing emphasized variable costs of raw materials and direct labor, that varied with volume of production.

In contrast, overhead costs can be variable or fixed, and may or may not vary with the volume of production. Today with advanced technology and automation, the manufacturing industry has become less labor intensive and incurred more overhead costs. While direct labor cost has decreased to 5% of total cost of a manufactured product, overhead cost has exceeded 55% [3]. In computer-integrated manufacturing (CIM), overhead cost often averages 35% of production cost [4]. Thus, advanced technologies, such as robotics, computer-aided design, and flexible manufacturing, have accelerated production activities or processes, which are complex to trace and measure in terms of costs, particularly overhead costs.

Since traditional cost accounting allocates overhead cost to products based on production volume, the types of products affect amounts of overhead cost incurred and profit earned. For instance, low-volume, customized products or services require special setup and handling, which use more resources and increase overhead costs. When varieties of specialized products or services increase, production and sales volume may not increase, and dis-economy of scale, not economy of scale, may prevail. Thus, traditional allocation of overhead cost can understate costs and overstate profits for low-volume, speciality products or services, and conversely, overstate costs and understate profits for high-volume, homogeneous products or services.

Due to the complexity of allocating costs to products and services, businesses experimented with various ways to measure performance and costs. In traditional costing, businesses experimented with additional cost bases beyond the three costs of raw material, direct labor, and overhead. For example, to allocate overhead costs to products or services, businesses expanded the single, predetermined plant-wide overhead rate based on budgeted direct labor hours or dollars; they added allocations based on floor space, number of employees, or other criteria. Others expanded the chart of accounts and number of cost pools for cost accumulation. Manufacturers used direct labor and other resources to 'convert' raw materials into finished products, and tried measuring performance using 'conversion costs' (direct labor and overhead costs).

In automated facilities, maintenance has become a major activity using resources. In advanced systems, computer-aided maintenance is technology-intensive and a major part of overhead costs. Yet, traditional cost accounting often treats maintenance cost as a final, one-line item in overhead, and thus may not accurately allocate maintenance cost to products and services.

In manufacturing, progressive operational managers became aware of the problems of traditional cost accounting. The disadvantages are that traditional accounting: (1) does not isolate and trace costs of necessary activities, (2) does not quantify non-financial, 'soft' benefits, such as quality, throughput, or flexibility, (3) penalizes overproduction, and (4) does not reflect justifications and effects of investments in advanced technologies [5].

The accounting profession has not accepted ABC as a part of generally accepted accounting principles. ABC meets neither financial reporting nor regulatory requirements. Businesses have not adopted ABC as an ongoing general ledger system in cost accounting. Most businesses consider routine collection of ABC information not worthwhile [6].

However, ABC has emerged as a powerful new tool for strategic planning, cost management, and decision making. ABC is a useful management tool, as businesses continue to reduce costs, 'downsize,' implement total quality management, or re-engineer business processes. ABC also complements other operational techniques, such as value chain analysis, capital investment analysis, and competitive benchmarking.

In cost management literature for almost a decade, activity-based costing appeared as a 'new' method to assign costs to products and services [7]. Many users of ABC reported successful applications. The Consortium for Advanced Manufacturing International is instrumental in promoting new cost management concepts for advanced manufacturing [8].

With advances in computers and data processing, all types and detailed breakdowns of cost data are available for management analysis. However, to maximize the accuracy of data and prevent information overload, businesses must input accurate, relevant data and weigh tradeoff between costs and benefits of obtaining additional information.

Because the benefits of ABC outweigh the costs of more precise cost information, we will compare the procedures of ABC to traditional cost accounting. As an alternative method to traditional costing, ABC provides more accurate details to assign overhead cost to products produced or services rendered. Between these two methods, the major similarity and difference are as follows:

- *Similarity*: for direct costs of raw materials and direct labor:
  - In the same exact way, both traditional cost accounting and ABC directly trace and assign these costs to products produced or services rendered.
- *Difference*: for indirect costs of overhead:
  - Traditional cost accounting allocates overhead costs, using bases that change with volume of production, such as machine hours, direct labor hours, or dollars.
  - ABC assigns overhead costs based on costs of relevant activities (processes) needed to produce products or render services.

9.2.2 ABC procedures

Based on a cause-and-effect rationale, ABC links resources, activities, and costs to products produced or services rendered. This rationale (Fig. 9.2) describes the following flow: use resources to perform activities that produce products or render services. Assign costs first to the activities in each activity center, and then the cost of these activities to the products or services.

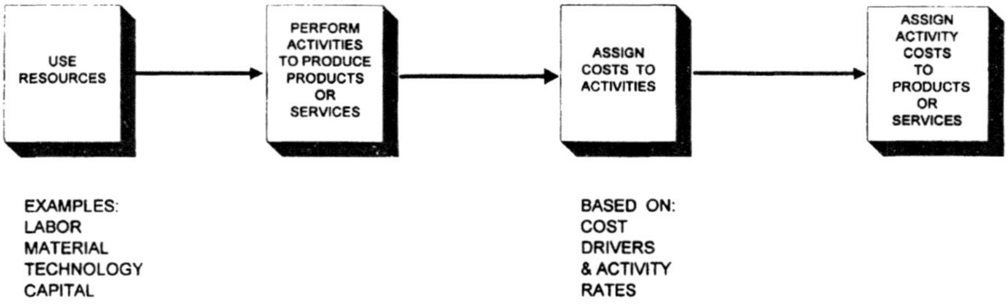

Fig. 9.2 Flow in activity-based costing.

We now apply ABC procedures to maintenance-related activities. These procedures consist of five major 'how-to' steps:

- define activities, resources, and activity cost centers
- combine similar activities
- select cost drivers
- calculate rate for an activity
- assign cost of an activity to a product or service.

Step 1: Define activities, resources, and activity centers

Activities are actions or processes that use resources to produce products or render services. Activities include work done within a partial or entire entity, such as an activity center or organization. ABC helps to identify and minimize non-value-added activities, which are wasteful and do not contribute to profit or customer satisfaction.

Resources include basically labor, material, technology, and capital. Companies must use money and other assets to acquire resources for activities to produce products or render services.

Activity centers can be cross-functional cost centers, for which management wishes to report costs. Depending on management's intent, a traditional department may have one or multiple activity cost center(s).

Activity (cost) centers serve as intermediate links for assignment of costs to products or services. An example is a two-stage method to assign costs. The first stage links and assigns costs to activities in activity centers, such as manufacturing and maintenance. The second (final) stage traces and assigns costs of activities in these centers to individual products or services.

Step 2: Combine similar activities

In an organization, the number of activities performed can be numerous. Although tracing cost to each activity would be ideal, the benefits derived from this detailed information may not be worth the costs to obtain it. Furthermore, detailed breakdowns of costs may not be available or practical. For these reasons, ABC combines similar activities (processes) into activity levels. Manufacturing industries can classify activities into three or more levels [9]. At each activity level, resources, such as labor hours for performing activities, have different connotations. For example, different levels of activities include:

- Activity at unit level for units of a product or service. Labor hours for assembling a unit of a product or service are unit-level resources. At this level, the activity can vary with volume of production, such units produced or direct labor hours used.
- Activity at product-line (service-type) level for some products or services. Labor hours in a machine setup for a batch of products are product-line level or service-type resources. At this level, the activity can vary with production requirements for one or more product lines or service types.
- Activity at organization level or activity-center level for all products. Labor hours for supervision of an entire plant or cost center are organization or activity-center level resources. At this level, the activity can vary with production requirements of an organization or activity center.

Step 3: Select cost drivers

Cost drivers are activities (actions or processes) that incur costs. Cost drivers are the bases for assigning costs to products or services.

For analyzing and selecting cost drivers, information comes from sources such as time-keeping or labor reports, work orders, interviews, surveys, observations, questionnaires, brainstorming, and story-boards. In maintenance, cost drivers can be the number of labor hours, training hours, work orders, breakdowns, component parts used, machine hours, computer hours, etc.

In a finished product, the costs of raw material and direct labor are direct cost drivers. In contrast, indirect overhead cost is not directly traceable to products or services. An example of overhead cost is non-cash depreciation of equipment. To link overhead costs to products or services, ABC managers would use cost drivers and activity rates. Cost drivers and activity rates may link activities to activities; activities to activity-centers; and activities or activity-centers to products or services. Depending on the level of activity, cost drivers may or may not be easy to identify. For example:

- At the *unit level*, activity costs are direct, variable costs, traceable to individual units of a specific product or service, and vary with volume of production (units produced). Examples of direct cost drivers include cost of raw materials used, number of units produced, cost of direct labor hours, and number of machine hours.

- At the *product-line or service-type level*, activity costs are mostly fixed for specific units of product or service, but vary with product lines or types of services. Cost drivers, such as number of product lines and number of machine setups, help to trace and assign costs to specific products or services in a product line or service type.
- At the *organization level or activity-center level*, activity costs are fixed for many products or services. At this level, assigning costs to specific products or services is most difficult; thus, cost drivers may be necessary to link costs to activities, and then assign activity costs to specific products or services. Examples of cost drivers are hours of maintenance, administration, and supervision. At this level, costs of activities may sometimes be based on arbitrary activity rates or percentages.

Step 4: Calculate rate for an activity (activity rate)

To calculate the rate for an activity, we will select a single cost driver for an activity, on-line computer usage. The information systems department logged a total cost of \$50 000 and a total of 500 hours for on-line computer use. The selected cost driver is hours of computer time; the activity rate is calculated as follows:

$$\text{Activity rate} = \frac{\text{total cost}}{\text{total activity}} = \frac{\$50\,000}{500 \text{ hours}} = \$100 \text{ per computer hour}$$

Step 5: Assign costs of an activity to a product or service

ABC assigns activity costs to 'cost objects,' such as a product, a service, a cost center, or an entire organization. From the above example, we apply the activity rate of \$100 per computer hour to the maintenance activity center. To diagnose the condition of forklifts, the maintenance activity center used 20 hours of actual computer time. The cost of on-line computer activity or use assigned to maintenance is as follows:

Activity cost assigned to a product or service
= Activity rate × units of actual activity or usage

$$= \frac{\$100}{\text{computer hour}} \times 20 \text{ computer hours} = \$2000$$

If management decides to assign this maintenance cost to specific products or services, additional cost drivers and activity rates would be necessary. For the remainder of this chapter, we will discuss and apply

ABC to maintenance and briefly introduce activity-based management (ABM).

9.2.3 Maintenance using ABC

Under traditional cost accounting, maintenance is most likely a single, final cost item in overhead. In contrast under ABC, maintenance can have three major types of costs:

- maintenance material cost
- maintenance labor cost
- maintenance overhead cost

ABC offers logical ways to analyze these three costs. Detailed costs of maintenance material and maintenance labor are easily traceable and obtainable. However, unlike costs of maintenance material and labor, maintenance overhead cost is not easily traceable and available. Yet the cost of maintenance overhead can exceed the combined costs of maintenance materials and labor.

The maintenance overhead cost is incurred, because maintenance activity centers use services of other activity centers to support maintenance activities. Today, with computer-aided maintenance systems, automated detailed analyses of the maintenance overhead cost are available. Thus, most of maintenance overhead cost can be analyzed in terms of costs for activities supporting maintenance, such as supervision, planning, scheduling, engineering, training, and documentation of monitoring and diagnostics. The computerized management systems can help create maintenance schedules, document maintenance work completed, and code and sort work orders by work area code, equipment code, equipment part code, and job code.

In addition to activities supporting maintenance, the maintenance activity itself can be subdivided and analyzed by categories. The following illustrates the five major steps of ABC procedures applied to maintenance in a computerized manufacturing operation.

Step 1: Define maintenance activities, resources, and other supporting activities

Maintenance activities act as a support function for manufacturing, and thus become a part of manufacturing overhead cost. Computer-aided maintenance consists of three major types of activities:

- preventive maintenance
- corrective maintenance
- predictive maintenance

Preventive maintenance activity is routine or scheduled maintenance. This type of maintenance activity includes cleaning (maintenance 'house-keeping'), lubrication, inspection, or parts replacement or adjustment. A work order can initiate preventive maintenance. Alternatively, as a part of work routine, empowered team members can start maintenance activities without work orders. Cost drivers for preventative maintenance include the amount of maintenance material and maintenance labor hours, number of machine hours, work orders, and routine maintenance transactions.

Corrective maintenance activity is emergency maintenance. When critical computers or machines break down and disrupt normal operations, managers would push 'panic buttons' and use resources from various activity centers or even outsiders. Due to the emergency nature of sudden computer or machine breakdowns, maintenance staff tend to spend down-time waiting for calls of parts, instead of doing other more productive work. Such unexpected and unscheduled events are costly and incur an opportunity cost. Thus, management would like to improve computer or machine reliability, reduce the frequency of breakdowns, and minimize unscheduled corrective actions with various techniques; their goal would be to achieve zero breakdowns in operation. Cost drivers for corrective maintenance include the amount of maintenance materials, maintenance labor hours, and computer usage time, number of machine hours, and breakdowns.

Predictive maintenance activity attempts to achieve zero defects and zero breakdowns in equipment and maintenance service and includes two types: (1) fault detection, and (2) fault prediction. Techniques for fault detection can monitor and diagnose machine conditions and uncover incipient or impending problems before breakdowns. From historical trends, techniques for fault prediction can estimate the remaining mean-time-before-failure. As an illustration, sample activities in predictive maintenance are shown in Fig. 9.3. Cost drivers for predictive maintenance include number of sensors, engineering hours, computer hours, and size of databases.

Maintenance activities use resources that incur costs. From process flow or expenses in the chart of accounts, management can analyze resources in terms of costs related to maintenance. Some examples include:

- Maintenance material – components and parts.
- Maintenance labor.
- Maintenance overhead – for other activities supporting maintenance:
  - plant or building floor space – for occupancy
  - computer time – for administration and diagnostics
  - off-site support – for time and travel
  - stockroom – for components and parts inventory
  - component repairs – at depot or supplier
  - recruitment

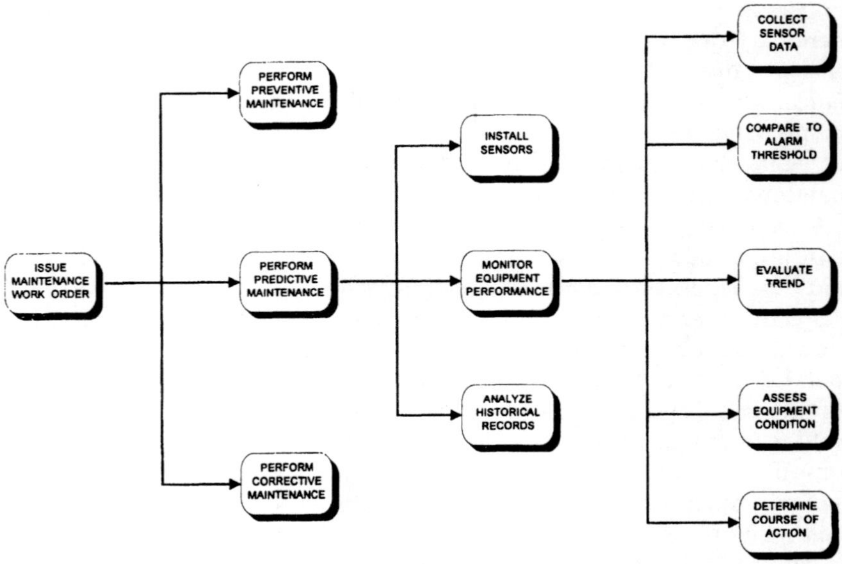

Fig. 9.3 Sample activities in predictive maintenance.

- team building
- training
- administration
- supervision
- rent, utilities, and communication
- depreciation – instruments.

Step 2: Combine similar activities

ABC combines related activities into relevant activity levels. Maintenance activities can be viewed at two activity levels:

- Maintenance activities at type-of-task level:
  - perform preventive maintenance
  - perform corrective maintenance
  - perform predictive maintenance.
- Maintenance activities at organization level:
  - establish maintenance schedules
  - issue work orders
  - maintain stockroom for maintenance components and parts
  - evaluate performance

- maintain standards and procedures
- conduct maintenance training.

Step 3: Select cost drivers

Selecting cost drivers is critical for assigning costs to products and services. Costs of maintenance materials and maintenance labor are direct, identifiable cost drivers. However, maintenance overhead cost relies on a number of cost drivers, which can be difficult to identify. However using historical data, management experience, process analysis, and Pareto's 80–20 rule, ABC managers can identify and select relevant cost drivers. Some sample cost drivers for maintenance activities are as follows.

Preventive maintenance activities

- Preventive maintenance material – from inventory records:
  - cost of maintenance components and parts used
  - number of maintenance components and parts used
  - cost of lubrication oil.
- Preventive maintenance labor – from work orders:
  - number of preventive maintenance hours
  - number of work orders
  - number of inspection hours
  - number of maintenance 'house-keeping' (cleaning) hours
  - number of hours for replacement or adjustment of parts.
- Preventive maintenance overhead:
  - number of maintenance training hours.

Corrective maintenance activities

- Corrective maintenance material – from inventory records:
  - cost of maintenance components and parts used
  - cost of overhaul of components and parts
  - number of maintenance components and parts used.
- Corrective maintenance labor – from work orders:
  - number of emergency maintenance hours
  - number of emergency maintenance calls
  - number of hours for idle (standby) hours
  - number of diagnostic hours.
- Corrective maintenance overhead:
  - cost of managing spare parts inventory
  - depreciation – allocation of cost of diagnostic equipment over a period of time.

Predictive maintenance activities

- Predictive maintenance material:
  - cost of sensors
  - number of sensors.
- Predictive maintenance labor – from work orders:
  - number of hours for fault detection
  - number of hours for fault prediction
  - number of total predictive maintenance hours
  - number of hours for data entry.
- Predictive maintenance overhead:
  - number of hours for engineering analysis
  - number of hours for software and database support
  - number of computer hours used
  - depreciation – allocation of cost of specialized equipment over a period of time.

General operational activities

General operational activities related to maintenance act as support functions for manufacturing, and thus become a part of manufacturing overhead costs. Although some costs of these activities are readily traceable, other costs are not directly traceable to products or services. With proper coding of payroll records, time sheets, and work orders, ABC can trace these activities, identify cost drivers, and assign activity costs to specific products or services. Some examples of general operational activities and cost drivers are:

- Management activities – from time sheet the number of labor hours or from payroll records the cost of labor for:
  - administration
  - supervision
  - computer log-on time – number of online computer hours
  - rent and utilities – number of square feet and number of watts used
  - stockroom activity – number of components and parts used.
- Human resource activities – from time sheets the number of labor hours or from payroll records the cost of labor for:
  - recruitment
  - team building
  - training
  - benefit administration.
- Other charges or cost pool – from time sheets, computer log, travel expense report, and work orders:
  - plant (building) occupancy – number of square feet

- computer time – number or cost of online computer hours
- off-site support – number or cost of hours for work and travel.

Step 4: Calculate rate for a maintenance activity

The following numerical example illustrates ABC with six tables. We will discuss Tables 9.1 through 9.6 in greater detail later. For each of the three maintenance activities, we selected a cost driver (Table 9.1) and calculated an activity rate (Table 9.2).

Step 5: Assign maintenance cost indirectly and manufacturing cost directly to products or services

In the following numerical example of ABC, we *indirectly* assign *non*-manufacturing overhead costs of maintenance and other activities to three manufacturing activity centers (Table 9.3). We then *directly* assigned costs of the manufacturing activities to two products (Table 9.5).

9.3 EXAMPLE

9.3.1 A numerical ABC example and comparison

RST Company manufactures two products (Product A and Product B) and would like to use ABC in two stages to assign costs of activities to these products. Specifically, RST management has selected a cost driver for each activity and would like to assign costs of nine activities to two products.

Traditional cost accounting recorded most costs and other data. The actual data included the following:

	Product A	*Product B*	*Total*
Units produced	100	100	200
Direct labor hours	30	70	100
Cost of raw material	$15 000	$15 000	$30 000
Cost of direct labor	$30 000	$70 000	$100 000

Activity center no.	*Non-manufacturing activity center*	*Manufacturing activity center*
110	Administration	
150	Information systems	
230	Material handling	
360	Preventive maintenance	
370	Corrective maintenance	
410		Fabricating
420		Machining
430		Assembly

Table 9.1 ABC: Activity centers, cost drivers, and all costs

(1) *Activity center no.*	(2) *Activity centre*	(3) *Cost driver*	(4) *Direct (raw) material cost*	(5) *Direct labor cost*	(6) *(Indirect) overhead cost*	(7) *Total cost*
	Non-manufacturing					
110	Administration	Assignment %			$50 000	$50 000
150	Information Systems	No. of transactions			60 000	$50 000
230	Material Handling	Cost of materials			60 000	$60 000
360	Preventive Maintenance	No. of direct labor hours	$5 000	$5 000	10 000	$20 000
370	Corrective Maintenance	No. of breakdowns	5 000	30 000	10 000	$45 000
380	Predictive Maintenance	Cost of maintenance	0	5 000	10 000	$15 000
	Manufacturing					
410	Fabricating		20 000	10 000	10 000	$40 000
420	Machining		10 000	60 000	20 000	$90 000
430	Assembly		0	30 000	10 000	$40 000
Total			$40 000	$140 000	$240 000	$420 000

Note: To simplify illustrations, the costs of direct materials and direct labor for some non-manufacturing activity centers are omitted.

This example illustrates the use of the aforementioned Tables 9.1 through 9.6. Under ABC, we:

- Indirectly assign non-manufacturing overhead costs (or costs of six non-manufacturing activity centers) to three manufacturing activity centers (Table 9.3). (The reasons for this indirect assignment are that the manufacturing activity centers used services of the non-manufacturing activity centers, and thus, incurred non-manufacturing overhead costs for these services.)
- Directly assign costs of manufacturing activity centers (or direct or raw manufacturing material, direct manufacturing labor, and manufacturing overhead cost) to two products, Products A and B (Table 9.5).

Table 9.1 lists all costs for six non-manufacturing activity centers and three manufacturing activity centers and selection of cost drivers for the six non-manufacturing activity centers. Under ABC, Table 9.1 provides the starting point and data for Tables 9.2 and 9.3. For instance, the cost drivers from Table 9.1 provide the bases for Stage 1 in both tables: (1) calculation of activity rates of non-manufacturing activity centers in Table 9.2, and (2) assignment of non-manufacturing overhead costs to three manufacturing activity centers in Table 9.3.

Table 9.2 starts Stage 1 under ABC and illustrates the calculation of activity rates for non-manufacturing activity centers: administration, information systems, material handling, and the three types of maintenance. To calculate the activity rate for a non-manufacturing activity center:

$$\text{Activity rate (Table 9.2: Column 3)} = \frac{\text{Total cost of each activity (Table 9.1: Column 7)}}{\text{Corresponding total activity (Table 9.2: Column 7)}}$$

In addition, Table 9.2 shows the amount of non-manufacturing services (activities) actually used by the three manufacturing activity centers (fabricating, machining, and assembly) to produce products.

Table 9.3 continues Stage 1 under ABC. Costs of six non-manufacturing activity centers represent non-manufacturing overhead costs, that are indirectly assigned to three manufacturing activity centers. To calculate the non-manufacturing overhead cost for each manufacturing activity center:

Non-manufacturing overhead cost (Table 9.3: Column 7, 8 or 9)
= calculated activity rate (Table 9.2: Column 3)
× Units of corresponding manufacturing usage (Table 9.2: Column 4, 5, or 6)

For example, Fabricating used computer services from Information Systems and incurred an overhead cost of \$10 000 (\$0.25/transactions times 40 000 transactions).

Table 9.2 ABC – Stage 1: Calculation of non-manufacturing activity rates

(1) *Activity center no.*	(2) *Non-manufacturing*	(3)	(4)	(5) *Manufacturing activity or usage*	(6)	(7)
	Activity center	*Activity rate (Total cost/Total activity) (Table 9.1/Table 9.2)*	410 *Fabricating (Amount of usage)*	420 *Machining (Amount of usage)*	430 *Assembly (Amount of usage)*	*Total activity or usage*
110	Administration	\$500 = 1%	40%	30%	30%	100%
150	Information Systems	\$.025/Transaction	40 000	60 000	140 000	240 000 transactions
230	Material Handling	\$1/Dollar of material	\$40 000	\$20 000	\$0	\$60 000
360	Preventive Maintenance	\$200/Direct labor hour	10	60	30	100 Direct labor hours
370	Corrective Maintenance	\$5000/Breakdown	1	3	5	9 breakdown
380	Predictive Maintenance	\$0.10/Dollar of maintenance	\$10 000	\$40 000	\$100 000	\$150 000

Table 9.3 ABC – Stage 1: Indirect assignment of non-manufacturing overhead costs to manufacturing activity centers

(1)	*(2)*	*(3)*	*(4)*	*(5)*	*(6)*	*(7)*	*(8)*	*(9)*	*(10)*
		Non-manufacturing overhead costs				*Manufacturing activity centers*			
Activity center no.	*Cost of product*	*110 Administration*	*150 Data processing*	*230 Material handling*	*360–380 Maintenance*	*410 Fabricating*	*420 Machining*	*430 Assembly*	*Total Activity cost*
Non-manufacturing overhead:									
110	Administration	$50 000				$20 000	$15 000	$15 000	$50 000
150	Information System		$60 000			10 000	15 000	35 000	60 000
230	Material Handling			$60 000		40 000	20 000	0	60 000
360	Preventive Maintenance				20 000	2 000	12 000	6 000	20 000
370	Corrective Maintenance				45 000	5 000	15 000	25 000	45 000
380	Predictive Maintenance				15 000	1 000	4 000	10 000	15 000
Manufacturing:									
Overhead						10 000	20 000	10 000	40 000
Raw materials						$20 000	$10 000	$0	$30 000
Direct labor						10 000	60 000	30 000	100 00
Total overhead cost						$118 000	$171 000	£131 000	$420 000

Manufacturing costs (Raw material + Direct labor + overhead)

Table 9.4 ABC – Stage 2: Cost drivers and activity rate for manufacturing overhead

(1) *Activity center no.*	(2) *Activity center*	(3) *Cost driver*	(4) *Produc t A* *Amount of activity* *(Machine or direct labor hrs)*	(5) *Product B* *(Machine or direct labor hrs)*	(6) *Total activity (Machine or direct labor hrs)*	(7) = (4)/(6) *Product A Activity rate [Amount of activity] [Total activity]*	(8) = (5)/(6) *Product B Activity rate [Amount of activity] [Total activity]*	(9) *Manufacturing overhead cost (Table 9.3)*
410	Fabricating	No. of machine hours	50	150	200	0.25	0.75	\$88 000
420	Machining	No. of machine hours	400	100	500	0.80	0.20	\$101 000
430	Assembly	No. of direct labor hours	30	70	100	0.30	0.70	\$101 000

Total manufacturing overhead cost \$290 000

In each manufacturing activity center (Fabricating, Machining, or Assembly), the subtotal manufacturing costs include:

- cost of direct (raw) material
- cost of direct labor
- cost of overhead from non-manufacturing activity center(s) (when non-manufacturing services are used)
- cost of overhead from a manufacturing activity center.

Table 9.4 starts Stage 2 under ABC for the three manufacturing activity centers and the two products. For each manufacturing activity center, Products A and B have corresponding, selected cost drivers and calculated activity rates. To calculate the final activity for each product

$$\text{Final activity rate for each product (Table 9.4: Column 7 or 8)}$$
$$= \frac{\text{Units of activity for each product (Table 9.4: Column 4 or 5)}}{\text{Corresponding total activity (Table 9.4: Column 6)}}$$

Table 9.5 continues Stage 2 under ABC, and illustrates the final, direct assignment of manufacturing costs to the two products. As stated previously, total manufacturing costs of a finished product generally include:

- cost of direct (raw) material
- cost of direct labor
- cost of overhead.

This example does not illustrate but assumes that *manufacturing* costs of raw material and direct labor from the three manufacturing activity centers (Table 9.1: Column 4 and 5) are *directly* traceable to Products A and B. However, in assigning overhead cost to a product, we must do some calculations:

Cost of overhead for each product (Table 9.5: Column 3)
= Calculated activity rate (Table 9.4: Column 7 or 8)
× Corresponding manufacturing overhead cost (Table 9.4: Column 9)

For instance, Fabricating activity center assigns to Product A an overhead cost of \$22 000 (50/200 times \$88 000) using machine hours. Likewise, the remaining manufacturing activity centers, Machining and Assembly, assign overhead costs to products using machine hours and direct labor hours respectively.

Table 9.6 compares two costing methods to allocated or assign costs to Products A and B: (1) traditional cost accounting, and (2) ABC. Under both methods, corresponding costs of raw material and direct labor are directly traceable and assigned to products and thus the same. The two costing methods treat manufacturing overhead cost differently. Table 9.6 illustrates:

Table 9.5 ABC – Stage 2: Direct assignment of total manufacturing costs to products

(1) *Activity center no.*	(2) *Activity center*	(3) *Manufacturing costs for products*	(4) *Cost of product A*	(5) *Cost of product B*	(6) *Total cost*
410	Fabricating	Direct (raw) material	$10 000	$10 000	$20 000
		Direct labor	3 000	7 000	10 000
Table 9.4: Column [(7) × (9)] or [(8) × (9)]		**Assigned overhead**	**22 000**	**66 000**	**88 000**
420	Machining	Direct (raw) material	5 000	5 000	10 000
		Direct labor	18 000	42 000	60 000
Table 9.4: Column [(7) × (9)] or [(8) × (9)]		**Assigned overhead**	**80 800**	**20 200**	**101 000**
430	Assembly	Direct (raw) material	0	0	0
		Direct labor	9 000	21 000	30 000
Table 9.4: Column [(7) × (9)] or [(8) × (9)]		**Assigned overhead**	**30 300**	**70 700**	**101 000**
Total cost			$178 100	$241 900	$420 000

Table 9.6 Comparison of two costing methods: Traditional versus ABC

Under traditional cost accounting:				
Manufacturing Costs for products	*Allocation rate*	*Cost of product A*	*Cost of product B*	*Total cost*
Direct (raw) materials	–	$15 000	$15 000	$30 000
Direct labor	–	30 000	70 000	100 000
Overhead (Table 9.3, 9.4)*	(30/100 or 70/100)	**87 000**	**203 00**	**290 000**
*No. of direct labor hours × Total overhead cost / Total direct labor hours				
Under traditional cost accounting Total cost of product	(Table 9.6)	$132 000	$288 000	$420 000
Under activity-based costing (ABC):				
Manufacturing Costs for products	*Assignment rate*	*Cost of product A*	*Cost of product B*	*Total cost*
Direct (raw) materials	–	$15 000	$15 000	$30 000
Direct labor	–	30 000	70 000	100 000
Overhead (Table 9.4, 9.5)	**Various**	**133 100**	**156 900**	**290 000**
(B) Under activity-based costing (ABC): Total cost of product	(Table 9.5)	$178 000	$241 000	$420 000
Comparison of traditional costing & ABC		*Cost of product A*	*Cost of product B*	
Difference in total cost		$46 100	$46 100	
Difference in percent		–35%	16%	

- Traditional cost accounting used only one base for all manufacturing activity centers – direct labor hours (Column 7 or 8 of Table 9.4) to allocate total manufacturing overhead cost to the products (Formula in Table 9.6). Specifically, Product A has a manufacturing overhead cost of $87 000 (30/100 times $290 000). Product B has a manufacturing overhead cost of the remaining $203 000 (70/100 times $290 000).
- ABC used a different base for each manufacturing activity center – three cost drivers to assign three separate manufacturing overhead costs to the products (Tables 9.4, 9.5, and 9.6).

As shown in this numerical example, traditional costing can overstate or understate overhead, and in turn, total costs of products or services. The inaccuracy stems from using only one base, direct labor hours, to allocate total manufacturing overhead cost, regardless of the number of manufacturing activity centers. In contrast, using a different base (cost driver) to assign corresponding manufacturing overhead cost of each manufacturing activity center, ABC can provide more accurate costs of products or services.

As previously mentioned, ABC is related to activity-based management (ABM). We will explore ABM and apply it to maintenance activities.

9.3.2 Activity-based management (ABM)

Besides assigning costs to products or services, ABC can become a proactive tool for cost management. The goals of cost management include: measure resources used, control non-value-added costs, evaluate performance, and implement improvements. Since ABC links resources, costs, and activities, ABC can give managers the tools to manage and monitor use of resources and optimize trade-offs among such criteria as cost, quality, time, service, and flexibility. Thus, when ABC shifts from cost assignment to cost management and planning, the ABC system evolves into an activity-based management (ABM) system.

The transition to ABM shows that ABM has a broader focus than ABC. Under ABC, the purpose of costing is specific, such as to reduce costs and improve productivity on selected activities, not across the board. In contrast under ABM, the purpose of costing is broader, such as to improve rate of return and capital budgeting. ABM can complement other activities related to productivity, such as re-engineering, continuous improvement, and benchmarking. Instead of relying mainly on traditional financial information, such as budgets, standards, and variances, managers can use ABM to optimize customer value and shareholder value [10].

Similar to ABC, ABM focuses on activities or business processes. Hammer and Champy describe a business process as 'a collection of activities that takes one or more kinds of inputs and creates an output that is of value to the customer' [11]. Porter views business processes as a 'value chains' of activities that convert resources into products or services and increases customer satisfaction [12].

Management often use ABM as predictive models for planning and managing business processes. Using ABM data, management can re-engineer business processes, identify wastes, and reduce cycle times to achieve profitability [13]. Two representative techniques are data flow model and temporal model, as described next.

Data flow model

Various models are available to analyze different aspects of business processes. System models describe the flow of information; re-engineering models analyze business operations or activities; and work-flow models link people and information together. For each of creating and manipulating graphical models, computer-aided software engineering (CASE) can replace tedious manual effort.

Data flow diagrams use graphical network to describe the flow of data and the activities or processes that change this flow. Each activity or process changes not only the data but also the data stores (databases and files). These diagrams can describe business activities of ABM and help analyze the value of each activity. System analysts also use these diagrams to design information systems.

Using data flow models, management can analyze business processes and obtain insights on 'what ifs' or alternatives to fine tune strategies. In operational audit, for example, management can analyze the effectiveness and efficiency of empowerment using throughput time as a criterion to identify non-value-added activities [14]. To achieve high productivity, ABM can help eliminate non-value-added activities and focus on time as an important resource.

Temporal model

As an illustration of time in manufacturing or service activities, the throughput time consists of: set up time, process time, inspection time, move time, storage time, idle (standby) time, and downtime.

- Setup time is the period spent to prepare machines for manufacturing or service.
- Process time is the period spent to manufacture products or to render services.
- Inspection time is the period spent for quality control of products or services.

- Move time is the period spent during various stages of producing a product line or rendering services, including travel time.
- Storage time is the period finished products are in storeroom, waiting to be shipped to customers.
- Idle (standby) time is the wasted period, waiting for repair parts, products, or crucial services.
- Down-time is the period when machines breakdown, and productions or services stop. Down-time includes actual repair, adjustment, maintenance time, and idle time.

Non-value-added activities can exist in inspection, move, storage, idle, and down-time. Some companies measure operational efficiencies by the ratio of process time divided by throughput time. The goal is to achieve a ratio close to 1:1. For many activities, the ratio is only 0.05:1 [15]. ABM can help management analyze and improve this ratio.

For ABC and ABM, critics contend that the complexity of collecting data exceeds the value of improved information. Others equate cost reduction with 'downsizing' and layoffs. Since the accounting profession is conservative by nature in maintaining fair representation of financial information, few accounting publications deal with the effectiveness of ABM. Few firms are willing to expose their ABM effort to internal or external scrutiny [16]. Other users of ABM encounter political and cultural resistance. However, non-financial professionals have continued research and experiments on ABC and ABM as a successful pioneering effort.

The success of ABM depends on conditions similar to those for total quality management. These conditions include the following:

- commitment from senior management to implement ABM,
- access to advanced technological and operational equipment and techniques,
- shared and common goals of management and employees,
- empowered work teams or groups,
- focus on processes and activities.

Therefore, most importantly for the success of ABM, senior management must commit and support the goals and implementation of ABM.

9.4 CONCLUSION

As alternatives to traditional cost accounting, activity-based costing and activity-based management (ABC and ABM) emphasize business activities or processes. By identifying the activities and measuring the costs of these activities to produce products or to render services, management can improve cost control and exploit new strategic alternatives to add value to operations. In addition, ABM can support continuous improvement and re-engineering of business processes.

With increasing acceptance of ABC and ABM, different cost modeling techniques are emerging [17]. A large selection of software packages is also available to facilitate using ABC and ABM for cost modeling and other complex analyses. These developments will further promote the implementation of ABC and ABM [18]. In the near future, we expect to see not only more applications of ABC and ABM but also more research and experimentation in these areas.

In maintenance, ABC and ABM can help operational management maximize value-added activities, reduce costs, and improve quality of maintenance services. ABC helps categorize overhead costs into meaningful details, based on activities in maintenance activity cost centers. Maintenance managers can apply ABC to analyze costs of activities, and use ABM to realign operations and thus minimize breakdowns and downtime.

Often, senior management has neglected and considered maintenance a 'necessary evil,' rather than a strategic opportunity. With more precise, relevant cost data, ABC and ABM will enable maintenance managers to develop innovations, improve performance, and achieve belated recognition.

REFERENCES

[1] New York Times (1995) Skilled workers watch their jobs migrate overseas. *New York Times*, 25 August, A-1.

[2] Mobley, K. (1990) *An Introduction to Predictive Maintenance*, Van Nostrand, NY, NY, 1.

[3] Kelly, K. (1990) Managing for quality costs: a bean-counter's best friend. *Business Week/Quality*, 42–3.

[4] Miller, J.G. and Vollmann, T.E. (1985) The hidden factory. *Harvard Business Review*, September/October, 142–50.

[5] National Center for Manufacturing Research (1990) *Competing in World-Class Manufacturing, American's 21 Century Challenge*, NCMR, Irwin, Chicago, IL, 159.

[6] Cooper, R. and Kaplan, R. (1991) *The Design of Cost Management Systems: Text, Cases, and Readings*, Prentice-Hall, Englewood Cliffs, NJ.

[7] Anthony, R.N. and Govindarajan, V. (1995) *Management Control Systems*, Irwin, Chicago, IL, 331–6.

[8] Berliner, C. and Brimson, J.A. (eds) (1988) *Cost Management for Today's Advanced Manufacturing*. Harvard Business School Press, Boston, MA.

[9] Anderson, L.K. and Sollenburger, H.M. (1992) *Managerial Accounting*, South-Western Publishing, Cincinnati, OH, 81.

[10] Cypress, H. (1994) Re-engineering. *OR/MS* Today, February, 18–29.

[11] Hammer, M. and Champy, J. (1993) *Re-engineer the Corporation: A Manifesto for Business Revolution*. Harper Business, NY, 200–213.

[12] Porter, M. (1985) *Competitive Advantages*. Free Press, NY, NY.

[13] Cokins, G., Stratton, A., and Helloing, J. (1993) *An ABC Managers' Primer*, Irwin, NY, 5.

[14] Tsai, A. (1992) Operational audit of empowerment. *Proceedings of the 18th Annual Advanced Control Conference*, Purdue University, 19–29.

[15] Kaplan, R.S. (1989) Management accounting for advanced technological environments. *Science*, 245, 25 August, 819–23.

[16] Celt, F. (1995) Implementing activity-based management. *Journal of Cost Management*, Summer, **10**(2), 36–49.

[17] Keys, D. (1994) Tracing costs in the three stages of activity-based management. *Journal of Cost Management*, Winter, **7**(4), 30–7.

[18] Albright, T. (1995) Software of activity-based management. *Journal of Cost Management*, Spring, **9**(3), 6–25.

CHAPTER 10

Life cycle maintenance management

Shozo Takata

10.1 CONCEPT OF LIFE CYCLE MAINTENANCE

10.1.1 Facility life cycle and needs for life cycle maintenance

Today's manufacturing has become increasingly dependent upon facilities with the advances of automation and integration of manufacturing systems. This leads to a growing concern about the effectiveness of the facility, because this is a critical factor for achieving high productivity and quality. The effectiveness of a facility should be assessed not only in terms of its functionality available at each point in time but also in terms of the available period of time. The life cycle effectiveness of a facility, F_E, can be defined by the following equation.

$$F_E = \int_0^{t_L} f(t)dt \tag{10.1}$$

where $f(t)$ and t_L represent available functionality at each point in time and the life of the facility, respectively. It is, therefore, important to maintain the facility in the best condition throughout the life cycle in order to improve life-cycle effectiveness as well as to design the facility with better functionality. Figure 10.1 shows an example of improvement of the rate of down-time in a factory where TPM (total productive maintenance) was implemented (Nakajima 1992). It shows how maintenance is effective and thus important for increasing the effectiveness of the facility.

Facility maintenance involves the following basic activities:

1. *Reliability and maintainability design:* improving design based on the evaluation of reliability and maintainability of the facility development phase and providing the design data for maintenance planning and maintenance task management.
2. *Maintenance strategy planning:* selecting the best maintenance strategy appropriate to each part of the facility.
3. *Maintenance task control*: planning and executing the maintenance task based on the selected strategy.

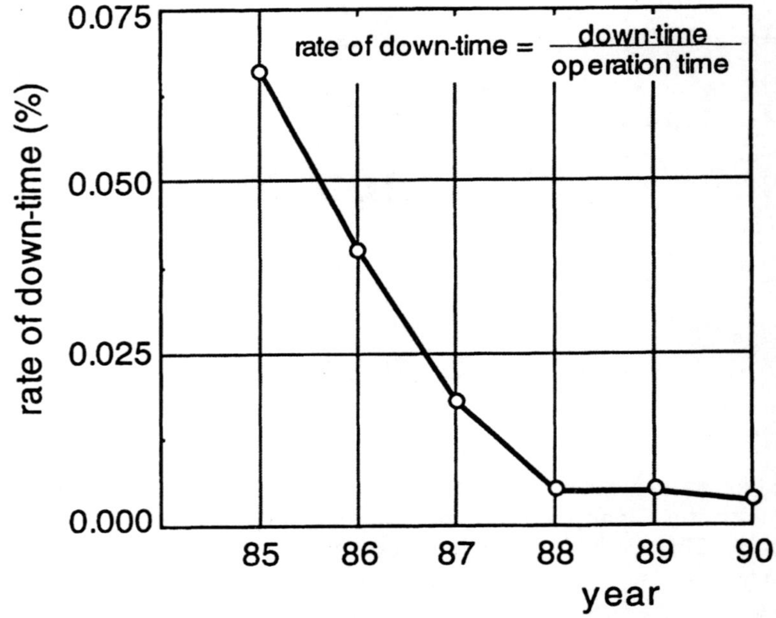

Fig. 10.1 An example of improvement in the rate of down-time in a factory where TPM (total productive maintenance) was implemented (Nakajima 1992).

4. *Evaluation of maintenance results:* evaluation of the results of maintenance as to whether the maintenance strategy planning and maintenance task control are appropriate.
5. *Improvement of maintenance:* improving maintenance strategy planning and maintenance task control based on the evaluation of maintenance results.
6. *Improvement of facility:* design improvement based on the evaluation of maintenance results.
7. *Dismantling planning and execution:* planning and executing dismantling at the end of the facility life cycle.

10.1.2 Mechanisms necessary for life cycle maintenance

The objective of life cycle maintenance management is to manage the above listed activities in an effective way throughout the facility life cycle. For this purpose, it is necessary to provide the following mechanisms for the maintenance management system.

1. *Adaptation to various changes in the facility life cycle:* During the facility life cycle, there could be various changes in the operating environment, operating conditions and the facility itself. The maintenance

should be adaptive to these changes in order to perform effective maintenance.

2. *Continuous improvement of the facility:* It is not feasible to design perfect facilities without any flaw especially in the case of manufacturing facilities, because they are usually one-of-a-kind products. Therefore, it is essential to provide the mechanism for continuous improvement based on the knowledge acquired during their life cycle.
3. *Data integration throughout the facility life cycle:* Where maintenance is concerned, most data necessary in the facility development phase and in the operation phase is common. In the development phase, it is essential to know the real operating situations and the problems encountered during operation. On the other hand, it is necessary to have exact design data including results of various analyses and the designers' intention for performing maintenance strategy planning and maintenance task control.

With regard to adaptation to changes and continuous improvement, it is important to execute the P-D-C (plan-do-check) cycle of maintenance effectively. Maintenance strategy planning plays a key role for this purpose. Concerning data integration, we need an information infrastructure which provides common data and tools in order to support maintenance management throughout the total life of the facility.

In the following, we will discuss maintenance strategy planning first. Then, we will discuss the information infrastructure for life cycle maintenance.

10.2 MAINTENANCE STRATEGY PLANNING

10.2.1 Maintenance strategy planning – a decision center for life cycle maintenance management

There are many ways to execute maintenance actions. They can be categorized, from various aspects such as opportunity of inspection or diagnosis, criteria for giving treatment, and type of treatment as shown in Fig. 10.2. Maintenance strategy planning selects the method of maintenance from among these options. No matter which maintenance action is taken, we have to determine the maintenance strategy beforehand.

In the past, most maintenance efforts have been directed at how to react to the occurrence of a problem. However, it is not feasible to prepare for every problem. Concerning monitoring and diagnosis, for example, it is obvious that you cannot install numerous sensors everywhere in the facility in order to be able to detect any kind of problems. If you do not know what to expect, you can hardly prepare for it. Only by anticipating potential problems can you devise countermeasures.

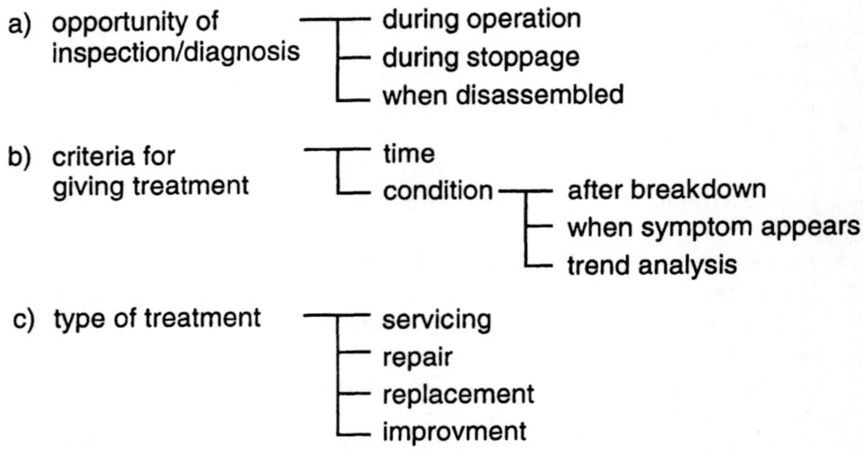

Fig. 10.2 Aspects for categorizing maintenance strategies.

Therefore, it is necessary to evaluate potential problems that may occur in the facility as the first step of maintenance strategy planning. Then the best maintenance strategy should be determined taking the characteristics of the problems and other conditions into account.

Figure 10.3 shows a role of the maintenance strategy planning in the frame of life cycle maintenance. It serves as a bridge between the facility development phase and the facility operation phase. It obtains design data and construction record from the facility development phase. The potential problems which may occur during operations are first evaluated, then the maintenance strategies for treating them are selected. They are given to the operation phase where a maintenance task planning module generated procedures and a schedule of maintenance tasks. After the maintenance tasks have been executed, the results are evaluated in terms of discrepancies between the anticipated and real conditions of the facility. If there are discrepancies, the information is fed back to the maintenance strategy planning, where the maintenance strategies are revised based on re-evaluation of potential problems taking the actual data into account. If the newly selected strategy is corrective maintenance, the information is further fed back to the development phase where design for improvement and modifications are performed.

As explained above, there are three feedback loops in Fig. 10.3. The first is the loop within the facility operation which consists of task planning, task execution and assessment of maintenance results. This is the loop for controlling routine maintenance work. The second loop is the one consisting of strategy planning and task control. By means of this loop, the maintenance strategies can be improved based on the

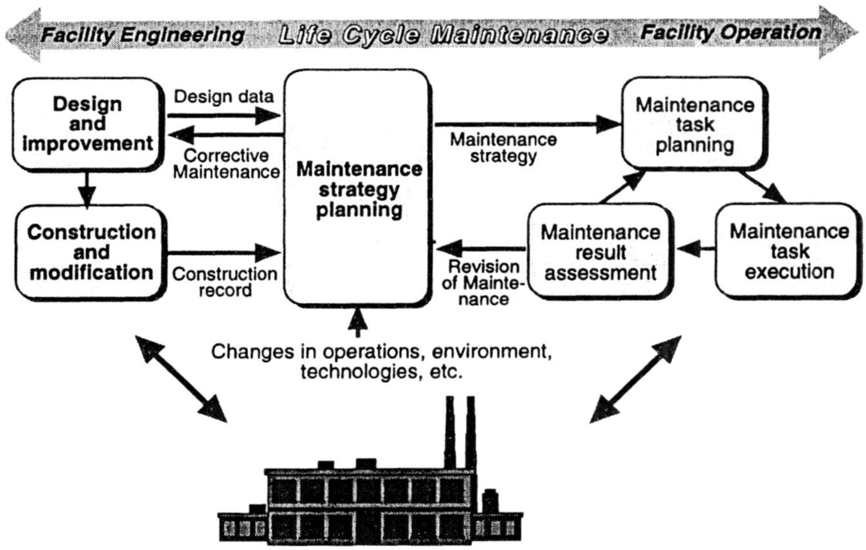

Fig. 10.3 Maintenance strategy planning – a decision for life cycle maintenance management.

observation of actual phenomena and knowledge accumulated during the facility life cycle. The second loop is effective in adapting the maintenance strategies to various changes such as those in the operation conditions and environment. The third loop is the one that includes the facility development. This loop is essential for continuous improvement of the facility during the life cycle.

These three loops provide effective mechanisms for adapting maintenance strategies to various changes and continuously improving the facility, required for the life cycle maintenance management as described in the previous section. Maintenance strategy planning plays a key role in these mechanisms by making, as it does, a basic policy of maintenance management. In the following sections, the evaluation of potential problems is first discussed. Then, criteria for the selection of maintenance strategies are considered.

10.2.2 Evaluation of potential problems

There are several causes of problems within a facility, such as improper design, defects in production, improper adjustment, improper operating conditions, and aged deterioration. Among these causes, deterioration is a matter of utmost concern for maintenance strategy planning. Eliminating the first three causes is, of course, an important part of maintenance activities. However, once they can be settled, it is not necessary to worry about them afterward. On the other hand, deterioration is essentially inevitable to a greater or lesser degree for the facility.

Deterioration is a physical and/or chemical process occurring at various parts of the facility such as fatigue, wear, and corrosion. Mechanisms which induce deterioration are called deterioration mechanisms. The resultant deteriorated states are identified by deterioration modes. The deterioration mechanism is activated by a certain set of conditions which we call causal factors. They are classified into the following categories:

1. Inherent characteristics such as geometry, material, and surface finish.
2. Exerted stress such as mechanical stress and thermal effects.
3. Relative motion between parts.
4. Operating environment such as in a gas, in a liquid, or in particles.

Deterioration occurring at certain area of parts may induce changes in the behavior of the facility. If these behavioral changes are related to required functions of the facility, they are recognized as functional degradation or failures. In this sense, we distinguish deterioration and failures. However, the causality of failures sometimes seems to be complicated because of the hierarchical structure of the facility and its functions. An example of failure causality is represented in Fig. 10.4. The hierarchy of the facility (in this case, it is a machine tool) is arranged vertically in the figure. At the bottom, deterioration occurs. The effect of deterioration is propagated to higher hierarchies and leads to functional degradation identified at each level. The characteristics of the functional deterioration or failure are recognized differently depending on the level at which they are observed.

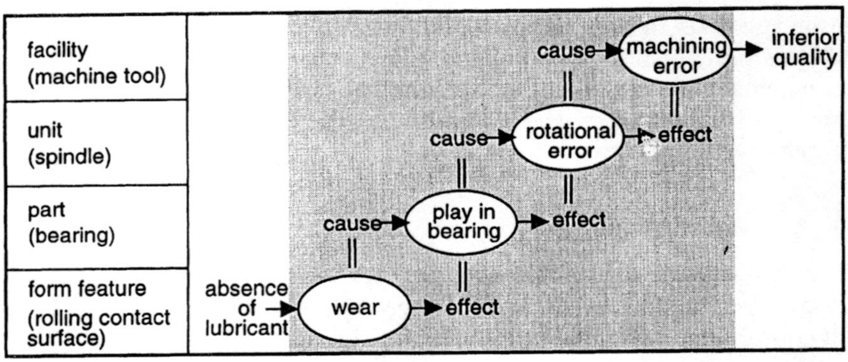

Fig. 10.4 An example of the failure causality.

From the above consideration, it may be realized that the evaluation of potential problems can be divided into several steps. They are:

1. Evaluation of causal factors existing at each part of the facility.
2. Evaluation of potential deterioration.
3. Evaluation of functional degradation or failures induced by the predicted deterioration.

Evaluation of potential problems does not always need to be precise. Certainty of the evaluation could vary depending on the amount of knowledge and experiences about problems. What is important is that you should recognize the certainty of the evaluation, because you can then select maintenance strategy taking the uncertainty into account. Usually, a lesser amount of knowledge is available at the beginning of the facility life cycle. In this case, the inspection cycle is, for example, set to short, making allowances for inaccuracy of evaluation. However, after experiences of operations are accumulated, more precise evaluation becomes possible, which may enable a longer inspection cycle. It is important for life cycle maintenance to make this sort of evolution possible.

10.2.3 Criteria for selecting maintenance strategies

As the second step of maintenance strategy planning, measures to deal with each potential problem should be determined. The following three criteria should be considered in this step.

1. Characteristics of deterioration and resultant functional degradation.
2. Severity of problems induced by potential deterioration.
3. Applicable maintenance actions and their effectiveness.

The most important characteristics of deterioration and functional degradation to be considered in the maintenance strategy planning are a progress pattern of deterioration or functional degradation. Figure 10.5 illustrate the progress pattern schematically. The curve shows the change in an observed parameter with time. The lower the parameter value becomes, the further the deterioration progresses. Two thresholds are defined in the figure, that is, a detection limit L_D and a functional limit L_F. L_D is a level from which the progress of deterioration or functional degradation can be identified. L_F is a level from which the facility becomes faulty. The pattern is divided into three periods associated with these two thresholds, that is, a normal state, a symptomatic state, and a faulty state. The pattern can be characterized by the time length and the predictability of the period of the normal state t_N and the period of the symptomatic state t_D, in comparison with the cycles of monitoring and/or diagnosis T_D and the cycle of treatment T_M. If t_N is unpredictable and t_D is very short, for example, the problem is recognized as random and sudden failure. It should be noted that the pattern could change

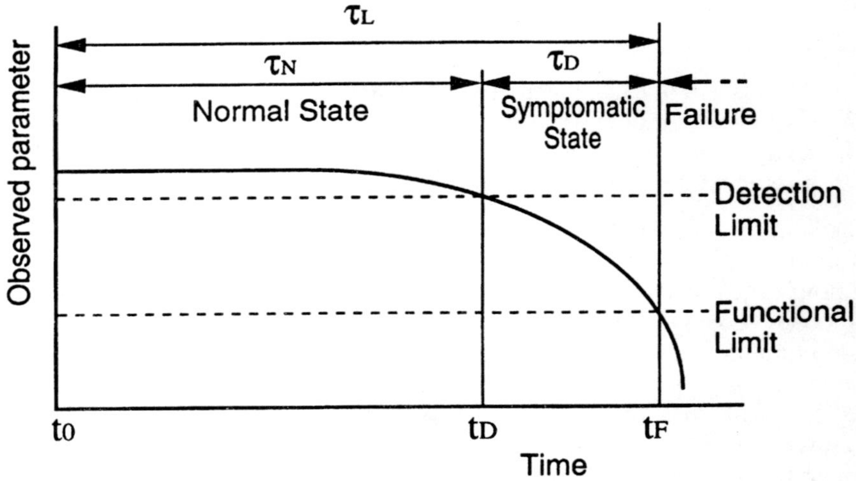

Fig. 10.5 Progress pattern of deterioration or functional degradation.

depending on what parameter is observed by what kind of sensing technology in which hierarchy level. Even though the deterioration progresses gradually, the failure induced by the deterioration could be recognized as a sudden one because of the difference of the hierarchy level observed.

Based on the characteristics of the progress pattern, maintenance strategies which are technically applicable to the problem can be determined. Figure 10.6 shows a logic tree to determine the maintenance strategies based on the length and predictability of t_N and t_D.

Severity of problems is another important criterion for maintenance strategy planning. Severity is usually defined as the magnitude of effects of the problem to the outside of the system concerned. (The effect of deterioration within the system is considered as functional degradation or failure.) The effects can be assessed from various aspects. Typical examples are (1) safety, (2) operational, and (3) economic. Another measure of severity often used is risk, which is defined as a product of the probability of occurrences of the problem and the magnitude of its effects. However, it is usually difficult to make a quantitative evaluation of the severity. In many practical cases, it is estimated qualitatively from various aspects and the results are combined to give a final rating.

The applicability of maintenance actions and their effectiveness should also be considered in selecting maintenance strategies. The effectiveness of maintenance actions is evaluated in conjunction with the characteristics of the problem concerned. If the problem is of a random nature, for example, periodic replacement is in no way effective. The evaluation of effectiveness of maintenance actions should also involve

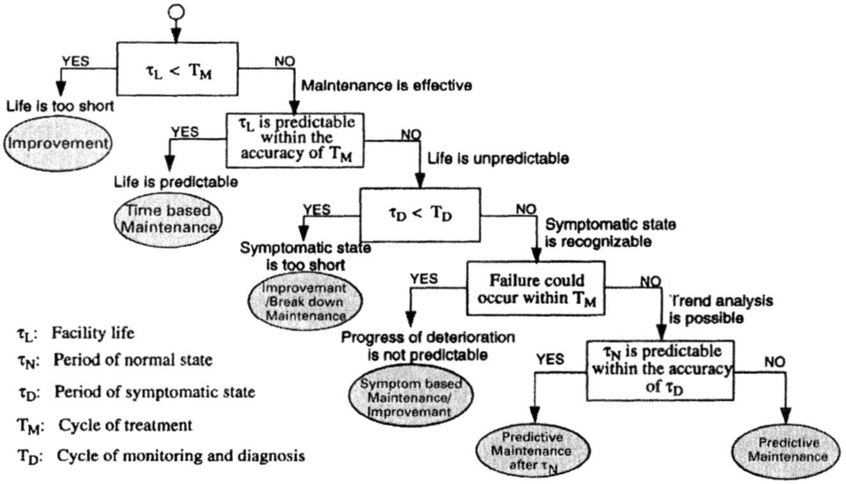

Fig. 10.6 Selection of maintenance strategies based on deterioration pattern.

evaluation of maintenance techniques. It is evaluated in terms of detectability of problems in the case of inspection methods, the rate of correct diagnosis in the case of diagnostic systems, and so on.

10.2.4 Methodologies for maintenance strategy planning

Study on systematic approaches to the maintenance strategy planning is a relatively new area of research. The most well-known methodology for this purpose is reliability centered maintenance (RCM) (Nowlan 1978). It was developed in the field of aircraft maintenance in the late 1970s. Recently it has been applied to other areas such as nuclear power plants and chemical plants. The general procedure of maintenance strategy planning by RCM is shown in Fig. 10.7. RCM provides a systematic procedure for maintenance strategy planning based on qualitative evaluation of failures. Failures are categorized in terms of severity first, then effectiveness of maintenance actions are evaluated.

In addition to RCM, risk based inspection (RBI) has attracted attention as a systematic method for maintenance strategy planning especially in the area of nuclear power plants in recent years (ASME 1991). While RCM selects the maintenance strategy based on categories of failure effects, RBI uses risk for prioritizing the potential failures. Risk is defined as the product of a failure probability for each item and its respective consequence. The basic concept of RBI is to focus inspection efforts on items with the highest risk.

With regard to methodologies for evaluating potential problems,

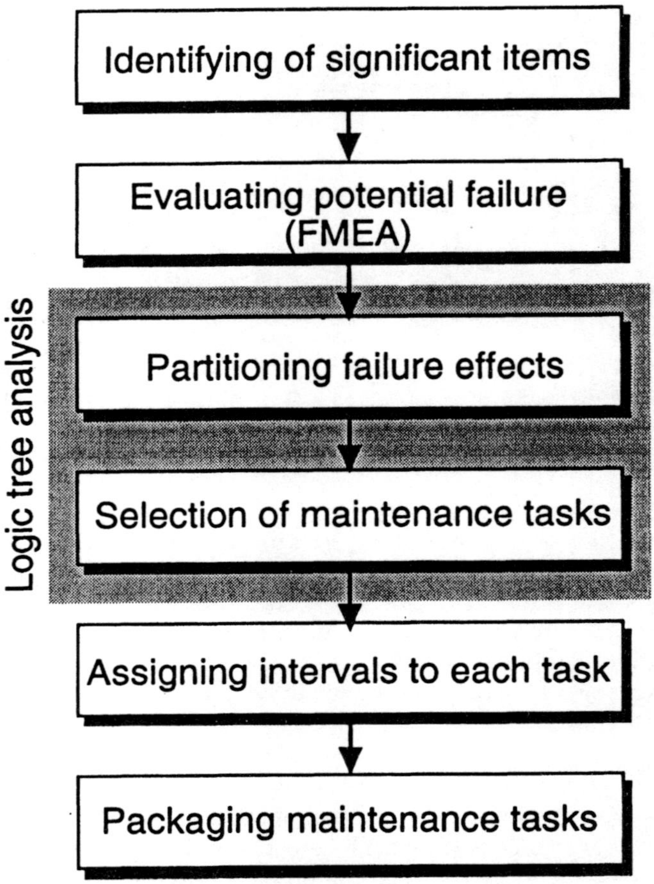

Fig. 10.7 Procedure of maintenance strategy planning by RCM.

FMEA (failure mode and effects analysis) is most widely used. It is used also in RCM and RBI. FMEA provides a systematic procedure for qualitative analyses of potential problems. However, there are some drawbacks from the aspect of maintenance strategy planning. The important one is that there is no clear distinction between failure and deterioration in the procedure, because analyses are carried out basically from the function point of view. Failures are defined from the functional point of view, whereas deterioration is defined as a physical/chemical process. FMEA does not provide any guide to identify the causal factors which induce deterioration. Another drawback of FMEA is that it needs an enormous amount of man-hours when comprehensive analyses are carried out, because each decision in the analysis procedure is performed by human experts based on their expertise. For evaluating potential problems repetitively during the facility life cycle, it is essential to have computer support to lighten the burden for the analysis.

10.3 INFORMATION INFRASTRUCTURE FOR LIFE CYCLE MAINTENANCE MANAGEMENT

10.3.1 Architecture of information infrastructure for life cycle maintenance management

We need a significant amount of data management and data processing for performing various activities of life cycle maintenance. It is, therefore, essential to have computer support for life cycle maintenance management. In maintenance strategy planning, for example, we need to refer to various data such as design data, operation, and maintenance history and knowledge about deterioration and failures. An effective data base system is indispensable for this purpose. Use of effective computer tools is also required for the comprehensive evaluation of the potential deterioration and resultant failures.

Figure 10.8 represents a basic concept of information infrastructure for life cycle maintenance management. It has two types of data bases: a facility model and a common data base. The facility model maintains all information specific to the facility. It includes the design data such as

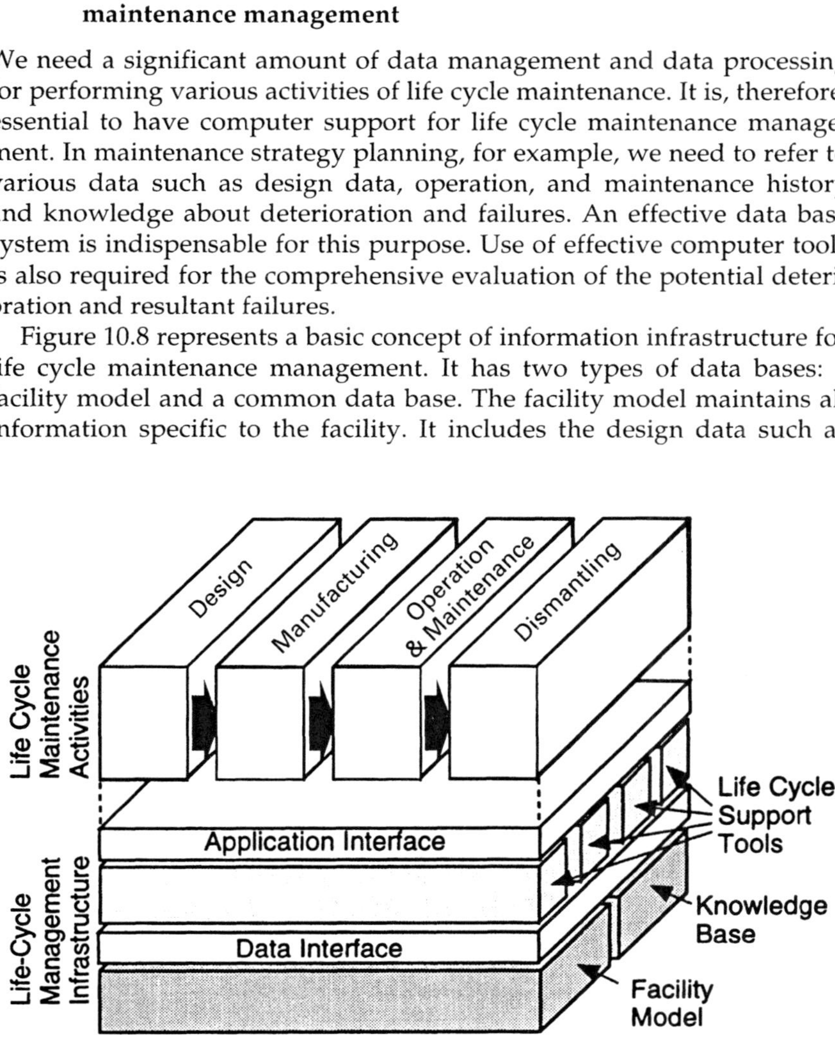

Fig. 10.8 Architecture of information infrastructure for life cycle maintenance management.

geometric and material data of parts, and assembly structures. It also includes records of construction, operating environment, and history of operation and maintenance. In the common data base, generic data which is independent of an individual facility is stored. Such data includes theoretical and empirical knowledge about deterioration and failures, knowledge about maintenance technologies such as inspection, monitoring, diagnosis, and repairs.

Various tools which support maintenance activities based on these data bases should also be implemented in the infrastructure. They include geometric modelers, various CAE tools such as stress, kinematics, flow, and heat transfer analysis, process simulators, tools for reliability analysis, and decision support tools.

The important feature which is required by the information infrastructure for life cycle maintenance management is to enable access to these data bases and tools from any phase of the facility life cycle, because life cycle maintenance requires repetitive evaluations of the facility in terms of potential problems, actual conditions of the facility, and effects of maintenance actions. Another reason for this is that new data could be generated any time in the life cycle. Knowledge about deterioration and failures should be enhanced based on experience during the facility life cycle. Design data could also be changed by improvements or modifications of the facility during the operation phase.

With regard to information management systems for maintenance, there has been a long history from the late 1960s. Most of the systems have been, however, devoted to the support of maintenance task control in the operation phase, which corresponds to the right side of Fig. 10.3. In recent years, as the concept of RCM has become popular, systems which support maintenance strategy planning have been developed. On the other hand, the information processing environment for product development has significantly improved recently. CAD and CAE systems provide the capability to create and manipulate product models in the computer and to make precise analyses based on these models. All data associated with product development can be managed by means of a PDM (product data management) system. The concept of information infrastructure for life cycle maintenance management could be realized by integrating such an advanced product development environment with recent maintenance management systems.

The following sections look at the facility model which serves as a central vault for life cycle management, and the deterioration evaluation system, which is one of the essential tools for the prediction of potential problems.

10.3.2 Facility model

The facility model is an essential part of the information infrastructure for life cycle maintenance management. The model is regarded as a data vault which is used for managing all data relevant to the facility in a consistent way throughout its life cycle. It should represent the basic information of the facility such as material and geometric data of parts, and assembly structures. It should also have the capability of representing additional information necessary for various applications, such as design intention and information related to operation and maintenance. Since the data relevant to the facility is changed or modified during the life cycle, the model should have a function for managing the history of these changes as well.

As a base of the model structure, the assembly structure of the facility is widely adopted. The representation of the assembly structure has been recently studied in association with product modeling (Sodhi 1994; Sugimura 1994). Figure 10.9 shows an example of a data structure of the facility model (Takata 1995). The model consists of *assembly items* and *assembly relations* between the items. Notations in the figure are based on EXPRESS-G (ISO 1992) where an attribute is shown as a thin line with its name on it and marked with a circle at the value end. A thick line shows a class hierarchical relation, with a circle attached to the sub-class side.

An assembly item represents a physical substance in the facility. Assembly items are classified into *part* and *form feature*. A part is an individual physical substance. In this model, an assembly is considered as a kind of part that can be divided into multiple parts. This concept allows us to represent the hierarchical structure of the facility in a flexible manner.

Form feature is a group of geometric elements that carries some functions or behavior. Note that it does not necessarily belong to a single part. An assembly feature is a form feature that mates with another form feature of a different part or sub-assembly to make an *assembly*, such as holes/pins and grooves/extrusions.

Two types of assembly relations are identified to represent assembly

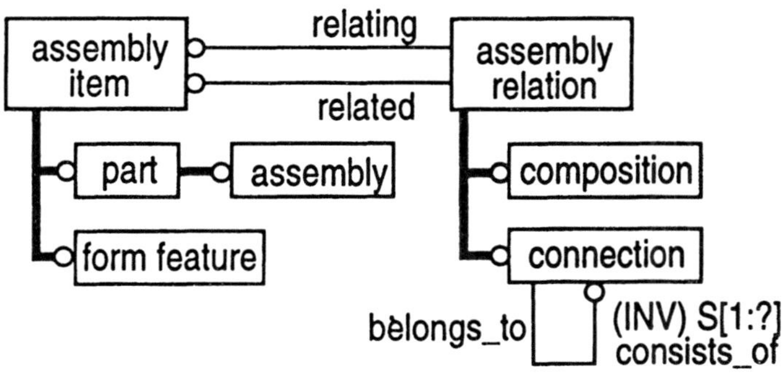

Fig. 10.9 Structure of the facility model.

structures. *Connection* is an assembly relation between two items that have no inclusion relations with each other, e.g. a part to a part, an assembly feature to an assembly feature. As hierarchical information, connection has a pointer to the connections in which it participates. For example, a connection between two parts may consist of several connections between assembly features. They have a pointer to the parent part-to-part connection.

Composition is an assembly relation between an assembly item and another assembly item that consists of it. Composition has a transformation that represents the position and orientation of the child item in the coordinate space of the parent item.

Both assembly items and assembly relations can have technical information in addition to configuration structures as attributes. Parts, for example, contain technical information, such as material and mass properties. Connection contains information that represents the type of the connection, such as the type of pairs, fastenings, and fittings. Any attribute can be added depending on the requirements from various simulations and analyses performed for the purpose of life cycle maintenance management.

As an example of the facility model, the model of a motor assembly with a gear shown in Fig. 10.10 is represented in Fig. 10.11. Note that Fig. 10.11 is an instance-level diagram. Only a part of the entire model is shown here. Attribute names have been omitted. Connections and compositions are represented as circles and triangles. The hierarchy of assembly items are arranged vertically. The motor assembly is related to the rotor-shaft sub-assembly by composition; and the rotor-shaft sub-assembly is also related to the rotor and the shaft in turn. In this example, you can see that any level of assembly items can be related using connection, thanks to object-oriented representation. The gear part is

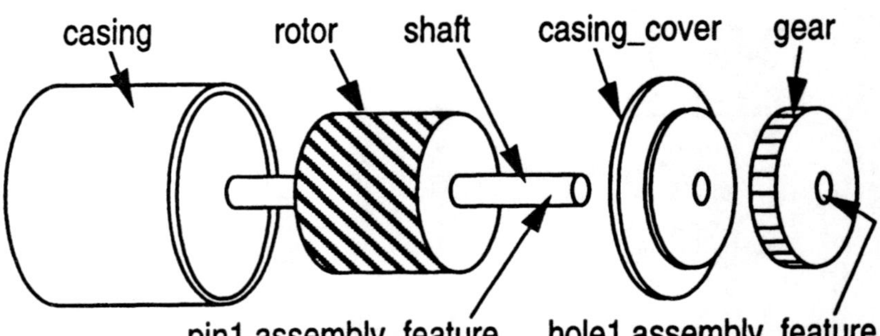

Fig. 10.10 Motor assembly with a gear.

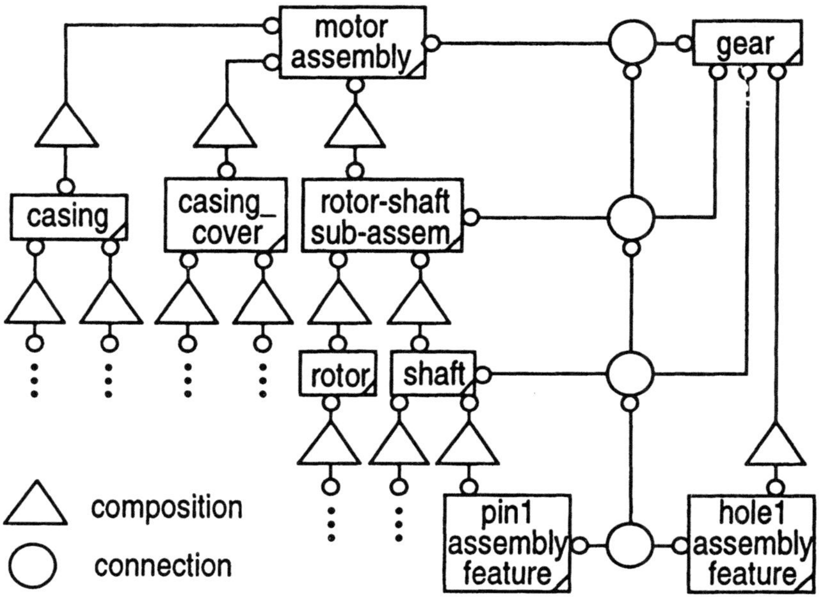

Fig. 10.11 Model representation for motor with gear.

related by connection to the shaft part, to the rotor-shaft sub-assembly and to the motor assembly.

10.3.3 Model-based deterioration evaluation

The role of life cycle maintenance is to maintain the facility in a proper condition to fulfill its required functions throughout the facility's life cycle. The reason we have to maintain the facility lies in the fact that the functions of the facility degrade mainly due to deterioration. Deterioration is a physical and/or chemical process occurring at various parts of the facility depending on operational and environmental stresses. To realize systematic maintenance, it is essential to predict potential deterioration. With this knowledge, proper maintenance strategies can be selected. Deterioration prediction is also required to design for reliability and maintainability. Therefore, we have developed a system for predicting the deterioration of facilities.

First, we describe the model for representing deterioration. Then, a deterioration evaluation system based on this model as well as the facility model is described.

Modeling of deterioration processes

Mechanisms, such as fatigue, wear, and corrosion, which induce deterioration at certain areas of parts or assemblies are called deterioration mechanisms. The resultant deteriorated states are distinguished by deterioration modes. The deterioration mechanism is caused by a certain set of conditions which we call causal factors. They are classified into four categories: (1) inherent characteristics such as geometry, material and surface finish; (2) exerted stress such as mechanical stress, thermal effects and electro-magnetic effects; (3) relative motion; and (4) operating environment such as in a gas, in a liquid, or in particles.

Although there seems to be an infinite number of phenomena recognized as deterioration of the facility, we can identify a certain set of deterioration mechanisms which are basic and common for many types of facilities. We call them fundamental deterioration mechanisms.

In many cases a chain of multiple fundamental deterioration mechanisms are related to failure. For example, fatigue failure could be initiated by a notch created by corrosion. In this way, one of the causal factors of a deterioration mechanism could be provided by other deterioration mechanisms. There is also a case where some of the causal factors are provided by mechanisms other than deterioration mechanisms, which we call causal factor formation mechanisms. An example of this type of mechanism can be seen when the rotation of a shaft with a radial load creates cyclic stresses which lead to fatigue at a stepped part of the shaft. The chain of deterioration mechanisms and causal factor formation mechanisms is termed a deterioration process. Figure 10.12, for example, schematically represents the deterioration process of fatigue for a spindle. In the figure, the rectangles represent the causal factors of the deteriorated state, and the shaded ovals represent the formation mechanism of the deterioration.

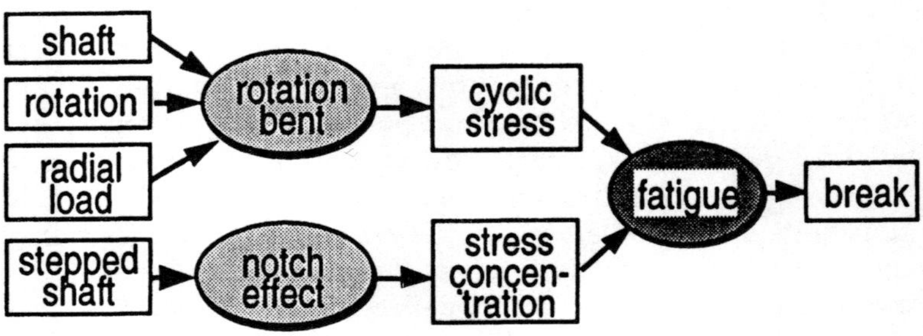

Fig. 10.12 Example of the deterioration process.

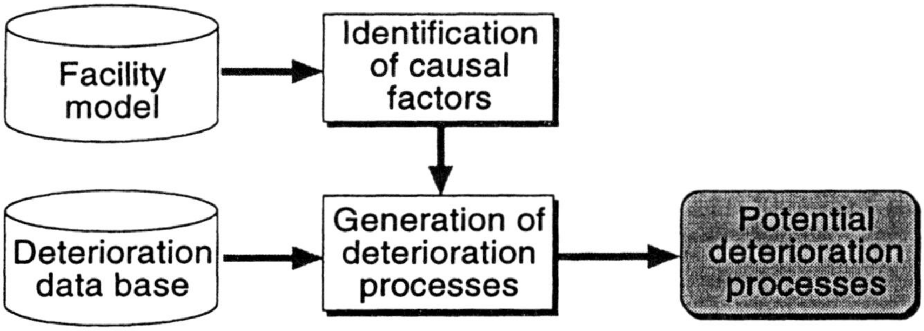

Fig. 10.13 Configuration of the deterioration evaluation system.

System for deterioration evaluation

The deterioration evaluation system carries out qualitative evaluations of potential deterioration based on the facility model and a deterioration data base (Takata 1994). It generates potential deterioration processes which may occur at particular parts of the facility. Figure 10.13 shows the configuration of the system. It consists of the facility model, the deterioration data base, and the inference mechanism.

1. *Facility model:* The facility model represents the structure and properties of a facility to be evaluated as described in section 10.3.2. The properties are represented in terms of values of attributes defined in assembly items and relations. Fig. 10.14 shows attributes currently defined in the model. The model has a mechanism to inherit the values of attributes from those of higher level items in the hierarchy, unless they are explicitly defined at their level.
2. *Deterioration data base:* The deterioration data base contains the fundamental deterioration mechanisms and the causal factor formation mechanisms. They are expressed in terms of a set of causal factors and the resultant deterioration modes or causal factors. This data base can be prepared independently from facilities.
3. *Inference of deterioration of form features:* This is conducted for every form feature of every part defined in the facility model, since different form features have different causal factors even in the same parts. The causal factors associated with a particular form feature are identified from the facility model, that is, the attributes of the form feature itself, the related form feature connection, and the part on which it is defined. The system searches mechanisms in the deterioration data base whose causal factors match with those identified in the model and generates potential deterioration processes which could occur at the form feature.

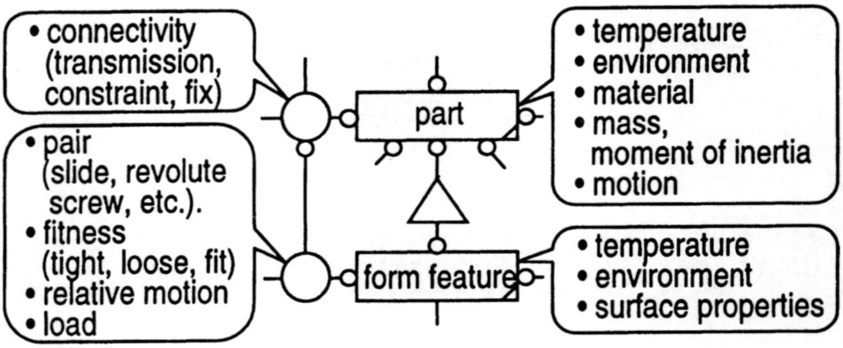

Fig. 10.14 Attributes defined in the facility model.

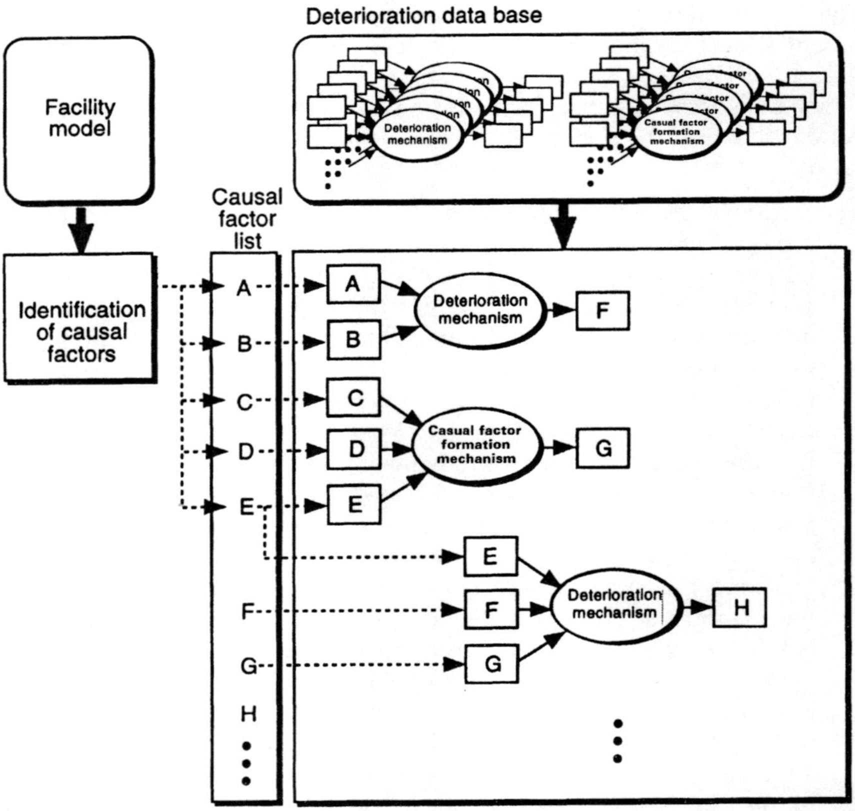

Fig. 10.15 Inference mechanism of deterioration evaluation.

Figure 10.15 illustrates how the system evaluates the component deterioration. The procedure of the evaluation is:

(a) A component to be evaluated is selected.
(b) The requisites provided by the component are extracted from the facility model and described in the requisite list.
(c) Deterioration mechanisms or requisite formation mechanisms whose requisites are satisfied with the current requisite list are searched. If there are no more mechanisms induced by the requisites on the list, then go to (e).
(d) The outputs of the selected mechanisms (requisites or deterioration modes) are appended to the requisite list, and go to (c).
(e) The deterioration process(es) is (are) formed by the selected mechanisms and displayed as the result of the evaluation.

4. *Inference of range of effects of deterioration:* In the above procedure, the effects of causal factors and deterioration modes are considered within the specified form feature. However, there are causal factors and deterioration modes which also have effects on other form features or parts. The following three cases have been identified:

(a) Some types of deterioration, such as loose bolt fastenings, have ranges of influence involving multiple parts or part connections.
(b) Effects of causal factors, such as vibration and heat conduction, propagate from one part to another.
(c) Effects of deterioration change environmental causal factors of other form features or parts. For example, leakage of coolant due to a damaged seal makes a bearing wet.

In the system, the first and the second cases are dealt with. Inference algorithms were developed for the first case. For the second case, the propagation rules for the corresponding causal factors were defined.

The inference process of the extent of the effect of a loose bolt, for example, consists of two steps: (a) inferring parts fastened by the bolt and (b) inferring the unstable parts among them. In the first step, the system searches for the mating part which is connected to the bolt with a thread, that is, for example, a nut. Then the system traces the mating assembly features of the parts fastened by the bolt and nut starting from the bearing surface of the bolt to that of the nut. The parts which have assembly features involved in the traced path are identified as loose parts. In the second step, the system checks whether any of these parts are fixed to a base part via connections other than those traced in the first step. If a part has such connections, the system identifies the part that is stable and excludes it from the loose part list.

Experimental system

The experimental system has been developed by using the object-oriented expert shell G2. An interface with a solid modeling system has been established. The assembly items and relations of the facility model are defined as object classes. The deterioration mechanisms and causal factor formation mechanism are also described as objects. The inference is executed by production rules.

Taking a centrifugal pump as an example, deterioration evaluation was performed. Twenty-four major parts were examined by the experimental system. The model of the impeller assembly is shown at the bottom of Fig. 10.16. The upper left part of the figure shows that the system identified the looseness of the bolt as the potential deterioration. The bolt fixes the impeller to the spindle. The result of inferring the extent of the effect of a loose bolt is shown at the upper right of the figure. The bolt-part list represents the inferred list of the parts fastened by the bolt. The loose-part list represents the inferred list of the parts fastened by the bolt. The loose-part list, on the other hand, shows the list of unstable parts among them. It is noted that SPINDLE is not included

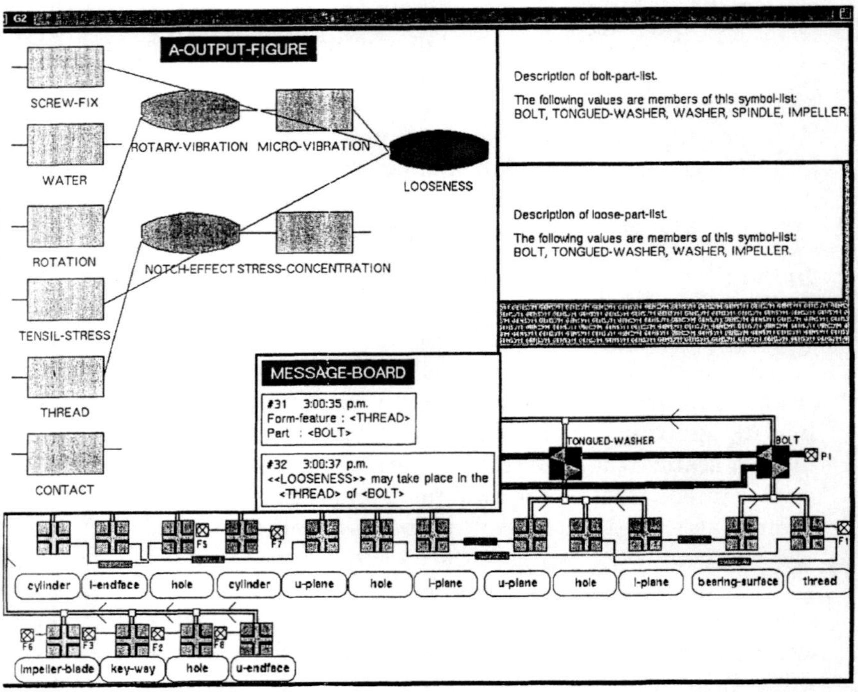

Fig. 10.16 The system output of deterioration evaluation for the bolt which fixes the impeller to the spindle.

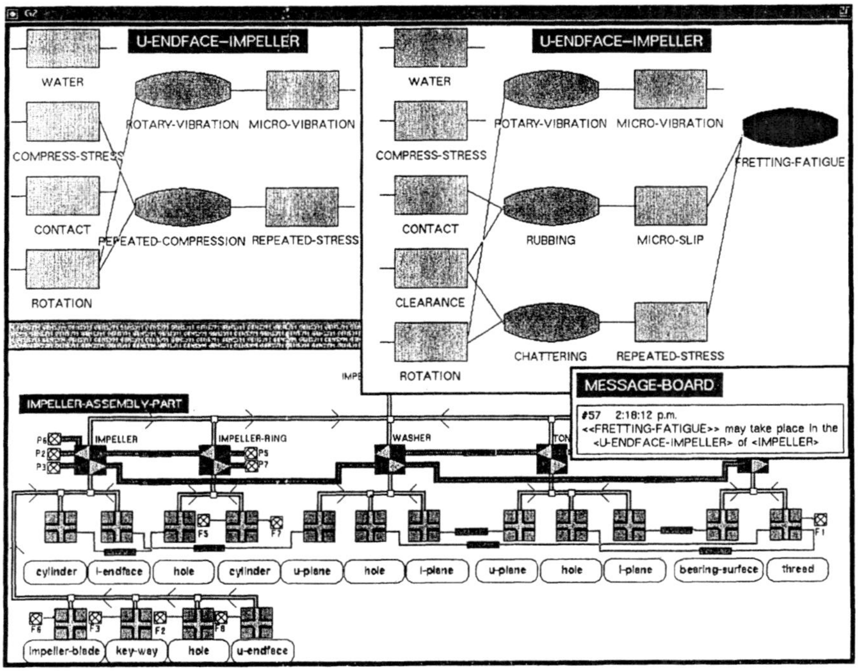

Fig. 10.17 The effect of the loose bolt on the deterioration of other parts.

in the loose-part list, because it is supported by a bearing which is fixed to a casing. The effect of the loose bolt on the deterioration of other parts is demonstrated in Fig. 10.17. It shows the results of the deterioration evaluation of the upper end face which is one of the assembly features of the impeller. The upper left part of the figure shows the result of the evaluation before the loose bolt has been identified. The system did not suggest any deterioration mode. However, when the loose bolt was identified, the clearance was added as one of the causal factors as shown at the upper right part of the figure. Then, the system identified fretting fatigue as the potential deterioration of this assembly feature.

REFERENCES

ASME (1991) *Risk-based Inspection – Development Guidelines*, ASME CRTD-20–1.

Nakajima, S. (1992) *Introduction to TPM*, Japan Institute of Plant Maintenance.

ISO (1992) *ISO CD10303-11 Product Data Representation and Exchange – Part 11, The EXPRESS Language Reference Manual*, ISO TC184/SC4/N151.

Nowlan, F.S., and Heap, H.F. (1978) Reliability centered maintenance. *Proceedings of Annual Reliability and Maintainability Symposium*, IEEE.

Sodhi, R., and Turner, J.U. (1994) Towards modelling of assemblies for product design. *CAD*, **26**(2) 85–97.

Sugimura, N. *et al.* (1994) A study on product model for design and analysis of mechanical assemblies based on STEP. *Japan-USA Symposium of Flexible Automation*, 495–98.

Takata, S., Yamaoka, N., Asama H., and Hiraoka, H. (1994) Model based evaluation of component deterioration. *Proceedings of 2nd International Seminar on Life Cycle Engineering*, 168–80.

Takata, S., Hiraoka, H, Asama, H. Yamaoka, N., and Saito, D. (1995) Facility model for life cycle maintenance system. *Annals of the CIRP*, **44**(1), 117–21.

CHAPTER 11

Life extension of operating machinery using the National Information Infrastructure (NII)

Shashi Phoha and Asok Ray

11.1 INTRODUCTION

Modern requirements of precision and quality are resulting in an alarming increase in the cost of maintaining complex machinery and the associated human and environmental safety risks. The cost of maintaining US Navy surface ships, aircraft carriers, and submarines is estimated to be in excess of $7 billion per year and at least as much is spent in logistics support and maintaining the maintenance infrastructure. The aging aircraft fleet in the civil sector and the increasing emphasis within the US on high performance manufacturing systems to meet global competition (Lee 1994; NRCS Report 1990) are two other examples which call for alternatives to time-based maintenance and run-to-fail maintenance policies. Information-based predictive maintenance using advanced sensor and model based prognostic maintenance policies is expected to significantly reduce costs (see Fig. 11.1). The electric power industry predicts a potential 50% reduction in maintenance costs due to prognostic maintenance (EPRI 1992 Annual Reports).

Advanced prognostic maintenance methods, however, are just emerging from research laboratories (Ray *et al.* 1994a; 1994b), pilot studies (ARL/PSU Technical Report, November 1995), and limited experimentation (Tangirala *et al.* 1995). Some of the recent research is summarized in section 11.2.1 of this chapter.

As the prognostication of damage in complex operating machinery gets beyond the reach of human senses and depends increasingly on electronic sensing devices and model outputs, the issues of real-time data availability, fault tolerance, interoperability, knowledge representation, and coordination of plant dynamics and operational planning gain significant technical and economic importance. The architecture of the

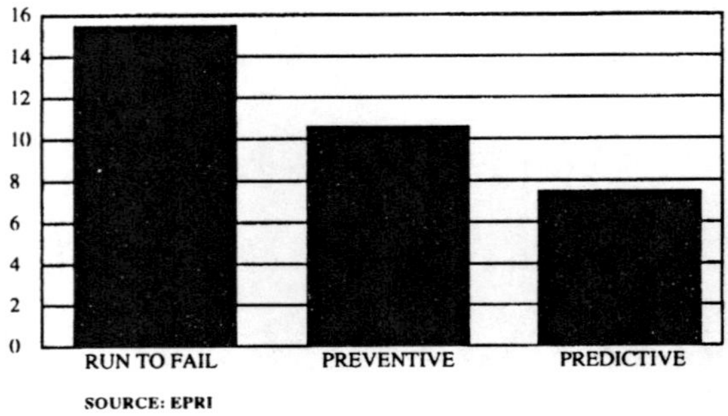

Fig. 11.1 Relative costs of maintenance policies.

maintenance infrastructure – its performance, acquisition, operation and maintenance – become major concerns in realizing the promise of life extension methods for operating machines. Section 11.2.2 discusses concepts for the electronic delivery of maintenance, diagnosis and prognosis (mdp) services for operating machinery.

The ubiquitous presence of complex operating machinery and the evolving need for electronic delivery of mdp services for prognostic maintenance provide the motivation for utilizing the emerging National Information Infrastructure (NII) in meeting the computational and economic challenges of equipment life extension methods. Electronic mdp services delivered over the NII from specialized and generic centers will provide major benefits including:

- Small and medium-sized businesses need not duplicate expensive diagnostic and prognostic maintenance support and expertise. Instead services may be shared from a common service center on a subscription or fee-for-service basis.
- A higher and consistent quality of service can be provided with experts and tools in mdp centers with access to historic databases, trends, performance analysis and sophisticated decision support tools.
- The availability of sophisticated in-process prognostic services to mobile platforms like ships, aircraft, tanks, etc., over wireless or other digitized battlefield networks is specially significant in warfighting or transportation planning applications.
- Flexible maintenance support architectures can be developed so that immediate alarm conditions can be handled locally and more extensive database access and powerful model calculations can be performed in the background in a seamless integrated environment.
- A major potential benefit is the technical feasibility and affordability of prognostic maintenance in a distributed operational environment. As opposed to performing equipment maintenance in response to

failure or according to a time schedule, information-based prognostic maintenance (IPM) is based on equipment condition, history, analysis and coordination of maintenance resources with operations. Results target avoiding catastrophic failures, avoiding unnecessary scheduled maintenance and minimizing downtime by coordinating resources and by interactive electronic access to maintenance manuals and support tools.

11.2 METHODOLOGY

11.2.1 Life extension methods for operating machinery

The life extension methods are based on stochastic modeling and control of accumulated damage at critical points in the plant. The damage prediction system, which is an essential ingredient of the life extending control system, is described in Chapter 7 of this book by Ray and Phoba. It consists of a plant model, structural model(s) of the critical component(s), and the respective damage model(s) as is depicted in Fig. 11.2. The component structural model in a typical operating machinery (e.g. a gas turbine in an aircraft engine or a heat exchanger in a chemical plant) consists of three parts, namely, heat transfer, thermoelastic and thermoviscoplastic where the time-dependent process variables such as temperature, pressure, and flow rate are external inputs. The damage model calculates both damage rate and accumulation as continuous functions of time. The ensemble of the component structural model and the damage prediction model in Fig. 11.2 is referred to as the structural damage model in sequel. Fig. 11.3 presents the flow diagram of a typical structural model.

The heat transfer model in Fig. 11.3 is derived from the classical diffusion equation for heat conduction to calculate the temperature distribution in a critical component. The thermoelastic model is formulated on the principle of finite-differences or finite-elements to calculate the stresses at discrete nodes. The model uses generalized Hooke's law, equilibrium and compatibility conditions and plastic strain relations to

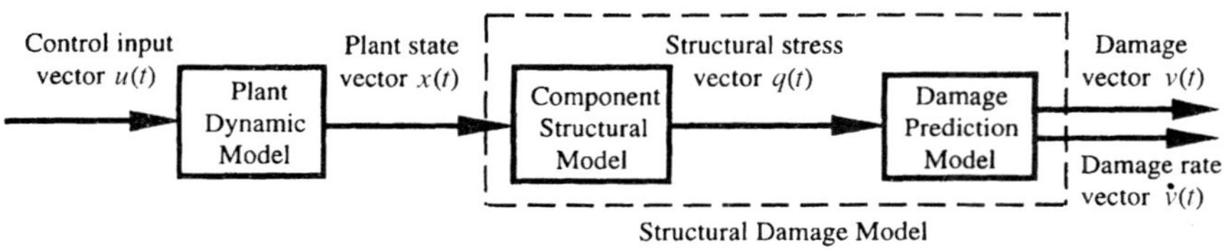

Fig. 11.2 The damage prediction system.

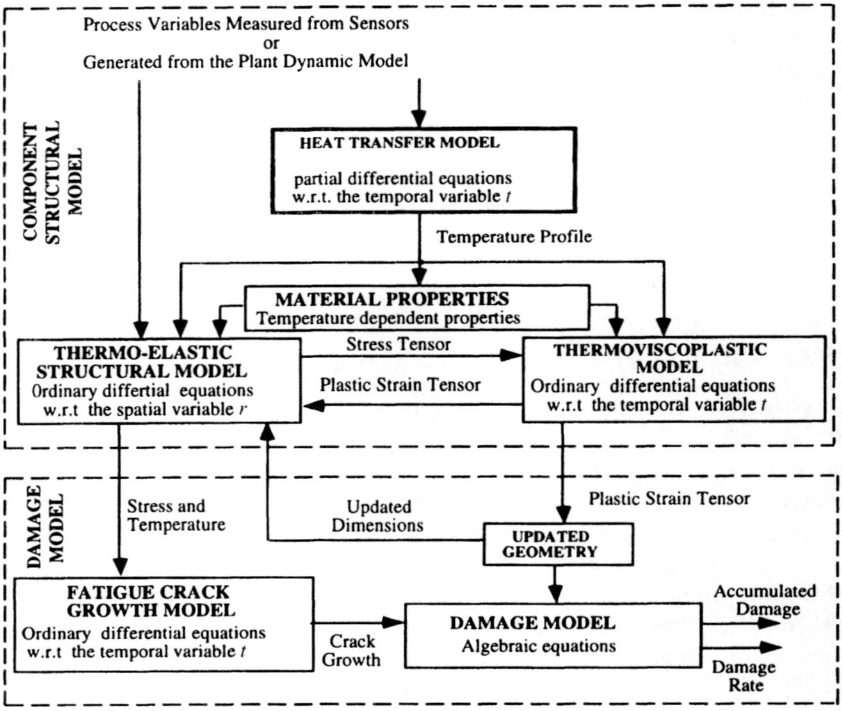

Fig. 11.3 Flow diagram of the structural damage model.

arrive at different equations with stress being the dependent variable with respect to the spatial variables. Plastic strain at each spatial node for each instant of time is generated by a unified thermoviscoplastic model of time-dependent creep/plasticity. The instantaneous plastic strain components at each node are inputs to the thermoelastic model as seen in the flow diagram of Fig. 11.3. The model inputs are stress components and temperature at each node generated by the structural model and the heat transfer model, respectively. Tensorial back stress components and the scalar drag stress and limit stress at each node are the internal state variables. The thermoviscoplastic model takes into account the effects of dimensional changes due to both rate-dependent creep and rate-independent plasticity. The model equations are in the form of time-dependent ordinary differential equations in the state-variable setting. A major advantage of separating the thermoelastic model from the thermoviscoplastic model is that the derivatives due to the spatial variables are decoupled from those due to the temporal variable.

The thermoviscoplastic model calculates the (irreversible) deformation in the component structure due to plastic strain. The crack growth part of the damage model is based on linear elastic fracture mechanics (LEFM) where the stress intensity factor range (ΔK) is the key to fatigue crack growth. The LEFM approach can be used for modeling fatigue crack growth phenomena at high temperatures provided that the material crack growth parameters are correctly obtained. In many applications, the rationale for using the ΔK approach as opposed to ΔJ is relatively mild plasticity and low creep rate of the structural materials based on the following observations: (1) the stress calculations in the component structural model take into account the effects of plastic strain; (2) the structural material partially loses its ductility as it strain-hardens after several stress cycles; (3) the component temperature, even under transient conditions, is less than half the melting point of the structural material; (4) the effects of cavitation crack growth are retarded by dynamic recrystallization because the structural load is often at high mean stress with low amplitude; and (5) microstructural analysis of cracked materials show the primary mode of fracture to be transgranular which is in agreement with the LEFM theory.

The fatigue crack growth model in Fig. 11.3 is built upon the concept of linear elastic fracture mechanics (LEFM). The fracture behavior of a linear elastic structure can be characterized by the stress intensity factor K, which is given as

$$K = C\sigma\sqrt{\pi a} \tag{11.1}$$

where C is the dimensionless geometric correction factor; σ is far field applied stress; a is characteristic crack dimension.

The fatigue crack process is assumed to belong to Mode I where the principal load, applied to the crack plane, is normal to the crack surface and tends to open the crack. The influence coefficient approach is used to account for the non-uniform stress in the main steam header. In this method remote stress normal to the crack plane (Mode I) is represented by a cubic equation as:

$$\sigma_{normal} = A_0 + A_1x + A_2x^2 + A_3x^3 = \sum_{j=0}^{3} A_j x^j \tag{11.2}$$

where x is a distance measured from the crack tip along the crack. Dimensionless stress distribution is given by

$$\sigma j = \left(\frac{x}{a}\right)^j \quad \text{for } j = 0, 1, 2, \text{or } 3 \tag{11.3}$$

Equation (11.2) can be rewritten in the following form:

$$\sigma_{normal} = \sum_{j=0}^{3} A_j a^j \left(\frac{x}{a}\right)^j \tag{11.4}$$

For semi-elliptic cracks, the stress intensity factor in the cubic polynomial stress distribution of Equation (11.4) is given as:

$$K = \sqrt{\frac{\pi a}{Q}} \sum_{j=0}^{3} G_j A_j a^j \tag{11.5}$$

where the crack is assumed to be semi-elliptic, Q is the crack shape parameter which is a function of a; G_js are pre-determined influence coefficients which are given by Anderson (1994) as functions of cylinder dimensions, crack shape, and crack dimensions.

Note that Equation (11.5) is a modification of Equation (11.1) where

$C \equiv \frac{1}{\sqrt{Q}}$ and $\sigma \equiv \sum_{j=0}^{3} G_j A_j a^j$.

The load history in this application is of varying amplitude where the effects of crack retardation/acceleration must be considered. Therefore, the crack closure model which is an extension of the Paris' model is adopted here. It assumes that residual displacements in the wake of a crack cause crack faces to contact or close before the tensile load is removed. This model predicts crack retardation in high to low loading sequence and acceleration in low to high loading. The cracks open only if the applied stress is greater than the crack opening stress, σ_{op}, which depends on the loading history. To this effect, the stress intensity factor range is calculated as:

$$\Delta K_{eff} = K_{max} - K_{op} \tag{11.6}$$

where K_{max} and K_{op} are calculated using σ_{max} and σ_{op} following Equation (11.5). The modified Paris' equation is:

$$\left(\frac{da}{dN}\right)_i = C_o(\Delta K_{eff})_i^n \tag{11.7}$$

where i corresponds to the i th cycle; C_o and ΔK_{eff} correspond to the same closure level and are defined as:

$$C_o = {}^{C}\!\Big/_{(U)^n}\,; \text{ and } U = {}^{\Delta K_{eff}}\!\Big/_{\Delta K} \tag{11.8}$$

and $\Delta K = K_{max} - K_{min}$ and K_{max} and K_{min} are calculated from Equation (11.5) using σ_{max} and σ_{min}.

The computation-intensive calculations of ΔK_{eff} from σ_{op} can be simplified by the block loading approach if the crack growth during any given loading block is not large enough to grow out of the plastic zone. Blocks can be created if there is no very large overload. Furthermore, if the stress amplitude is small compared to the mean stress ratio for each cycle within the blocks, it is assumed that

$$\sigma_{op} = \sigma_{min} \quad \text{if } \sigma_{min} > 0 \tag{11.9}$$

which is often the case in pressure vessels. This assumption allows calculation of crack growth per cycle by first calculating K_{max} and K_{op}, and then using Equations (11.8) and (11.9) to evaluate crack growth per cycle in a block. The crack growth equation is then transformed into a differential equation following the procedure of Ray *et al.* (1994a).

Life extension methods for several critical components of the Eddystone Power Plant Unit 2, using the above damage prediction system, are illustrated in section 11.3.

11.2.2 The maintenance infrastructure for equipment life extension

Paralleling recent advances in equipment life extension methods is the need for evolving a maintenance infrastructure to help realize the benefits of economies of scale and consistent high quality of service at affordable cost. This maintenance infrastructure incorporating a multiplicity of users, service providers and intelligent agents (human or computer program) can be classified into three layers as shown in Fig. 11.4. The Maintenance Enterprise incorporates both the maintenance infrastructure, and operating equipment and maintenance services.

The first layer of the maintenance infrastructure is composed of (1) the physical pathways used to transmit information, (2) information services which provide the enabling technologies for interoperation on the pathways like File Transfer Protocols, World Wide Web, Gopher, and (3) data storage. The second layer consists of intelligent agents which broker the information and service access needs between the users and service providers. Much recent work has been done on intelligent brokers (Crowston and Malone 1988; Cypher 1993; Dent *et al.* 1992; Malone *et al.* 1988; Maes and Kosierok (1993)) in similar contexts. The third layer consists of mdp specific applications which customize user interfaces,

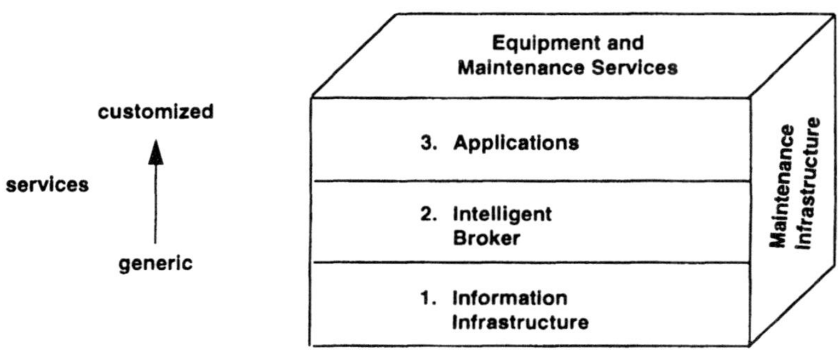

Fig. 11.4 The maintenance enterprise.

provide data and integrity checking, package distributed services and support commercial aspects of the enterprise.

The evolution of the National Information Infrastructure (ARL/PSU Technical Report 1995) provides the underpinnings for integrated maintenance enterprises of increasing scope and magnitude up to national or global levels. The following section provides a more detailed concept of operation for electronic mdp services over the NII. The next section discusses research issues and necessary advancements in information technology for a technically and economically feasible maintenance infrastructure. Finally current research in these areas is discussed and future research directions explored.

Electronic mdp services over the NII: concept of operations

The object model for electronic mdp services over the NII involves three sets of players: users, service providers and service brokers. A very simple process model of mdp services over the NII involves requesting a service, finding a service, selecting a service provider, establishing a contract, developing a service interface and electronically executing a service.

A functional architecture for electronically accessing mdp services is depicted in Fig. 11.5. The major components of this architecture include a

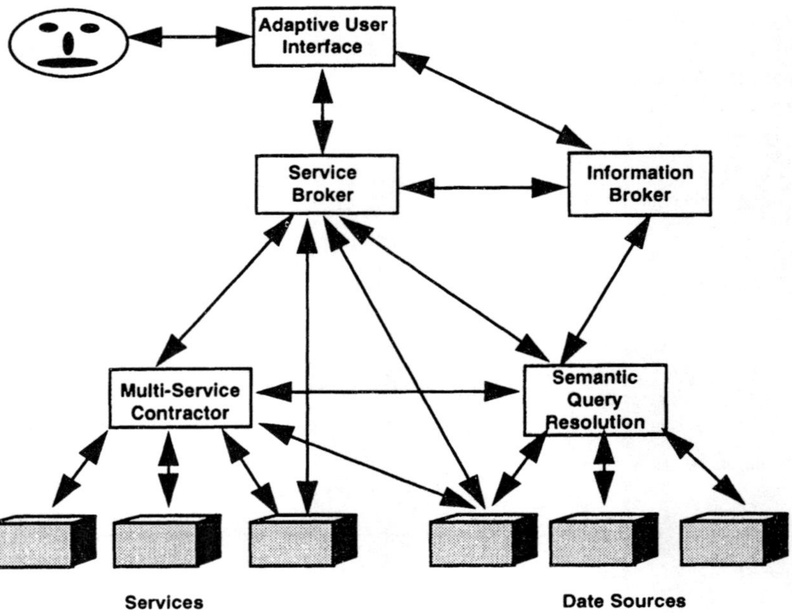

Fig. 11.5 Functional architecture of service access components.

web-type open user interface to request mdp services and to accept results, a service brokering function to find and invoke services or offer clarification dialogue with the user, and an mdp thesaurus for semantic resolution of multiuser terms. The service broker may invoke multi-party service contracts in response to a user query. These functional components of the maintenance infrastructure hide, as much as possible, the details of the locations of required services and their specific access terms and invocation procedures from the user, allowing him to focus on defining his needs in a standardized request for service. More technical details on this concept are given in the ARL/PSU Technical Report 1995. Major technical challenges of this architecture are addressed in the next section.

Technical challenges in electronic delivery of mdp services

Real-time interoperation of mdp service components and databases under dynamic conditions of thermomechanical and environmental stress requires complex interactions of internal representations of sensor data, operational requirements, resources and equipment models, with rich semantics. The problems of differences in data representation, heterogeneity of platforms, software and access terms, global constraints, global query processing and security represent the well-studied and visible surface of much deeper and more complex problems of dynamic conflicts in model autonomy, lack of standardization of requirements, control mechanisms and concurrency. The heterogeneity of the information infrastructure and other issues like data access mechanisms are best resolved by standardizing data and communication protocols, message formats and procedures. However, the timely and accurate resolution of global queries with differing and overlapping information needs, spatial scales and temporal assumptions which draw upon autonomous sources of transient information, requires more intelligent methods of retrieval and coordination of service access mechanisms.

In the context of an ubiquitous maintenance infrastructure, rapid advances in computing and communications technology have not, so far, triggered corresponding advances in the time-critical manipulation of dispersed and transient services and data for real-time analysis and decision support. Technology challenges abound in the intelligent access, retrieval and presentation of services and data in the context of a particular application and from a particular user's perspective. To a user with electronic access to multiple sources of information, anywhere in the enterprise, current technology provides two retrieval options:

- a fixed context application – like a menu system, which severely limits the usability of valuable data, or
- a query language access which requires extensive training and

specific knowledge of the structure, reference terms and system access terms for each target database.

The technical problem is to formulate the concept and prototype of an enterprise memory model which (1) presents itself to different intelligent agents (programs or humans) from dispersed parts of the enterprise in the context and perspective of that user, and (2) automatically keeps itself aware of the changes in information and changes in usage patterns. Furthermore, this enterprise memory model does not disturb the autonomy of the services and data sources which are locally maintained. There is also the need to not require retraining of local users to learn global access terms in order that they may continue using their local services and data sources after incorporation into the enterprise memory.

Various partial mechanisms dealing with the structure of an electronic enterprise-wide memory have been explored in the literature using neural networks and inductive learning algorithms (Hurson *et al.* 1992; Johnson and Zaulkernan 1991; Zaulkernan *et al.* 1992). Concepts like ANSWERGARDEN (Ackerman, 1990) and OBJECTLENS (Malone *et al.* 1988) developed at MIT also incorporate ideas of self-adaptive electronic memories for sharing of transient and informal information. Also, a Summary Schemes Model (Bright and Hurson 1990; Dash *et al.* 1994) developed at The Pennsylvania State University (PSU) develops a semantic formalism for formulating imprecise queries to such an enterprise memory without knowledge of system access terms and target database structures, addresses or reference terms. An intelligent information broker prototyped by the Applied Research Laboratory at Pennsylvania State University (Dash *et al.* 1994) incorporates the self-adaptive features of ANSWERGARDEN (Akerman 1990; Malone *et al.* 1988) and an integrated self-learning point-and-click user interface to retrieve precisely and imprecisely formulated information from anywhere within the enterprise. Thus an integrated collaborative infrastructure is created for invoking mdp services over the NII. A constructivist method for enterprise mdp development is given below.

Then, there is also the problem of server interoperability at the application layer of the enterprise. Tools, standards and protocols which deal with the heterogeneity of services and facilitate interactive exchange are required to freely enable interoperation and packaging of services in the application layer. The plant associated applications must be able to communicate with the diagnostic service applications, and sensor data must be in a standard format. Additionally, the analysis results and the maintenance instructions must be standardized. Although the initial system will include man in the loop, the eventual goal is to be able to close the loop in a completely automated fashion for mdp and automated reconfiguration if necessary. The data acquisition standard will require the Machinery–Health Message (MHM) standard, Remote

Process Control language for monitoring and diagnostic services (RPC), security standard, and Maintenance Message Prescription standard. The MHM will specify both steady state and dynamic signals from a complete spectrum of sensors and system output. The RPC will provide for the remote execution of one M&D service by another, allowing one software package to interrogate another for data, or activate its processing algorithms. The Maintenance Message Prescription will standardize the prescribed maintenance actions required as a result of the mdp analyses.

Other technology challenges include real-time coordination of optimal maintenance policies under dynamic operating conditions and operational requirements.

Constructivist methods for mdp enterprise memory

Ongoing and future research efforts in resolving technical challenges in electronic invocation of services in an mdp enterprise are described below.

The adaptive user interface is a complexity-hiding interface which learns and provides the user-specific view of information in a dynamic work environment, and does not require the user to know the specific structure or access terms for the data he seeks needs to be developed. A point-and-click graphical user interface is being developed at the NII Testbed at Penn State (Dash *et al.* 1994) which provides all the power of SQL, and automatically adapts to the user context, and to changing information in the databases. High-level classes hide information details and present the user with a tree capturing his perspective of the information. Furthermore, the user actions tracked by inductive learning mechanisms (based on NASA's IND algorithms (Buntine 1993)) are incorporated into the user context to modify the presentation of information to him/her. Additions and deletions of nodes on the tree are also automatically depicted.

Other research directions revolve around developing standardized data models for mdp services. Much research work has been ongoing in this area. STEP – the Standard for the Exchange of Product model data (Nell 1995; Smith 1995), under the auspices of the International Organization for Standardization (ISO) in Geneva – is the premier activity in the formal standards arena for describing the configuration and characteristics of industrial products, and how they may be deployed within an operating plant. Closely related to STEP are the parts library (PLIB), manufacturing management data definition (MANDATE) activities, and standards to facilitate enterprise integration. Collectively, these efforts are defining standardized data models that are widely applicable within industry, and also to mdp activities. Consortia, such as the Machinery Information Management Open Systems Alliance (Mitchell

1995; Bever 1995), are developing standardized data models for specific types of mdp services, such as (initially) machinery vibration analysis.

Advances in the techniques used to define and implement standardized data models for mdp are ongoing within many different consortia and formal standards activities. Areas for current advanced work include: electronics representation standards for boards and components, open-electronic data interchange approaches, criteria for the formal definition of data elements, standard representations for software objects, techniques for information and process modeling, and standard formats for the exchange of such models.

Coordination tools for electronic invocation of mdp services: beyond the above enabling technologies of the Information Age which promote information integrated enterprises, recent work in coordination sciences is essential for coordination of packaged electronic maintenance services over the NII. Evolving fundamental techniques, aimed at coordinating autonomous service invocation and delivery in an enterprise, are generalizations of coordination mechanisms in a digital computer architecture: time-sharing, compiler optimization, memory-mapping, exception handling, etc. Concepts for automated tools like User Request Coordination and Prioritization, Service Coordination Mechanisms and Conflict Resolution Mechanisms are being formulated to support an effective maintenance infrastructure. Several coordination mechanisms relevant to mdp applications have been described (Finkelstein 1994; Klein 1991, 1992, 1995; Malone and Crowston 1991; Mark and Dukes-Schlossberg 1994; Brown 1985; Birnbaum *et al.* 1990; Winograd 1986).

The integration of advanced real-time mdp methods with a supporting maintenance infrastructure thus provides a basis for revolutionary improvements in both availability and affordability of operating equipment.

11.3 EXAMPLES OF ADVANCED LIFE EXTENSION METHODS FOR OPERATING MACHINERY

The following research experiments and pilot studies demonstrate technical feasibility of equipment life extension in distributed computing environments using advanced prognostic methods which monitor material damage characteristics during machine operation.

11.3.1 Distributed maintenance infrastructure for vibration damage mitigation

Experimental verification of equipment life extension was conducted at The Pennsylvania State University in various experiments during 1993 to 1995 under grants from DARPA (DE-FC36-9560100069), NSF

(DMI9424587, ECS-9216386, MSS-9112609), (NASA (NAG3-1240, NAG3-1673 and NAG3-1799) and EPRI (RP8030-5). A laboratory testbed, shown in Fig. 11.6, was established incorporating:

1. a two-degrees-of-freedom mechanical system excited by a computer controlled shaker table, representing an operational plant (Fig. 11.6), and
2. a remote mdp laboratory providing electronic mdp services and containing plant models, stochastic damage estimation algorithms, life prediction algorithms and a dynamic structural durability analysis module for achieving optimized system performance.

Test results demonstrate (see Fig. 11.7):

1. fatigue life extension up to three and one half times with no significant performance degradation, and
2. verification of analytical prediction of damage with experimental observations. Technical details of these experiments are given below and in Tangirala *et al.* (1995).

Description of the laboratory testbed

The testbed incorporates a plant subsystem and a remote mdp system. The plant subsystem consists of a simple two-degree-of-freedom (DOF) mass-beam system excited by the vibratory motion of a shaker table. One DOF is directly controlled by the shaker table whereas the other DOF is not controlled, but it is an observable DOF via position measurements of the vibrating masses. The test specimen, subjected to fatigue, is representative of mechanical damage in a plant component. The mechanical

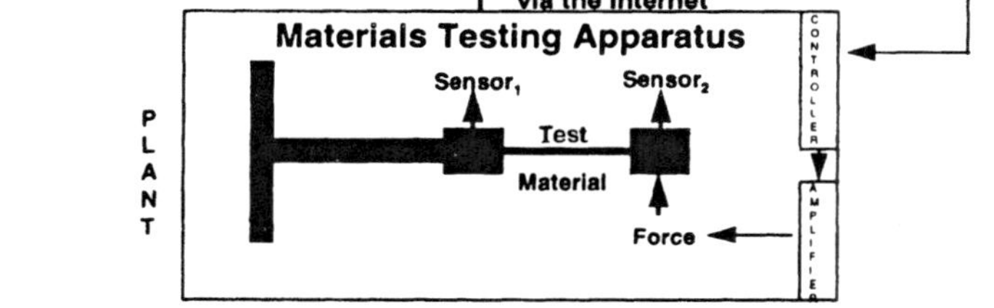

Fig. 11.6 Real-time monitoring for equipment life extension.

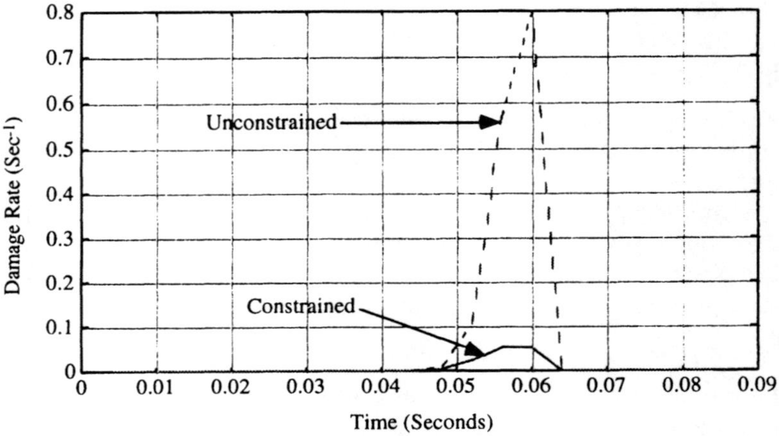

Fig. 11.7 Life extension verification by experimental observation.

system consists of two discrete masses supported on cantilever beams. The second mass is connected to the shaker table which acts as the actuator of the system through a flexible pivot. The flexible pivot rigidly links the translating mass in the actuator to the mass on the beam in the direction of motion. The translational positions of the masses are measured using linear variable differential transformer (LFDT) sensors which are located at the midpoints of the masses M1 and M2.

Since the objective of this testbed is to investigate the influence of different control policies on specific modes of fatigue failure in a dynamic setting, a structurally weakened element which is representative of a critical plant component is introduced to ensure the occurrence of an observable failure. In the two-mass configuration, two parallel failure sites are introduced by drilling a 0.312 inch hole located 2.66 inches from the center of mass M1. As the specimen is subjected to cyclic stresses, cracks initiate on the upper and lower surfaces of the beam and propagate through the thickness towards the neutral axis.

The remote mdp system consists of multi-computer based data acquisition, processing, and analysis tools and equipment life extension models shown in Fig 11.6 and described in the next section. The plant subsystem represents an experimental simulation of an operating plant and the mdp subsystem emulates a remote mdp center's services.

Advanced prognostic and life extension models

Mechanical structural stress model

A structural model of the mechanical system described above is formulated using a cubic interpolation polynomial to approximate the lateral displacement of the beams and a linear approximation for the lateral displacement of the masses.

The displacements of the beams 1 and 2, and the mass M_1 are approximated by the following relationships:

$$y_i(x) = C_{1i}\frac{x^3}{6} + C_{2i}\frac{x^2}{2} + C_{3i}x + C_{4i};\ i = 1,2 \text{ for the beams 1 and 2}$$

$$y_3(x) = C_5 x + C_6; \text{ for the mass } M_1$$

The ten constants in the above equations are evaluated on-line by applying standard boundary and compatibility conditions and using the physical dimensions to yield equations of motion.

The stresses at the failure sites on beam 2 shown in Fig. 2.2 are given by

$$\sigma_{FS} = 5.47 \times 10^5 \bar{y}_1 - 2.58 \times 10^5 \bar{y}_2 \text{ psi}$$

This accounts for the increased local stress at the critical point due to the reduction in the area moment of inertia caused by the hole drilled in the specimen. The stresses computed via this equation show excellent agreement with experimentally measured stresses. Time-domain frequency-sweep excitation (Ljung 1987) and experimentally measured data (Tangirala *et al.* 1995) were used to identify a sixth order voltage (actuator signal)-to-voltage (position sensor signal) transfer function.

Fatigue damage model

The fatigue damage model in the continuous-time setting, reported by Ray and Wu (1994a), is first derived based on linear damage accumulation using a combination of Coffin-Manson and Basquin relationships (Bannantine *et al.* 1990). Then, this linear damage model is modified following the damage curve approach (Bolotin 1989) to account for the dependence of the damage rate on the current level of accumulated damage. For details see Tangirala *et al.* (1995).

Using a crack propagation model such as the Paris model (Paris and Erdogan 1963), it is shown that the crack growth rate is dependent not only on the stress amplitude but also on the current crack length.

Structural durability analysis

Structural durability analysis consists of optimizing a cost functional subject to physically motivated constraints, such as actuator limits, and those based on damage rate and accumulation in the critical components (Ray *et al.* 1994c, 1994e).

Minimization is carried out subject to constraints. These constraints take the form of actuator limits and limits on the damage rate experienced by the material at the failure site during operation. Reducing the damage rate in a cycle of operation results in the reduction of the damage accumulated during that cycle, leading to an increase in the life of the system. The weight on the control inputs is included to avoid large fluctuations of the input signal as this could lead to greater fatigue damage and actuator chatter. The relative weights on the two components of the cost functional subject are selected by the design engineer to emphasize performance, transient plant behavior, damage dynamics or control effort as required. A judicious selection of the constraints and weights can result in, within limits, different levels of damage mitigation with an associated loss of performance.

In this work, a general purpose nonlinear programming software package NPSOL (Gill *et al.* 1991) has been used to solve the open loop constrained optimization problem. The details are reported by Ray and Wu (1994b) and Ray *et al.* (1994c).

On-line damage mitigation algorithm

The problem is to generate, for the mass-beam structure, an on-line optimal algorithm that will make a trade-off between the conflicting requirements of performance and damage mitigation over the time interval $[t_0, t_f]$. This optimization problem is represented in a discrete time setting to find an optimal control sequence (uk) that will maneuver the plant from a known initial condition to a prescribed final state subject to the following constraints:

1. The discretized version of the continuous-time dynamic system.
2. Preassigned tolerance limits on damage rate to assure structural durability.

The optimization is carried out over one half cycle of motion with the displacement of M2 chosen as the performance variable. The nominal output profile is determined by the closest approximation to a half square wave that the displacement trajectory of M2 can achieve on the actual dynamic system over one half cycle of operation. This profile and the corresponding nominal excitation are shown in Figure 11.8. Resonance of M1 causes fatigue damage of the test specimen. This is a direct consequence of the high local stresses at the failure site caused by the large displacements of M1. This configuration is, therefore, representative of a real physical situation where the specification of performance requirements in one subsystem of a dynamic system may lead to fatigue damage in other subsystems.

Experimental verification of life extension

The optimization task described in the previous section was carried out over one half cycle of motion of the mass-beam structure. It should be noted, however, that the initial damage of the specimen is a random variable (Ray *et al.* 1995) which depends on several factors like the machining quality, the heat treatment, size of the notch, etc. The level of

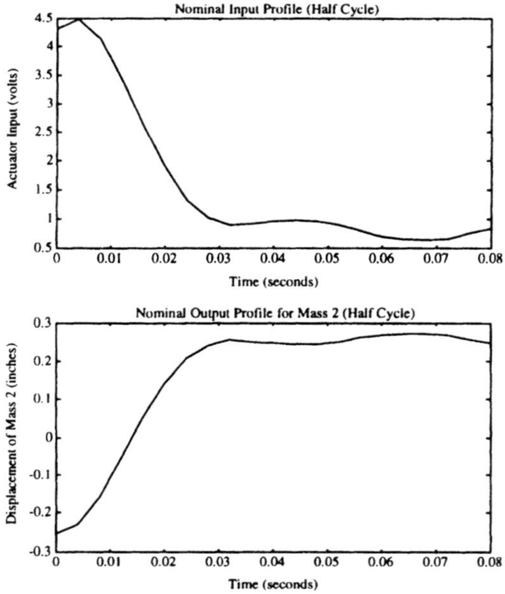

Fig. 11.8 Nominal input and output profiles for optimization.

initial damage used in the optimization has a significant effect on the damage rate and accumulation due to nonlinearities in the damage dynamics (Tangirala *et al.* 1995).

The constrained optimization yields the input waveform shown in Figure 11.9. The corresponding output profile is plotted along with the unconstrained (nominal) output in. This plot shows that the loss of performance, as defined by the deviation from the nominal profile, is quite small. Figure 11.10 shows the Fast Fourier Transform (FFT) of the constrained and unconstrained input waveforms. The FFT of the constrained waveform shows a marked reduction in the magnitude of the third harmonic component at ~18 Hz relative to the unconstrained waveform. As mentioned earlier, this third harmonic excites resonance of the mass, M1, and is the major factor causing fatigue damage at the failure site. Simply removing this component from the input excitation, however, leads to a considerable loss of performance which is unacceptable. Therefore, the magnitude of this third harmonic component is substantially reduced while some of the other (apparently harmless) harmonics are moderately increased to minimize the loss of performance. Figure 11.11 compares the damage rates obtained from the simulation of the dynamic system using the unconstrained and constrained input waveforms. A considerable reduction in the damage rate due to

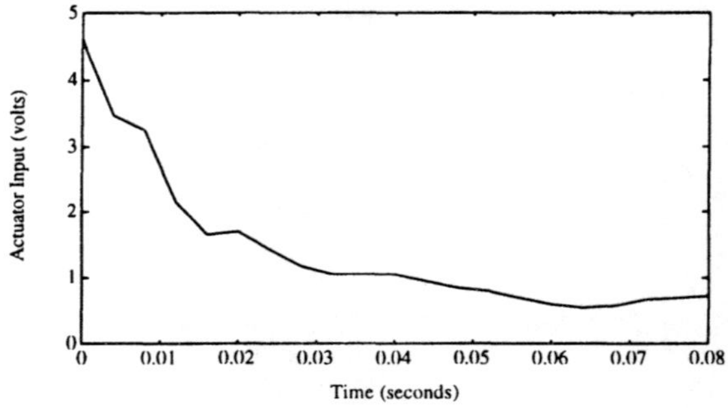

Fig. 11.9 Constrained actuator input.

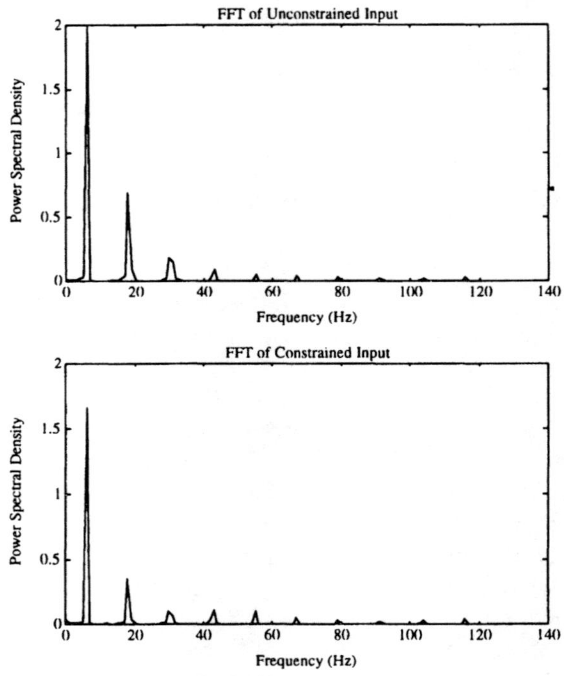

Fig. 11.10 FFTs of constrained and unconstrained inputs.

the constraint is clearly observed. The next two paragraphs present simulation and experimental results to demonstrate how the constrained optimization procedure achieves a substantial increase in the specimen life with only a marginal loss of performance.

Simulation of the dynamic system to the point of failure of the specimen

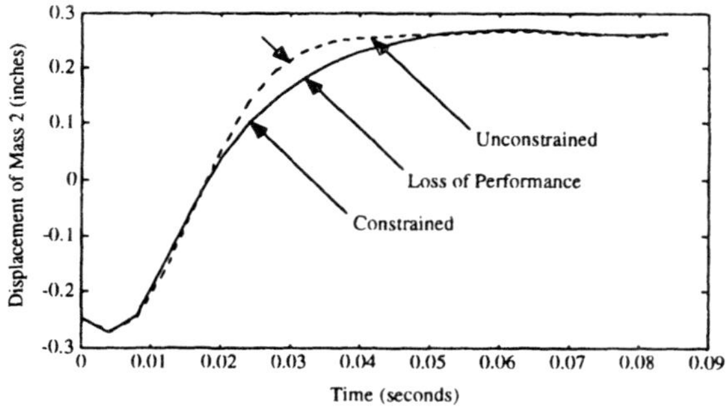

Fig. 11.11 Comparison of constrained and unconstrained outputs (simulation).

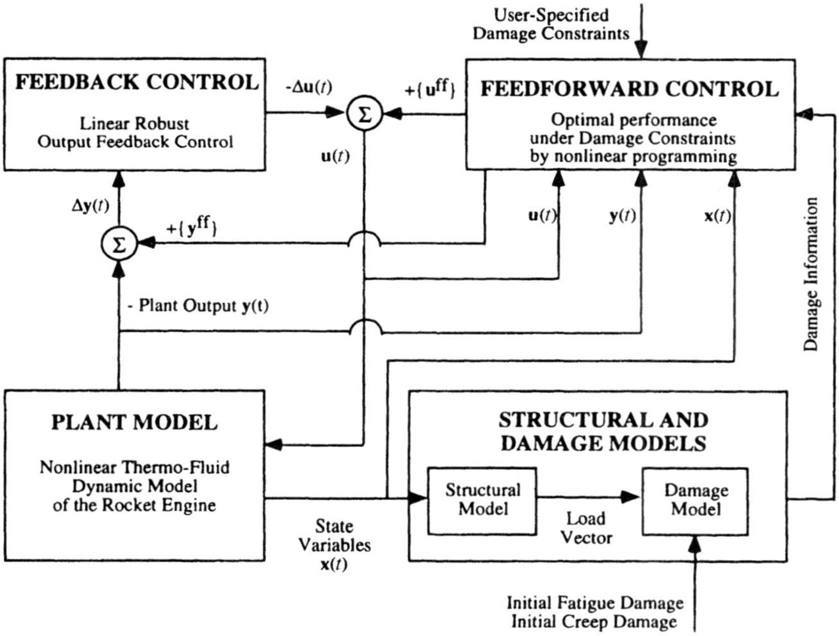

Fig. 11.12 Schematic of damage-mitigating control system.

(i.e. when the damage accumulation reaches unity) yields a predicted life of 3639 cycles for the unconstrained case and 11 814 cycles for the constrained case. These simulation results show an increase in life by a factor of ~3.25 while the loss of performance is insignificant, as seen in Figure 11.12. It should be noted, however, that the predicted life is based on an expected value of 0.0005 for the initial damage and therefore this approach may at most represent the specimen life in a statistical sense (Ray *et al.* 1995).

Table 11.1 Number of cycles to failure

Specimen number	*Cycles of failure*
1	3 934 (Unconstrained)
2	3 812 (Unconstrained)
3	3 554 (Unconstrained)
4	11 222 (Constrained)
5	13 960 (Constrained)
6	12 120 (Constrained)

The constrained and unconstrained input sequences, generated earlier for the simulation, were also used to excite the shaker table and mechanical structure in the testbed. These experiments were conducted using identical specimens that were machined from the same stock of 1018 steel. Fig. 11.1 compares the output signals (i.e. the displacement of the mass M2) obtained from the simulation and experimental results when excited by the unconstrained and constrained input signals. The mean value of specimen life is 3767 cycles for the unconstrained case and 12 434 for the constrained case. These results show an increase in life by a factor of ~3.30 which agrees very closely with the corresponding gain of ~3.25 predicted by simulation. Table 11.1 shows that there is some scatter in the number of cycles to failure for different specimens. This is due to the stochastic nature of the damage process and the randomness in the initial flaw size in apparently identical specimens.

11.3.2 Life extension of mechanical components of a reusable rocket engine

Under a grant from NASA Lewis Research Center (NAG3-1240, NAG3-1673 and NAG3-1799), the Pennsylvania State University has developed damage prediction algorithms for a reusable rocket engine such as the Space Shuttle Main Engine (SSME), and demonstrated through computer simulation the incorporation of these algorithms into a closed-loop controller to achieve an optimal trade-off between system performance and structural durability of the engine. An overall schematic of the Damage Mitigating Control (DMC) system is shown in Figure 11.12.

The simulation is carried out in a distributed fashion as follows:

- a nonlinear, state space model of the plant resides at the Mechanical Engineering laboratory representing plant operations,
- the structural models, damage models, and feedback controller reside on Collaboratory Internet worked computers at the Information Systems Collaboratory which emulates a remote mdp center operation, and,

- plant state variables and control inputs are passed between the two facilities over the Internet, emulating the first phase of NII implementations. The second phase of NII implementations is planned to introduce ATM switching and SONET technologies.

The simulation is configured so that the plant may run under open-loop control, in which case the mdp center computers monitor the damage in critical components in real time. This arrangement is representative of how such a system would be implemented in real life, since for most applications it would be undesirable to invest the extra money (and in the case of the rocket engine, incur the extra weight) involved in installing the computing power necessary to monitor the damage on-site. In addition, this arrangement allows the structural and damage models to be easily upgraded as better algorithms are developed, with the new models becoming available to all users simultaneously. Also, long-term load and performance histories can be maintained and used on-line at the mdp center for maintenance decision support. The simulation may also be run under closed-loop control, in which case the DMC is active, generating a feedback control policy in real time to limit the damage to critical plant components by user-defined constraints.

Technical details of the simulation setup are given below.

Description of plant model

In order to make the damage prediction calculations computationally tractable, while maintaining sufficient accuracy for the purpose of control synthesis, a reduced order nonlinear model of the thermo-fluid dynamics of the rocket engine was formulated. A standard lumped parameter approach was used to approximate the partial differential equations describing mass, momentum, and energy conservation by a set of first-order differential equations, with time as the independent variable. The resulting 20th order model proved to be valid for the operating conditions of main thrust chamber pressure between 1200 psi and 3000 psi, as demonstrated by comparison with the engine manufacturer's own high order nonlinear model.

Further details of the plant model are available in NASA Contractor Report 194470 and NASA Contractor Report 4640 (Dai and Ray 1994; Freed 1988).

Description of damage models

While the rocket engine contains hundreds of components, it was deemed both impractical and unnecessary to attempt to monitor the damage in every single component, since the severity of operating conditions varies significantly between different sections of the engine. Following consultations with NASA personnel, three components were identified as being critical to the safe operation of the plant, as well as being subjected to extreme thermomechanical stresses resulting in high rates of damage accumulation:

- blades in the oxygen turbopumps,
- blades in the hydrogen turbopumps, and
- coolant channel ligaments in the main thrust chamber.

For each component it was necessary to identify the relevant mode(s) of damage, develop appropriate damage algorithms, and formulate a structural model to capture the relevant material behavior. The turbine blades were assumed to suffer primarily from high-cycle fatigue cracking due to low amplitude vibratory stresses induced by unsteady aerodynamic loading. The stress amplitude at the blade root is obtained via a linear elastic finite element analysis, and crack length is determined using the Paris crack propagation model.

Due to the high gas temperatures and high coolant pressures, the ligaments of the main thrust chamber cooling channels experience both creep and creep ratcheting. The structural behavior of the ligaments is represented by an equivalent sandwich beam model (Dai *et al.* 1994), consisting of two thin faces with identical thickness, separated by an incompressible core. A unified viscoplastic theory (Freed 1988) was selected to model the damage dynamics because of its ability to represent both rate-dependent creep and rate-independent plastic behavior.

Control system formulation

Formulation of the DMC system is accomplished in two steps:

1. Synthesis of a feedforward control policy with the objective of achieving structural durability and high performance.
2. Synthesis of a feedback control policy for intelligent decision-making such as damage prognosis and risk analysis via on-line feedback of the plant and damage monitoring information.

The feedforward control policy is obtained via nonlinear programming by optimizing a specified cost function of the plant dynamic performance without violating preassigned constraints on the damage rate and accumulation. The damage dynamics are incorporated in state space form into the complete dynamic model of the engine, thus allowing the instantaneous damage rate to be dependent on the current level of accumulated damage. This is consistent with physical observations that show damage rate to be strongly dependent on initial conditions.

In order to account for deviations from the nominal plant trajectory due to modeling uncertainties, sensor noise, and external disturbances, a feedback controller was synthesized to maintain the trajectories of the plant output variables close to the respective nominal trajectories. The robust feedback control policy was formulated via μ-synthesis (Packard *et al.* 1993).

Simulation results

The test case chosen to evaluate the DMC was to bring the pressure in the main thrust chamber from equilibrium at 2700 psi to a new equilibrium state at 3000 psi in 300 ms. The control inputs available to accomplish this maneuver are the positions of two preburner oxidizer valves. Simulations were carried out for a wide variety of initial conditions and constraints on the allowable amount of damage. Figure 11.13 gives representative results for the oxidizer turbine, showing the increment in damage accumulation for two different damage constraints, as well as results for the unconstrained case. In this particular example, the main damage constraint is on the maximum allowable rate of creep damage in the cooling channel ligaments, thus demonstrating the need to consider the effects of plant operation on all of the critical components simultaneously, since changes made to reduce damage in one component may have a beneficial or adverse effect on other critical components.

Simulation results detained in Dai and Ray (1994) demonstrate on-line monitoring of plant operations is needed for accurate damage assessment because:

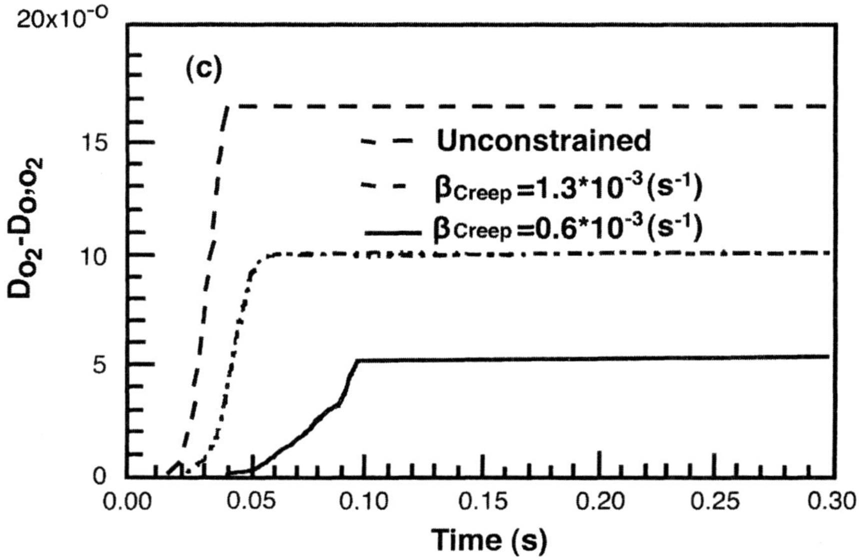

Fig. 11.13 Increment in accumulated fatigue damage in the oxygen turbopumps.

- damage accumulation rates are significantly higher during transient operation, as compared to steady-state operation, and
- high performance need not be sacrificed to achieve substantial increases in component life.

11.3.3 Pilot demonstration of life extension of the Eddystone Power Plant over the NII testbed

Remote on-line prognostics for the fossil-fueled power plant at Unit 2 Eddystone, PA are being developed in a pilot demonstration of the National Information Infrastructure testbed for remote equipment mdp. The purpose of the pilot demonstration is to:

1. Demonstrate multiuser access to remove mdp services requiring standardized access and invocation mechanisms, semantic query resolution for a heterogeneous user group.
2. Predict the remaining life of critical power plant components, including the main steam header, in operational conditions, using current equipment health status inputs from sensor data, load history and planned operations.

The distributed maintenance infrastructure

The pilot demonstration incorporates two NII nodes, the Pennsylvania State University node incorporating the power plant structural and damage models, and life prediction algorithms using input data obtained by an interface computer from the Eddystone Power Plant, Unit 2. The input data is collected at the power plant using various sensors. Other potential user and service nodes are emulated on various computers at the Applied Research Laboratory of Penn State University.

The physical pathways of the distributed maintenance infrastructure for the pilot demonstration consist of a variety of networking and internetworking systems. These include Ethernet LANs, desktop ATM systems connected by a SONET backbone, Internet connections provided by commercial Internet service providers as well as direct connections via modem and ISDN. These systems are reconfigured for testing the performance of available mdp maintenance services and protocols under a variety of conditions.

The enabling technologies or information services allow the heterogeneous conglomerate of inter-networks to act as a single, homogeneous path for information. File transfer protocols, Telnet protocols and the World Wide Web are all examples of technologies that allow users to directly access a service over the NII.

Applications and user services are distributed across the networks. In addition to the mdp services, the testbed consists of software modules called intelligent brokers to support the interaction between users and service providers. Brokers provide many of the functions of middleware in a client/server environment.

The NII testbed uses standardized remote access and invocation of services. This allows third parties to develop and deploy services independently. A service broker provides aid to users in locating the service and selection of service providers. The service broker will help integrate mdp services and data sources on the NII testbed by providing a mechanism to generate queries based on a model of the data needs of the service.

The NII testbed provides for multi-user access by a heterogeneous user community. There are many ways for users to access the applications and services in the Distributed Maintenance Infrastructure. Initially most people will use a small basic set of widely available access software such as a WWW client, a Telnet client, an FTP client and a news and email reader. Eventually users will have sophisticated software agents who interact with the applications and needed mdp services provided on the network.

mdp services

Remote mdp services are provided to the Eddystone Power Plant, located south of Philadelphia. In particular, the analysis focuses on the main steam header of Unit 2. This component is a large pipe which directs high pressure, high temperature steam from the boiler to the high pressure turbine. The continuous high pressure, high temperature operating environment causes the main steam header to experience long-term creep damage, resulting in a thinning of the pipe walls. In addition, fluctuations in steam temperature and pressure experienced during transient plant operations induce fatigue crack formation and propagation, and may also cause additional plastic strains to accumulate in the pipe.

The prognostic analysis requires the continuous, on-line monitoring of plant sensor data, including the steam pressure, temperature, and mass flow rate through the main steam header, as well as the overall plant output and fuel consumption. This data will be forwarded to the Information Systems Collaboratory at ARL/PSU, where structural and damage models have been formulated for the steam header. Specifically, both time-independent inelastic strains and time-dependent creep damage are modeled using a unified viscoplastic theory, while fatigue crack growth is calculated using linear elastic fracture mechanics. For determining the current damage state, sensor data provides sufficient input to the structural and damage models. However, for component remaining-life predictions, the plant macroscopic variables (i.e. steam temperatures and pressures, etc.) need to be generated based on the planned future operation of the plant. For this purpose, ARL/PSU has developed a dynamic model of the Eddystone plant. The model is a finite-dimensional, 27th order state space representation of the plant, with a standard lumped parameter approach used to approximate the partial differential equations describing mass, momentum, and energy conservation by a set of first order differential equations with time as the independent variable; a detailed description of this model, configured to represent a generic powerplant, is given in Ray *et al.* (1994). Using output from the plant model as input to the structural and damage models, ARL/PSU is able to predict, within a specified confidence level, the remaining life of the main steam header.

The testbed will provide operators at Eddystone Unit 2 with the current damage state of the main steam header. At first, off-line data will be used to test the system and to tune and validate the structural and damage models. As the project proceeds, real-time plant sensor data will be forwarded to ARL/PSU on a regular basis, where the damage calculations will be conducted automatically. Damage information will be returned to Eddystone, both on a regular schedule, and in response to a plant operator's request. The operator decision process will be modeled to determine an upper bound on response time.

In addition, the system will predict, within a specified confidence level, the remaining life of the main steam header, based upon expected future operating conditions. This will not only be of use in maintenance scheduling; it will also aid in operations planning, allowing the user to run several 'what if' scenarios for different operating conditions, and use the resulting prognostic analyses to identify the most economical operating schedule.

The pilot tests must demonstrate the technical and economic feasibility of life extension services over the NII. This must include the demonstration of timely and reliable responses to a user request for service and invocation and integrity of services delivered. Furthermore the plant dynamic model and creep model must be validated by the experimental demonstration.

11.4 CONCLUSIONS

This chapter establishes distributed system architectures for on-line monitoring, diagnostics and prognostics (mdp) of equipment. The mdp information is used for predicting remaining life, and for real-time constraining of operational parameters for the life extension of operating machines. Both technical and economic feasibility issues are examined. Technical feasibility analysis is supported by laboratory experiments and simulations of a reusable rocket propulsion engine and electric power plants. These experiments demonstrate substantial reductions in total accumulated damage by slight adjustments in the process variables to reduce system stress. Furthermore the results lead one to devise life extension methods for operating machinery by constraining the process variables based on load history, operational requirements and total accumulated damage of critical components.

Typically, the health and monitoring of equipment is done on-site, historic and diagnostic databases are archived, logistics and schedule data are available from plant management, and the dynamic models of damage and operating equipment performance under thermomechanical fatigue, creep, corrosion and wear are often proprietary and available only from vendors of equipment or research organizations. Concepts for utilizing the emerging National Information Infrastructure in meeting the computational and economic challenges of equipment life extension methods are developed. The technical vision supporting these concepts is that of electronic monitoring, diagnostic and prognostic services for electromechanical systems from specialized and generic centers over a ubiquitous information infrastructure. Advances in information technology and an operational concept for supporting such a maintenance infrastructure are presented.

REFERENCES

Ackerman, M. (1980) ANSWERGARDEN: a tool for growing organizational memory. *Proceedings of the ACM Conference on Office Information Systems*, 31–9.

Anderson, T.L. (1994) *Fracture Mechanics: Fundamentals and applications*, 2nd edn, CRC Press.

Applied Research Laboratory/Pennsylvania State University and EPRI M&D Center (1995) *Interoperability Testbed for Diagnostic and Prognostic Maintenance of Equipment and Processes: Concept of Operations*, Technical Report, Cooperative Agreement No. DE-FC36-95G010064, November.

Argote, L. and Epple, D. (1990) Learning curves in manufacturing. *Science*, **247**, 920–4.

Bannantine, J.A., Comer, J.J. and Handrock, J.L. (1990) *Fundamentals of Metal Fatigue Analysis*, Prentice-Hall, Englewood Cliffs, NJ.

Bever, K. (1995) MIMOSA – machinery information exchange standards. *Sound and Vibration*, September.

Birnbaum, L., Collins, G. Freed, M. and Drulwich, B. (1990) Model-based diagnosis of planning failures. *Proceedings of AAAI-90*, 318–23.

Bogdonoff, J.L. and Kozin, F. (1985) *Probabilistic Models of Cumulative Damage*, Wiley Interscience, New York.

Boller, C. and Seeger, K.T. (1987) *Materials Data for Cyclic Loading*, Elsevier, Amsterdam.

Bolotin, V.V. (1989) *Prediction of Service Life for Machines and Structures*, ASME Press, Warrendale, PA.

Bright, M.W. and Hurson, A.R. (1990) *Summary Schemas in Multidatabase Systems*, Penn State Computer Engineering Technical Report, TR-90-076.

Bright, M.W. and Hurson, A.R. (1992) Issues in multidatabase systems. *IEEE Computers*, **25**(3), 50–60.

Bright, M.W., Hurson, A.R. and Pakzad, S. (1993) Automated resolution of semantic heterogeneity in multidatabases. *ACM Transactions on Database Systems.*

Brown, D.C. (1985) *Failure Handling in a Design Expert System*, Butterworth and Co.

Buntine, W. (1993) Tree classification software. *NASA Conference Publication*, **1**(3189), 289–98.

Crowston, K. and Malone, T.W. (1988) Intelligent software agents. *Byte*, 267–71.

Cypher, A. (1993) in *Watch What I Do: Programming by Demonstration*, MIT Press.

Dai, X. and Ray, A. (1994) Damage mitigating control of a reusable rocket engine, Part 1: Life prediction of the combustion coolant wall. *ASME Winter Annual Meeting*, Chicago, November.

Dash, K., Hurson, A.R., Phoha, S. and Chehadeh, C. (1994) Summary schemas model – a scheme for handling global information sharing. *Proceedings of the ISMM International Conference on Intelligent Information Management Systems.*

Dent, L., Boticario, J., McDermott, J., Mitchell, T. and Zabowski, D. (1992) A personal learning apprentice. *Proceedings of AAAI 92*, 96–103.

Etzioni, O. and Weld, D. (1994) A softbot-based interface to the Internet. *Communications of the ACM*, July.

Finkelstein, A. (1994) Requirements engineering: a review and research agenda. *Proceedings of the First Asian & Pacific Software Engineering Conference*, IEEE CS Press, 10–19.

Freed, A.D. (1988) Structure of a viscoplastic theory. NASA TM-100749, Lewis Research Center, Cleveland, Ohio, prepared for the Summer Annual Meeting Symposium: Constitutive Equations and Life Prediction Models for High Temperature Applications, sponsored by ASME, Berkeley, California, June 20–22.

Gill, P.E., Murry, W., Saunders, M.A., and Wright, M.H. (1991) *NPSOL Version 4 06*, Software Distribution Center, Stanford University, Palo Alto, CA.

Hopcroft, J.E., and Ullman, J.D. (1979) *Introduction to Automata Theory, Languages and Computability*, Addison-Wesley, Reading, MA.

Hurson, A.R., and Bright, M.W. (1991) Global information access for microcomputers. *Journal of Microcomputer Application*, **10**(2), 73–81.

Hurson, A.R., and Bright, M.W. (1991) Multidatabase systems: an advanced concept in handling distributed data. *Advances in Computers*, **32**, 149–200.

Hurson, A.R., Bright, M.W., and Pakzad, S. (1993) *Multidatabase Systems: An Advanced Solution for Global Information Sharing Process*, IEEE Computer Society.

Hurson, A.R., Jin, B., Pakzad, S.H., and Miller, L.L. (1992) Artificial neural systems in decision-aiding for large incomplete databases. *Heuristics, The Journal of Knowledge Engineering, Special Issue: Neural Networks*, **5**(3), 16–40.

Hurson, A.R., Pakzad, S., and Cheng, J.B.R. (1993) Object-oriented database management systems: evolution and performance issues. *IEEE Computer*, **26**(2), 48–60.

Johnson, P.E., and Zaulkernan, I.A. (1991) Preserving the effects of organizational learning. In *Proceedings of the INSEAD Conference on Organizational Learning and Strategic Renewal.*

Klein, M. (1991) Supporting conflict resolution in cooperative design systems. *IEEE Systems Man and Cybernetics*, **21**(6).

Klein, M. (1992) DRCS: An integrated system for capture of designs and their rationale. In *Proceedings of Second International Conference on Artificial Intelligence in Design*, Pittsburgh, PA.

Klein, M. (1993a) Capturing design rationale in concurrent engineering teams. *IEEE Computer*.

Klein, M. (1993b) Supporting conflict management in cooperative design teams. *Journal on Group Decision and Negotiation: Special Issue on the 1992 Distributed AI Workshop*, **2**, 259–78.

Klein, M. (1995) Integrated coordination in cooperative design. *International Journal of Production Economics: Special Issue on Integration and Collaboration Systems*.

Lee, J. (1994) Machine performance monitoring and proactive maintenance in computer-integrated manufacturing: review and practice. *International Journal of Computer-Integrated Manufacturing*, Vol. 8, No. 5, pp 370–80, 1995.

Ljung, L. (1987) *System Identification: Theory for the User*, Prentice Hall, Englewood Cliffs, NJ.

Maes, P. and Kozierok, R. (1993) Learning interface agents. In *Proceedings of AAAI-93*, 459–65.

Malone, T. and Crowston, K. (1991) *Towards an Interdisciplinary Theory of Coordination*. Technical Report 120, MIT Center for Coordination Science.

Malone, T., Lai, K. and Yu, K. (1988) OBJECTLENS: a spreadsheet for cooperative work. *ACM Transactions on Office Information Systems*, **5**(4), 297–326.

Mark, W. and Dukes-Schlossberg, J. (1994) A system for supporting engineering negotiation. *Concurrent Engineering Research and Applications: Special Issue on Conflict Management in Concurrent Engineering*, **2**(3), 173–82.

Mitchell, J.S. (1995) MIMOSA – Building the foundation for 21st century optimize asset management. *Sound and Vibration*, September.

National Research Council (1990) *The Competitiveness Edge: Research Priorities for US Manufacturing*, National Academy Press.

Nell, J. (1995) *STEP on a Page*. World Wide Web site, National Institute of Standards and Technology, http://elib.cme.nist.gov/nipde/stds/stepage. html.

Packard, A. and Doyle, J. (1993) The complex structured singular value. *Automatica*, **29**(1), 71–109.

Paris, P.C., and Erdogan, F. (1963) A critical analysis of crack propagation laws. *ASME Journal of Basic Engineering*, **D85**, 528–34.

Petrie, C.J. Jr. (ed.) (1992) Characterizing the need for enterprise integration. In Group Report. *Proceedings of the First International Conference on Enterprise Integration Modeling*, 114–30.

Phoha, S. (1986) The digital circuit switch and multilevel security. *Proceedings of the 6th Annual AFCEA Meeting*, Philadelphia, PA.

Phoha, S. (1987) Network management and control. *Proceedings of the Workshop on Computer Networking for Manufacturing Systems*, sponsored by National Science Foundation, Grant No. DMC-8714929, 4–6 November.

Phoha, S. (1988) Service Access Procedure (SAP) for a transport layer protocol. *Telematics and Informatics*, **5**(1), 184–6.

Phoha, S. (1989) Research directions in computer networking for manufacturing systems. *ASME Journal of Engineering for Industry*, May, 109–15.

Phoha, S., Sircar, S., Ray, A., and Mayk, I. (1993) Computational grammars for hierarchical command and control automata. In *Technical Proceedings of the 1993 Symposium on Command and Control Research*.

Phoha, S., Sircar, S., Ray, A., and Mayk, I. (1994) 'Discrete Event Control of Combat Dynamics in the C2 Setting', in *Proceedings of the 1994 American Control Conference*.

Ray, A., Dai, X., Wu, M.-K., Carpino, M. and Lorenzo, C.F. (1994) Damage-mitigating control of a reusable rocket engine. *AAAI Journal of Propulsion Power*, **10**, 225–34.

Ray, A., Tangirala, S., and Newman, J.C. (1995) Stochastic modeling of damage dynamics for failure prognostics and risk analysis. In *Proceedings of the American Control Conference*, Seattle, WA.

Ray, A., and Wu, M.-K. (1994a) Fatigue damage control of mechanical systems. *Smart Material Structure*, **3**, 47–58.

Ray, A., and Wu, M.-K. (1994b) *Damage-mitigating Control of Space Propulsion Systems for Structural Durability and High Performance*. NASA Lewis Research Center Contractor Report 194470, March.

Ray, A., Wu, M.-K., Carpino, M. and Lorenzo, C.F. (1993) Damage-mitigating control of mechanical systems: Part I – Conceptual development and model formulation. *Proceedings of American Control Conference*, San Francisco, CA.

Ray, A., Wu, M.-K. Carpino, M. and Lorenzo, C.F. (1994a) Damage-mitigating control of mechanical systems: Part I – Conceptual development and model formulation. *ASME Journal of Dynamic Systems, Measurements and Control*, **116**(3), 448–55.

Ray, A., Wu, M.-K., Carpino, M. and Lorenzo, C.F. (1994b) Damage-mitigating control of mechanical systems: Part I – Conceptual development and model formulation. *ASME Journal of Dynamic Systems, Measurements and Control*, **116**(3), September, pps 437–447.

Rychlik, I. (1993) Note on cycle counts in irregular loads. *Fatigue Fract. Eng. Mater. Struct.*, **16**, 377–390.

Smith, B. (1995) *An Overview of the ISO TC184/SC4 Committee*. World Wide Web site, National Institute of Standards and Technology, http://elib.cme. nist.gov/nipde/stds/stepage.html.

Sobczyk, K. and Spencer, B.F., Jr. (1992) *Random Fatigue: Data to Theory*, Academic Press, Boston, MA.

Stanfill, C., and Waltz, D. (1988) Toward memory-based reasoning. *Communications of the ACM*, **29**(2), 1213–28.

Suresh, S. (1991) *Fatigue of Materials*, Cambridge University Press, Cambridge.

Tangirala, S., Ray, A. and Carpino, M. (1995) Damage mitigating control of mechanical structures: experimental verification of the concept. *Smart Material and Structures*, **4**(2), 139–46.

Vidyasagar, M. (1992) *Nonlinear Systems Analysis*, 2nd Edn, Prentice Hall, Englewood Cliffs, NJ.

Walsh, J.P. and Ungson, G.R. Organizational memory. *Academy of Management Review*, **16**(1), 57–91.

Winograd, T.A. (1986) Language/action perspective on the design of cooperative work. In *Proceedings of CSCW '86*.

Zaulkernan, I.A., Das, A. and Johnson, P.E. (1992) Reuse of existing knowledge for decision support. *Proceedings of the 12th International Conference on Decision Support Systems*, 305–55.

PART TWO

CASE EXAMPLES

CHAPTER 12

Case Example 1: Motor incipient fault detection using artificial neural network and fuzzy logic technologies

Mo-yuen Chow, Yuan-Shin Lee and H. Joel Trussell

12.1 INTRODUCTION

Motor monitoring, incipient fault detection and diagnosis are very important and difficult topics in the engineering field. Applications range from a small DC motor used in an intensive care unit to the huge motors used in nuclear power plants. With proper machine monitoring and incipient fault detection schemes, early warning for preventive maintenance, improved safety and reliability can be achieved for different engineering system operations. The importance of incipient fault detection can be found in the cost savings which are realized by detecting potential machine failures before they occur. For example, in manufacturing areas, machine tools have been used extensively to shape objects to suit their needs. Recent studies indicate that as many as 90% of the failures of machine tools occur because of a malfunction of some elements in the machine tools (Lee 1992; Lee 1995). Such failures are found typically in machine components such as motors, bearings, ball screws, limit switches, pumps, solenoids, etc. There are literally a host of components on a machine tool which can cause, or threaten to cause, a shutdown. These machine malfunctions need to be detected and corrected before quality is degraded.

Many engineers and researchers have focused on incipient fault detection and preventive maintenance, which aim at preventing motor faults from happening. Different invasive and non-invasive approaches for motor incipient fault detection have been reported (Timperley 1983; Pengelly and Ast 1988; Reason 1988; Tavner and Penman 1989; Protopapas *et al.* 1990; Alguindigue and Uhrig 1991; Chow and Yee 1991;

Maier 1992; Trutt *et al.* 1993). With proper monitoring and fault detection schemes, the incipient faults can be detected in their early stages. Thus, maintenance and down-time expenses can be reduced while also improving safety. Non-invasive, inexpensive, and reliable fault detection techniques are preferred by engineers. Many of the inexpensive and non-invasive techniques available for fault detection and diagnosis in motors are based on mathematical models of the system of interest, such as parameter estimation (model-based techniques) (Isermann and Freyermuth 1991a; 1991b). However, since most machine dynamics are non-linear and stochastic, many assumptions must be made regarding the system in order to arrive at a simple and reasonable mathematical model of the machine. In other words, the fault detection or diagnosis system is not robust enough in the presence of noise and perturbations because the underlying mathematical model of the system is not well represented.

The emerging artificial neural network (ANN) technology has been applied successfully to perform fault detection and control of different engineering systems, such as motors (Miller *et al.* 1990; Narendra and Parthasarathy 1990; Chow and Nenozzi 1993) and power distribution systems (Chow *et al.* 1993; Chow and Taylor 1994). The demand for the use of artificial neural networks to solve engineering problems is expected to increase significantly in the next ten years, due to several breakthroughs in this field and also to the limitations of conventional engineering problem-solving techniques. Artificial neural network configurations, training data requirements, robustness issues, and design considerations of motor fault detection have been investigated (Chow, Sharpe *et al.* 1993). Results to date have demonstrated their significant performance advantages relative to the conventional methods. Although neural networks can provide the correct input–output fault detection relation, it is essentially a 'black box' device, that is, it does not provide heuristic reasoning about the fault detection process. Fuzzy logic could be a solution to this problem. Fuzzy logic can easily and systematically transfer heuristic, linguistic, and qualitative knowledge (preferred by humans), to numbers and quantitative knowledge (preferred by computers), and vice versa. This provides a simple method to heuristically implement fault detection principles and to heuristically interpret and analyze their results. Unfortunately, the major drawback of fuzzy logic is the difficulty in giving the exact solution to the problems, while an exact solution is essential for motor fault detection. Given the technologies of artificial neural networks and fuzzy logic, we can construct hybrid neural/fuzzy systems that can take advantage of the powerful features of both technologies while minimizing the effects of their drawbacks. The motor fault detection hybrid neural/fuzzy systems can be used to solve motor fault detection problems with great accuracy and robustness and also provide a heuristic explanation of the fault detection process.

This chapter focuses on motor incipient fault detection using neural network and fuzzy logic technologies and will give some overview of the advantages, challenges, and results obtained from the motor fault detection aspects. Parts of the research results obtained by the authors are highlighted. The detailed description and theory of most of the results presented in this chapter can be found in the references cited within the chapter. The first part of this chapter describes the use of artificial neural network technology to perform motor fault detection. The second part will describe a novel neural/fuzzy motor fault detection system that is a non-invasive, economical, accurate and reliable technique to perform motor fault detection. This system can produce heuristic explanations of faults that humans can easily understand. The third part of the chapter will describe the high-dimensionality challenges of motor fault detection and point out some approaches to solving the problem.

An induction motor is used in this chapter to illustrate the methodologies and issues in the use of neural network and fuzzy logic technologies for motor incipient fault detection and preventive maintenance. Usually, induction motors are protected by devices such as fuses, overload relays, and circuit breakers. Research has focused on different motor failure mechanisms, causes of stator and rotor failures, analyses of these failures, methodologies to determine whether a motor is suitable for extended service, test methods, the test equipment needed, application and limitations of these test procedures, data gathering, specific benefits, and costs. On-line monitoring of induction machines in critical applications has been increasingly necessary to improve their reliability and to minimize catastrophic failures. Microprocessor-based monitoring systems are of particular interest because they can be used for regular analysis of machine variables and to predict possible fault conditions, so that preventive maintenance can be organized in a cost-effective manner. It will be unrealistic at the present time to expect a technique that is able to detect all types of possible faults that may happen in a motor under any conditions. However, it is worthwhile to examine those common incipient faults encountered in daily motor operating conditions. With better understanding of motor incipient fault causes and consequences, their symptoms under different operating conditions, and with technology advancement, more accurate, reliable, and economical fault detection schemes can be developed for our motors.

Insulation failure of the motor windings is a common incipient fault in electric motors. The windings of a motor are insulated to prevent shorting of the winding turns. This insulation is categorized into several classes which are rated for different maximum operating temperatures. Often, however, these maximum temperatures are violated due to the operating environment, load requirements, etc. This violation leads to cracks, thinning, and eventual loss of insulation in certain locations of the stator windings. This effectively decreases the number of stator turns by causing inter-turn shorting of the windings. Different techniques have been developed for insulation fault detection: for example, measuring the resistance of the insulation is a common method used to determine the insulation condition, partial discharge testing, and a technique based on effective-negative-sequence impedance as a predictor of incipient failure.

Bearing wear is another common fault in electric motors. As the motor ages, the bearing will change shape due to an imbalance of the motor or just fragmentation of the bearing itself. These deformations cause vibrations in the motor as the rotor turns and eventually contribute to failures in other parts of the motor. Different bearing wear detection schemes have been developed by different researchers. An inexpensive means of determining bearing wear is by direct inspection. This method is much more accurate in determining the condition of the bearing; however, it requires down-time for the motor. A similar procedure determines the bearing condition based upon a particle analysis of the oil used to lubricate the motor. Frequency analysis of the vibration signals caused by the aging bearing, based upon the premise that different types of faults occur in different frequency spectra, can be done to monitor the bearing condition.

To demonstrate the concepts, this discussion will concentrate on two of the most common types of incipient faults in a single-phase squirrel-cage induction motor. Stator winding fault and bearing wear are used as illustrations of the application of the artificial neural network technology to perform incipient fault detection in motors. The motor is assumed to operate under a known load torque at steady-state condition. A turn-to-turn insulation incipient fault in the main winding of the motor will cause a change in the corresponding number of equivalent turns, K, of the main winding. Likewise, bearing wear of the motor is reflected in the damping coefficient of the motor, B. From analysis, the steady-state current, I, and the rotor speed, w, of the motor can be related to the main winding equivalent turns N and the damping coefficient B by a nonlinear equation:

$$g(l, w, K, B) = 0 \tag{12.1}$$

where $\mathbf{g} = [g_1, g_2]^T$. The stator current I and the rotor speed w are found to be good indicators to the changing conditions of the stator winding and hearings and can be used to indicate the stator winding and bearing conditions (Chow, Sharpe *et al.* 1993). Moreover, the current and speed are easily accessible and can be measured quite accurately. Therefore, for the detection of turn-to-turn insulation faults and bearing wear in single phase induction motors, the inputs to the neural network based fault detector are (I, w), and the desired network outputs are (K_c, B_c), the qualitative conditions of the winding insulation and bearing wear.

The conditions of the main winding and the bearing of the motor are reflected in the numerical values of the main winding equivalent turns K and the damping coefficient B. For the specific induction motor application under study, the conditions of K_c and B_c are then quantized into three condition levels (good, fair, bad) based on the values of K and B. Define M_1 as the mapping from (I, w) to (K, B):

$$M_1 : \{I,w\} \Delta \{K,B\} \tag{12.2}$$

and M_2 as the mapping of the quantization of (N, B) to (K_c, B_c):

$$M_2 : \{N,B\} \Delta \{K_c,B_c\} \tag{12.3}$$

As a result, the mapping M from (I, w) to (K_c, B_c) can be written as a composition of M_1 and M_2:

$$M = M_1 \_ M_2 : \{I,w\} \Delta \{K_c,B_c\} \tag{12.4}$$

The mapping M is very complex due to the high degree of nonlinearity of the induction motor dynamics and of the discretization involved; thus obtaining an accurate analytical expression for M for a given induction motor is rather difficult. The complexity of the mapping M nevertheless can be avoided by training a neural network to learn the desired mapping of M, based solely on input–output examples which can be obtained accurately.

An artificial neural network with input–output mapping N learns the mapping M by adjusting the network weights, w, in the weight space, W, through training algorithms such as backpropagation (Werbos 1974). The training process can be represented as:

$$\min_{w \in W} \| N(w,x) - M(x) \| \tag{12.5}$$

where input pattern $x \in X$, the input space, the corresponding output pattern $y = M(x) \in Y$, the output space, and $\| \cdot \|$ is some appropriate functional norm measure. Generally, m input–output measurements are taken from the mapping M and are used as the training examples for the artificial neural network to train the network. The training process is then approximated as:

$$\min_{w \in W} \| N(w,x) - M(x_i) \| , \mathrm{i} = 1,\cdots,\mathrm{m} \tag{12.6}$$

where $x_i \in T$ is the ith input pattern, $y_i = M(x_i) \in Y$ is the corresponding output, and $T \tilde{A} X$ is the training pattern space. Let $\hat{y}_i = N(w, x_i)$ be the corresponding network output given input x_i. Denote e_i, the error between the network output $\hat{y}_i$ and the actual output y_i, as:

$$e_i = \hat{y}_i - y_i = N(w,x_i) - M(x_i). \tag{12.7}$$

The network training is stopped, that is the network is sufficiently trained, when the training error E, defined as:

$$E = \sum_{i=1}^{m} \| e_i \| \tag{12.8}$$

is less than or equal to the preset tolerance value e_T. Again, $||\cdot||$ is some appropriate norm measure, two-norm being a popular choice.

12.2 DESIGN AND TRAINING CONSIDERATIONS OF THE MOTOR FAULT DETECTOR ARTIFICIAL NEURAL NETWORK

12.2.1 High-order net

In the posed problem, the values of K_c and B_c for each combination of (I, w) are obtained by experts' heuristic interpretation of the measured data. This set of input–output patterns can then be used to train the network to perform satisfactory fault detection and achieve 98% accuracy. Sometimes, if *a priori* information about the relationship of the inputs and outputs is available and used correctly, then the network structure and training time can be reduced and the network accuracy can be significantly improved. Based on the motor dynamics, as shown in Fig. 12.1, the bearing condition is basically a quadratic function of I and w. A high-order net, with network inputs $(I, w, I^2, w^2, I*w)$, is used to give faster training and better accuracy, as shown in Fig. 12.2. The performance of the high-order neural network and conventional neural network on bearing fault detection for different numbers of hidden nodes is shown in Fig. 12.3. In the figure, 2-*-B,trn stands for conventional two input neural network configuration with inputs (I, w) for bearing fault detection for training data set, while 5-*-B,tst stands for high-order five input neural network configuration with inputs $(i, w, I^2, w^2, I*w)$ for bearing fault detection for testing data set, etc. The increase in performance is attributed to the fact that the training error to be minimized is more convex in the high-order input space than in the

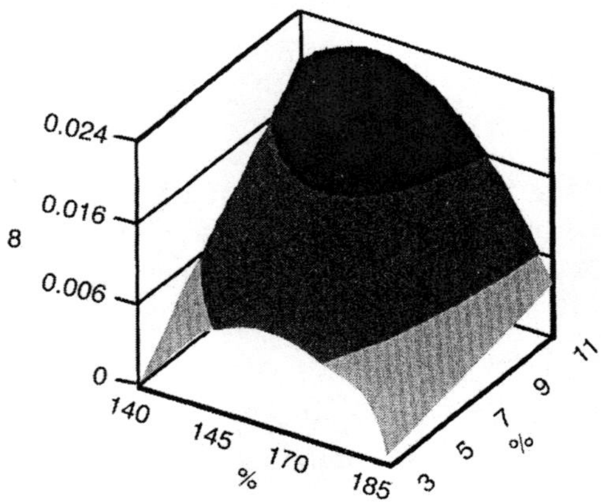

Fig. 12.1 Surface plot of bearing wear as a function of motor current, I, and rotor speed, w.

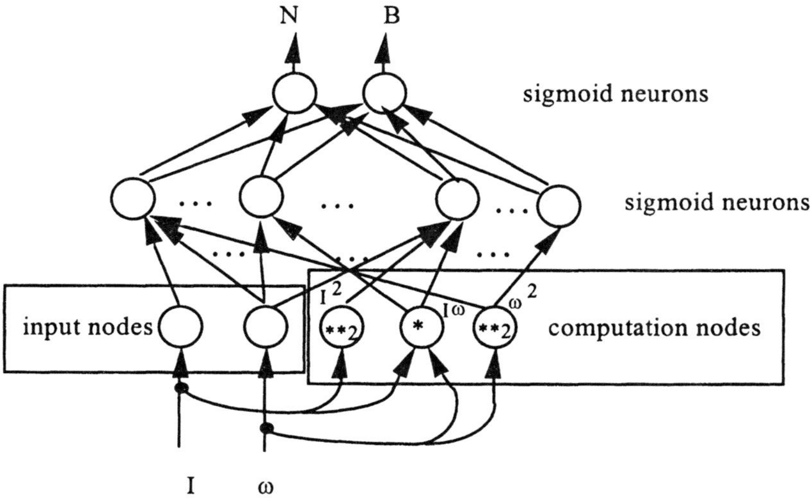

Fig. 12.2 Configuration of a high-order neural network for incipient fault detection.

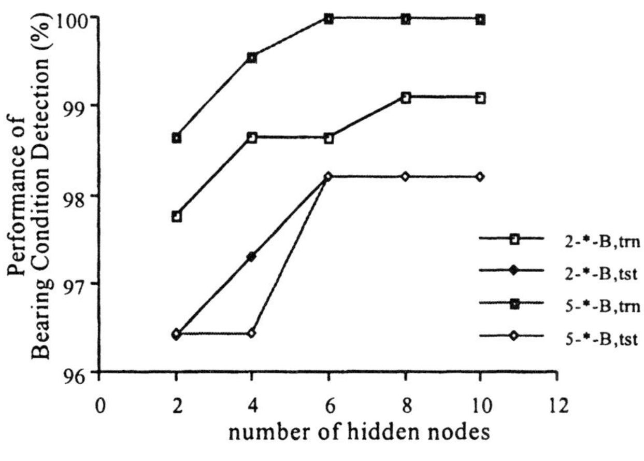

Fig. 12.3 Performance of the high-order neural network and conventional neural network on bearing fault detection for different numbers of hidden nodes.

normal input space. More detailed discussion on high-order net for fault detection can be found in Chow and Yee (1991).

12.2.2 Number of training data required – learning theory

The performance of a neural network, including its generalization ability, in a specific application depends on many factors including the structure of the neural network, the number of training patterns required, and the stopping criteria of the training process. Much of the concern in the design of neural networks to meet desired specifications and performance levels is centered around the training process of the networks. It is desired to be able to determine, to a certain degree of accuracy, (1) the appropriate size of the set of training examples as well as (2) the criterion to stop the training process. These factors are important for efficient applications and/or implementations of artificial neural networks.

In real world problems, actual training data is usually expensive to obtain. Therefore, it would be helpful if we could estimate, *a priori*, the number of training patterns required to train a network to meet some desired performance level, rather than collect the needed number of training examples by trial and error.

In order to achieve the goals of reasonably determining (1) the training set size and (2) the training stopping criteria, the use of learning theory is proposed. Learning theory attempts to understand the average behavior of a given network architecture at various stages of the training process, for various sized training sets (Chow, Bilbro *et al.* 1991). Although learning theory does not give the exact values of the network weights, it can nevertheless provide a prediction of the feasibility of successful training and an estimate of the performance of a specific network trained with different number of training examples.

In Chow, Bilbro *et al.* (1991) learning theory is specifically applied to the training process, Incipient Fault Detector for Induction Motors Artificial Neural Network (IFD-ANN). It was originally thought that the larger the number of training examples and training cycles used, the more accurate the fault detector neural network would be. Simulation results from learning theory soon contradicted this assumption. An increase in the number of training cycles might as well increase chances for overtraining the network, which results in a decrease of the network's overall accuracy. Moreover, the training process is long and inefficient with a large number of training examples and training cycles. With the aid of learning theory, the training of the IFD-ANN can be made faster as more information concerning the training process of the network in question is available before actually beginning the training procedure.

For the given three-layer IFD-ANN architecture with five inputs, ten hidden nodes, and two outputs, denoted as $N^3_{5,10,2}$ as shown in Fig. 12.2, we used learning theory and Monte Carlo estimation to predict the performance of the network as a function of the number of training examples and subject to different training stopping criteria. Fig. 12.4 shows the *learning theory's* estimate of the average correct prediction probability density of the IFD-ANN for the incipient fault condition, as functions of the number of examples used to train the network and of different error stopping criteria *e*. The average correct prediction probability density is directly proportional to its generalization ability. We can see that the relationship between the generalization ability of the network (y-axis) is *not* a linear function of the number of training examples used (x-axis). We can also see from Fig. 12.3 that beyond approximately 80 training examples the generalization ability of the neural network for predicting the bearing condition (i.e. B_c) saturates. So we can reduce the size of the training set to about 80 examples, which, compared to the 224 examples previously used, implies a great reduction in the number of training examples. This reduction, in turn, results in a decrease in the network's training time.

In order to confirm that the results provided by the curves generated by learning theory agree with the actual training results, extensive computer simulations were performed. The network IFD-ANN was trained using m = (5, 10, 15, 20, 30, 40, 60, 80, 100, 150, 224) training examples, successively. For each m, 10 different trials – each trial using m different examples from the training data set and the same initial network weights – were performed and their results were averaged. The computing time spent for these simulations was approximately 500 hours in a micro-VAX workstation. The performance of the IFD-ANN for

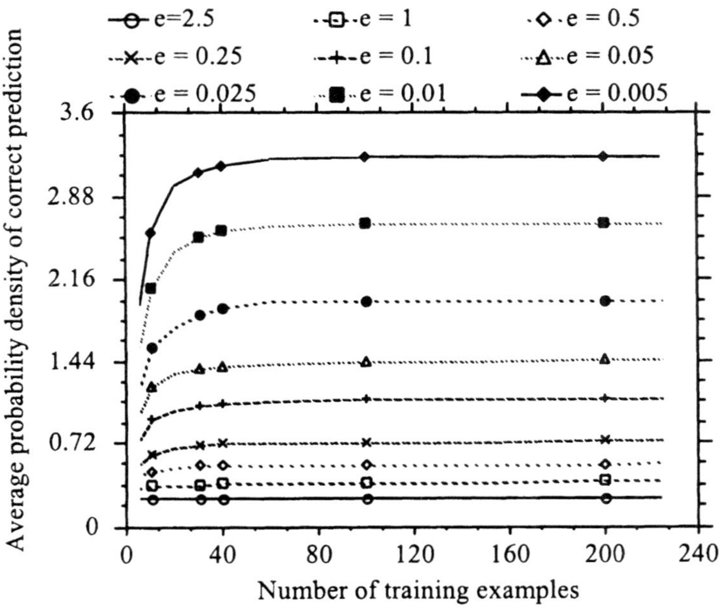

Fig. 12.4 The IFDIM-ANN's average probability density of correct prediction of B_c, as a function of the number of training patterns and of different stopping criteria.

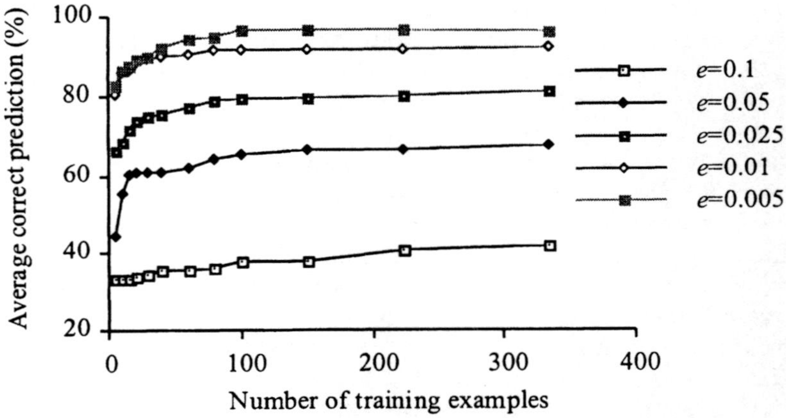

Fig. 12.5 Performance of the IFD-ANN with respect to the number of training patterns for different stopping criteria.

the correct prediction of B_c is plotted in Fig. 12.5, again, as a function of number of training patterns for different stopping criteria. The ordinate scale of Figs 12.4 and 12.5 are different due to the learning theory computing the proportionality measure of correct prediction instead of the actual prediction measure. However, the scaling factor does not affect the purpose in obtaining the sufficient data required to train the network.

12.2.3 Robustness

Many neural network configurations can solve the same motor fault detection problem satisfactorily. However, different configurations can have different performances in noisy conditions. In Chow and Yee (1992) a relative robustness measure, based on sensitivity analysis, is derived and proposed to be used as a measure of neural network robustness subject to small perturbations. The relative robustness of the network with respect to input x with trained weights w is expressed as:

$$R(x \mid w) = 1 - \frac{\| S(x \mid w) \|_2^2}{S_{max}} = 1 - \frac{\sum_{i=1}^{N_L} \sum_{j=1}^{N_I} s_{ij}^2 (x \mid w)}{S_{max}} \quad (12.9)$$

where

$$S(x \mid w) = \begin{bmatrix} \frac{\partial \hat{y}_1(x \mid w)}{\partial x_1} & \frac{\partial \hat{y}_2(x \mid w)}{\partial x_1} & \cdots & \frac{\partial \hat{y} N_L(x \mid w)}{\partial x_1} \\ \cdot & & & \cdot \\ \cdot & & & \cdot \\ \frac{\partial \hat{y}_1(x \mid w)}{\partial x_{N1}} & \frac{\partial \hat{y}_2(x \mid w)}{\partial x_{N1}} & \cdots & \frac{\partial \hat{y} N_L(x \mid w)}{\partial x_{N1}} \end{bmatrix} \quad (12.10)$$

is the network sensitivity matrix, and N_I and N_L are the number of inputs and outputs of the motor fault detector artificial neural network respectively, S_{max} is the maximum value of $\|S(x \mid w)\|_2^2$ for all x, is used as a measure of the overall network relative robustness subject to input noises with input x, and $\hat{y}$ is the output of the network as defined in section 12.3. S_{max} is a constant after the network has been trained. The choice of R in Eq. 12.10 is bounded between [0, 1]. The relative robustness measure has been used to demonstrate that a Multi-Sample Incipient Fault Detector Artificial Neural Network (MS-IFDANN) uses tapped-delay lines to increase the number of inputs to the neural network (Chow and Yee 1992). This modification increases the motor fault detection artificial neural network performance in noisy conditions, which in turn improves its reliability.

The input layer of the MS-IFDANN model is basically an expansion from the input measurements at time t_k to n consecutive measurements at times t_k, t_{k-1}, ..., t_{k-n+1}, obtained with the aid of a tapped delay line. The sampled input signal is applied to a string of delay boxes, each delaying the signal by one sampling interval. At any given instance of time, n sets of input measurements, where n is the sampling window size, are sent to the MS-IFDANN. Fig. 12.6 shows an overview diagram of the

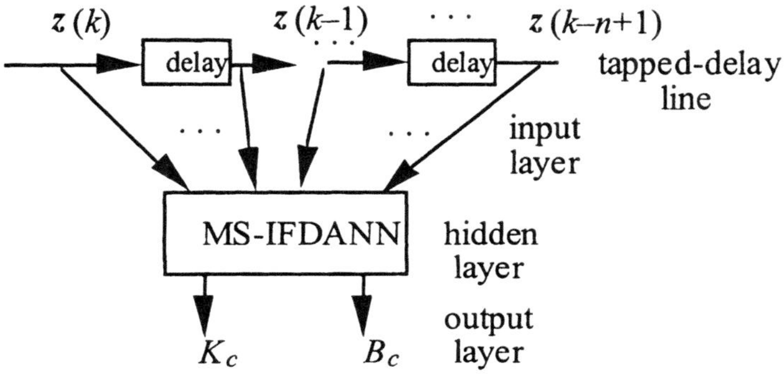

Fig. 12.6 Block diagram of MS-IFDANN.

MS-IFDANN model. The hidden layer and output layer structures remain the same as discussed before in the IFD-ANN model (eight hidden and two output units). The MS-IFDANN is denoted as $N^3_{n*5,8,2}$. The value $n = 1$ gives the original IFD-ANN of structure $N^3_{5,8,2}$.

This scheme increases the robustness of the fault detector neural network subject to noisy inputs, because the condition of the motor is now determined based not only on the most recent measurements $z(k)$ but also on the past $(n-1)$ measurements $(z(k-1), \ldots, z(k-n+1))$. The noise effect of each measurement sent to the detector is suppressed by the multiple inputs to the neural network.

With extensive simulations and using a nominal network for different MS-IFDANNs of structure $N^3_{n*5,8,2}$, the corresponding relative robustness measures were calculated and are plotted with respect to n for six different input patterns (p1–p6) in Fig. 12.7. The results clearly indicate that the relative robustness of the neural network increases as n, or correspondingly the number of input units, increases.

The accuracy of MS-IFDANNs with respect to different noise levels and sampling window size is shown in Fig. 12.8. By comparing Figs 12.7 and 12.8, we can see the relationship between the accuracy of the networks with noisy inputs is highly correlated with the relative robustness measure of the same networks. Note that as the sampling window size n increases, the relative robustness and accuracy of the networks

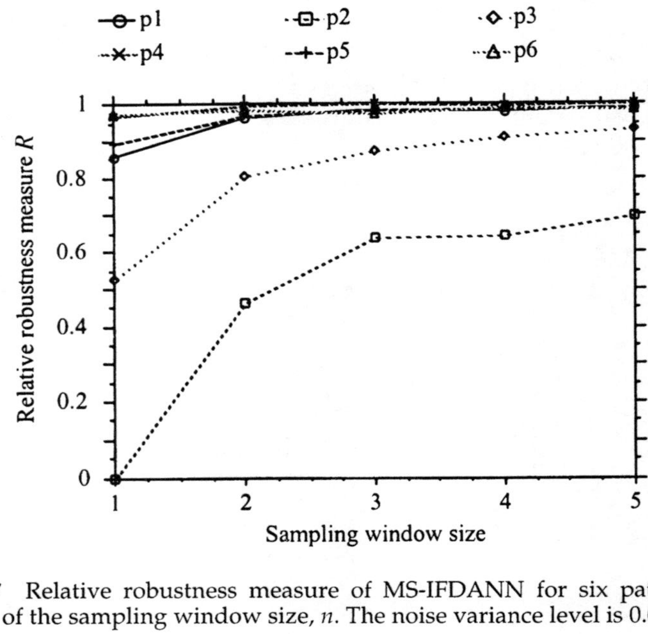

Fig. 12.7 Relative robustness measure of MS-IFDANN for six patterns, as a function of the sampling window size, n. The noise variance level is 0.01.

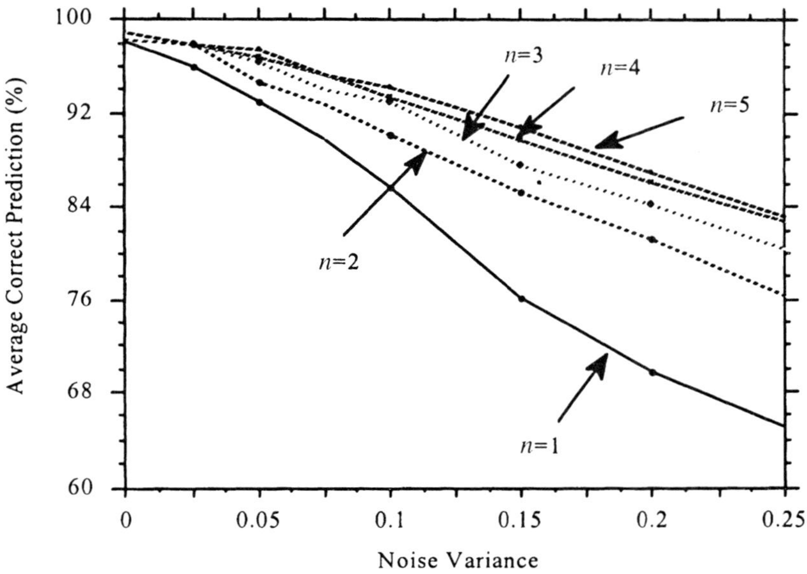

Fig. 12.8 Accuracy of MS-IFDANN $N^3_{n*5,8,2}$, where $n = 1, \ldots, 5$, as a function of noise variance.

increase, even in the presence of large noise variance. Based on this relationship, the relative robustness measure can be used as an important indicator of the accuracy of different neural network configurations in noisy environments.

12.3 MOTOR FAULT DETECTION NEURAL/FUZZY SYSTEM

The usual motor incipient fault detection procedure requires engineers/researchers to devote a significant amount of time and energy to investigate the motor system they are working on. They use their knowledge and intelligence to identify appropriate variables to be monitored, and design appropriate fault detection schemes and parameter settings for their tasks. They may rely upon machine theory or system theory to design their fault detection schemes. The end result is a highly complex motor incipient fault protection scheme that is not transparent to other people. Furthermore, the detection schemes and parameter settings are very specific to the problem under investigation.

As mentioned in section 12.1, the major drawback of using an artificial neural network alone to perform motor fault detection is its black box approach. In this section, we will present some preliminary results of the combination of the artificial neural network and fuzzy logic technologies to perform motor fault detection as well as give heuristic reasoning of the fault detection process. The artificial neural network motor fault detector is configured in a module form based on fuzzy rules and membership functions as shown in Fig. 12.9. Therefore, after training, the motor fault detection NN/FZ system will not only give satisfactory fault detection accuracy, but also provide heuristic reasoning behind the fault detection process (Goode and Chow 1995a; 1995b).

The fuzzy membership function module performs the mapping F_μ, which maps the input space, (I, w), to the fuzzy membership function space, M. Each output of the neural network in this module represents a fuzzy membership function set $\mu_i \in M$. Through back-propagation training, the mapping F_μ is manipulated to provide the correct membership function representation of the input space. The fuzzy membership space is linearly mapped to the fuzzy rule module in the fuzzy rule space, R. The mapping F_r is also performed by a fully connected feedforward structure. Through back-propagation training, the weights adjust the mapping F_r which is manipulated to provide the correct fuzzy rules necessary for proper fault classification. The rules are determined by a principle of excitatory and inhibitory weights which perform the mapping F_r.

Bearing wear fault detection is used once again as an illustration. As

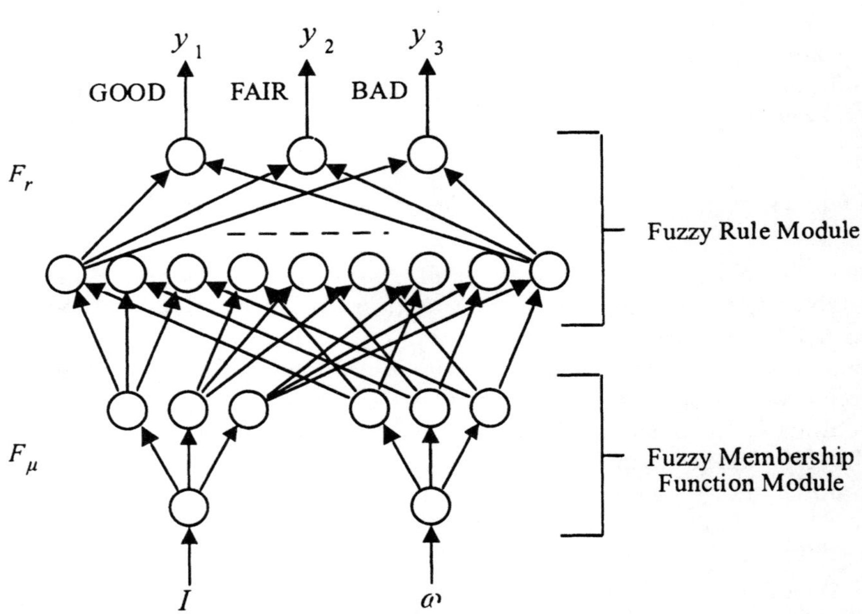

Fig. 12.9 Motor fault detection neural/fuzzy system structure.

bearing wear increases, the coefficient of friction and motor current also increase while the rotor speed decreases. Bearing wear generally contributes 5–10% of all rotational power losses experienced by a healthy condition motor. Considering this, a rule base for bearing wear can be easily constructed based upon the following heuristics:

$$\begin{aligned}
&P_{\text{loss}} = (P_{\text{stator loss}} + P_{\text{rotor loss}} + P_{\text{rotational loss}}) \\
&\text{If } P_{\text{rotational loss}} < 0.05\,(P_{\text{loss}})\ \textit{good} \\
&\text{If } 0.05\,(P_{\text{loss}}) < P_{\text{rotational loss}} < 0.10\,(P_{\text{loss}})\ \textit{Then fair} \\
&\text{If } 0.10\,(P_{\text{loss}}) < P_{\text{rotational loss}}\ \textit{then bad}
\end{aligned} \tag{12.11}$$

A preliminary rule base (for off-line training to initialize the fuzzy/neural system) of the neural/fuzzy fault detection system is constructed and shown in Table 12.1. The inputs, I and w, are evaluated as LOW, MEDIUM, or HIGH (L, M, H). The resulting rule decision is classified as either BAD, FAIR, or GOOD (B, F, G) or 'don't care' (d). The 'don't care' term means the corresponding rule has no effect on the fault detection scheme. One of the examples is that the motor will not operate in that input space due to physical constraints.

Through backpropagation training, the fuzzy rule base was modified by adaptation of the mapping F_r to provide the proper classification output. This adaptation modified the fuzzy rule base by eliminating three unnecessary rules. The modified fuzzy rule base is shown below in Table 12.2 which agrees with heuristic reasoning.

Given the initial fuzzy membership functions (three random fuzzy membership function sets per input), the neural/fuzzy system modified these fuzzy membership functions through backpropagation training by adapting the mapping F_μ. The change in the membership functions provide knowledge to determine the actual range of 'high' current, 'low' speed, etc. for safe operation of the motor in question. The initial membership functions for bearing wear detection are shown in Fig. 12.10. These modified membership functions (after appropriate training)

Table 12.1 Fuzzy rule base for bearing wear

	w			
		L	M	H
I	L	G	F	G
	M	F	F	G
	H	B	B	G

Table 12.2 Reduced Fuzzy rule base for bearing wear

	w			
		L	M	H
I	L	d	F	G
	M	F	F	G
	H	B	d	d

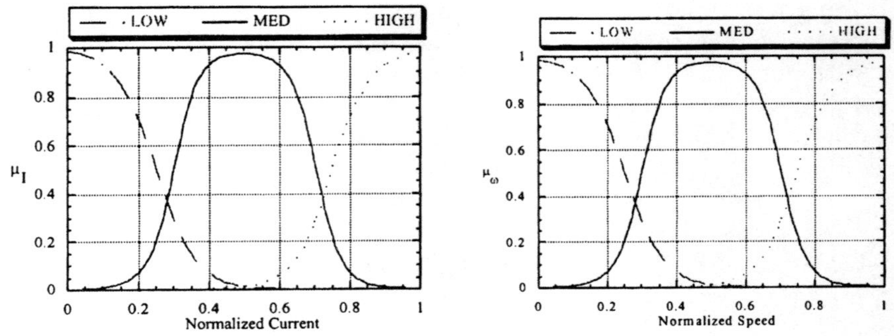

Fig. 12.10 Initial membership functions of the stator current and motor speed for motor fault detection (before training).

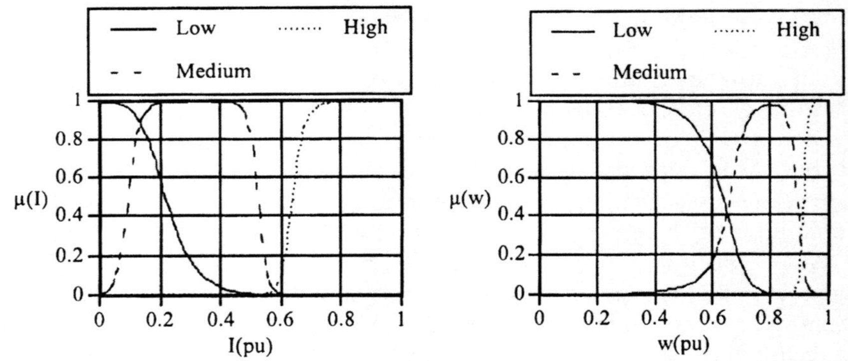

Fig. 12.11 Membership functions of the stator current and motor speed for motor fault detection (after training).

are shown below in Fig. 12.11, which gives the heuristic description of the membership functions of bearing wear detection.

12.4 CONCLUSION

This chapter has presented the advantages, challenges, and highlights of the research results of the application of the artificial neural network and fuzzy logic technologies to performing motor fault detection. This chapter has also presented some preliminary results on a motor fault detection neural/fuzzy system which not only gives highly satisfactory performance as a non-invasive, inexpensive, and robust motor fault detection technique but also provides valuable heuristic reasoning of the fault detection approach. The results obtained from the project so far indicate a promising future of using artificial neural network and fuzzy logic technologies to solve machine fault detection problems.

ACKNOWLEDGEMENTS

The authors of this paper would like to acknowledge the support of National Science Foundation, for Grant No. ECS-9521609.

REFERENCES

Alguindigue, I.E. and Uhrig, R.E. (1991) Vibration monitoring with artificial neural networks. Proceedings of *SMORN VI Symposium on Nuclear Reactor Surveillance and Diagnostics*, Gatlinburg, TN.

Chow, M.-y., Bilbro, G. and Yee, S.O. (1991) Application of learning theory to a single phase induction motor incipient fault detection artificial neural network. *International Journal of Neural Systems*, **2**(1&2), 91–100.

Chow, M.-y. and Menozzi, A. (1993) Design methodology of an intelligent controller using artificial neural networks. *Proceedings of IECON '93*, Maui, Hawaii.

Chow, M.-y., Sharpe, R.N. and Hurf, J.C. (1993) On the application and design consideration of artificial neural network fault detectors. *IEEE Transactions on Industrial Electronics*, **40**(2), 181–98.

Chow, M.-y. and Taylor, L.S. (1994) Application of information technology to distribution system operations. *NSF Electric Power Systems Infrastructure Symposium*, Pullman, WA.

Chow, M.-y. and Yee, S.-O. (1991) Methodology for on-line incipient fault detection in single-phase squirrel-cage induction motors using artificial neural networks. *IEEE Transactions on Energy Conversion*, **6**(3), 536–45.

Chow, M.-y. and Yee, S.-O. (1991) Using neural networks to detect incipient faults in induction motors. *Journal of Neural Network Computing*, **2**(3), 26–32.

Chow, M.-y. and Yee, S.O. (1992) A measure of relative robustness for feedforward neural networks subject to small input perturbations. *International Journal of Neural Systems*, **3**(3), 291–99.

Chow, M.-y., Yee, S.O. and Taylor, L.S. (1993) Recognizing animal-caused faults in power distribution systems using artificial neural networks. *IEEE Transactions on Power Delivery*, **8**(3), 1268–74.

Goode, P. and Chow, M.-y. (1995a) Using a neural/fuzzy system to extract knowledge of incipient fault in induction motors: Part I – Methodology. *IEEE Transactions on Industrial Electronics*, **42**(2), 131–38.

Goode, P. and Chow, M.-y. (1995b) Using a neural/fuzzy system to extract knowledge of incipient fault in induction motors: Part II – Application. *IEEE Transactions on Industrial Electronics*, **42**(2), 139–46.

Isermann, R. and Freyermuth, B. (1991a) Process fault diagnosis based on process model knowledge – Part I: Principles for fault diagnosis with parameter estimation. *Journal of Dynamic Systems, Measurement, and Control*, **113**, 620–6.

Isermann, R. and Freyermuth, B. (1991b) Process fault diagnosis based on process model knowledge – Part II: Case study experiments. *Journal of Dynamic Systems, Measurement, and Control*, **113**, 627–33.

Lee, J. (1992) *In Process Machine Degradation Monitoring and Fault Detection Using a Neural Networks Approach*, Department of Mechanical Engineering, George Washington University.

Lee, J. (1995) Modern computer-aided maintenance of manufacturing equipment and systems: Review and perspective. *Computers and Industrial Engineering*, **28**(4), 793–811.

Maier, R. (1992) Protection of squirrel-cage induction motor utilizing instantaneous power and phase information. *IEEE Transactions on Industry Applications*, **28**(2), 376–80.

Miller, W.T., Sutton, R.S. and Werbos, P.J. (1990) *Neural Networks for Control*, The MIT Press, Cambridge, MA.

Narendra, K.S. and Parthasarathy, K. (1990) Identification and control of dynamical systems using neural networks. *IEEE Transactions on Neural Networks*, **1**(1).

Pengelly, B.W. and Ast, G.E. (1988) A computer-based multipoint vibration system for process plan rotating equipment. *IEEE Transactions on Industry Applications*, **24**(6), 1062–73.

Protopapas, C.A., Kaminaris, S.D., Machias, A.V. and Papadias, B.C. (1990) An expert system for fault repairing and maintenance of electric machines. *IEEE Transactions on Energy Conversion*, **5**(1), 79–83.

Reason, J. (1988) Boost availability with up-to-date maintenance techniques. *Electrical World*.

Tavner, P.J. and Penman, J. (1989) *Condition Monitoring of Electrical Machines*, Research Studies Press Ltd./John Wiley & Sons Inc.

Timperley, J.E. (1983) Incipient fault identification through neutral RF monitoring of large rotating machines. *IEEE Transactions on Power Apparatus and Systems*, **102**(3).

Trutt, F.C., Cruz, C.S., Kohler, J.L. and Sittake, J. (1993) Prediction of electrical behavior in deteriorating induction motors. *IEEE Transactions on Industry Applications*, **29**(4), 1239–43.

Werbos, P.J. (1974) *Beyond Regression: New Tools for Prediction and Analysis in the Behavioral Science*, Harvard University, Cambridge, Massachusetts.

CHAPTER 13

Case Example 2: Data analysis for diagnostics and process monitoring of automotive engines

Bruce D. Bryant and Kenneth A. Marko

13.1 INTRODUCTION

This case example describes software for analysis of data from cold tests of fully and partially assembled engines. Cold testing of engines (testing without fueled running) at engine assembly plants has attracted considerable interest for its potential to reduce test costs and improve product quality. Cold testing of newly manufactured engines offers several advantages over traditional 'hot' tests (fueled, running tests):

- Speed: fewer hookups and detachments, shorter overall test time.
- Cost: less expensive test and facility hardware; cold testing also offers improved safety and environmental conditions.
- Quality: extremely sensitive to mechanical deviations which affect performance and therefore much lower probability of missing a failure.
- Coverage: wide range of potential defects can be detected and identified.

The primary motivation for embarking upon this work was to apply methods we developed to make improvements in the areas of engine testing and process control. Conventional procedures have traditionally relied upon ad hoc feature extraction and classification based upon visual inspection of signals by trained technicians, or limit-checking by automated test equipment. While experienced technicians can inspect and analyze the data traces and render rather accurate diagnoses of many problems, their ability to sustain high performance over long periods of time on the same task is questionable. Our goal was to develop data analysis tools which surpassed these approaches in required development time, sensitivity, accuracy, consistency and maintainability.

In the cold test, data is collected from a partially or fully assembled IC engine which is being cranked without fuel or ignition at low speeds by an external electric drive. Physical measurements such as torque, intake manifold pressure, exhaust manifold pressure, oil pressure, and vibrational activity are collected. Data is collected as a function of crankshaft rotation angle, typically taking one sample per degree of crankshaft rotation for one complete engine cycle (two rotations of four-stroke engine). The IC engine is connected to an electric motor/gearbox through a mechanical coupling which includes an in-line torque sensor. Other instrumentation includes pressure sensors on the intake manifold and exhaust manifold, and a high-bandwidth oil pressure sensor connected to the oil pressure sensor port. Additional instrumentation may include vibration sensors mounted at various places on the engine, in-cylinder pressure transducers, and microphones.

Cold testing can also be performed on partially assembled engines, and since these tests occur before the assembly is complete, they are referred to as in-line (IL) tests. In this case, the engine consists of the engine block plus all or some of the rotating components – crankshaft, piston assemblies, camshaft, etc. Its crankshaft is being turned by an electric motor which is connected to it through an in-line torque transducer. Accelerometers are attached directly to the engine block to monitor vibrations.

We will discuss the nature of the data collected and the processing schemes we developed to extract salient information from that data. A review of several of the methods we analyzed to make decisions related to the product quality will be presented, as will the results from pilot runs which have been completed. It should be noted that the methods we have developed are intended to be broadly applicable to many testing operations in the manufacturing plants, and therefore, the matters of interest to the reader should be the general approach and generic methods we have developed, rather than the specific details related to automotive engines.

Finally, we note that the methods developed here, applied to all production engines will generate large volumes of data, valuable for process control diagnostics and quality monitoring. Relentless competitive and regulatory pressures impose the need to improve quality and reduce unit-to-unit variability. It is evident that the information necessary to help meet these challenges lies in the data collected, but efficient methods to extract that information from the data must be available. We will demonstrate that the techniques we have developed for data compression and analysis are also applicable to the task of retrospective analysis or data-mining in the accumulated database and therefore add further value to the methods we advocate for process monitoring and quality improvement.

13.2 METHODOLOGY

A proper definition of diagnostics is the detection of a problem or failure and the simultaneous or consequent identification of its cause. The approach we have adopted for process monitoring and diagnostics separates these two features in the definition of diagnostics and establishes formal methods to help decide whether a product shall be classed as normal or faulty and then uses trainable systems to handle the somewhat more complex task of fault identification. The goal is to provide both (1) rigorous procedures for developing decision rules which can assess product quality and correctly identify the root cause of problems and (2) the means to develop these rules from parsimonious samples of real data. The procedures we have developed are outlined in step-wise fashion as follows.

13.2.1 Data collection

The process of data collection is frequently the most time-consuming and costly part of developing a diagnostic system. In order to minimize the efforts required to create effective tools for process monitoring and diagnostics, the data collection must be consistent (stable over long periods of time), comprehensive, and non-destructive. In addition, during the development of a diagnostic system, data which represents all foreseeable circumstances (i.e. faults) under all uncontrollable environmental conditions should be available to ensure that the decision analysis is robust with respect to all expected variations. Since one cannot be certain that diagnostic procedures can be developed for any given set of problems using a given set of sampled data, the task of data collection is often undertaken as a risky, speculative venture. It has been our good fortune to have the support of skilled and supportive engineers who helped collect the data that provided the foundations for the development of the methods we describe.

13.2.2 Processing

Raw data as it has been collected from a system under test is noisy, redundant, and usually not organized in a manner to expedite development of simple decision rules. Pre-processing tactics like averaging, feature detection, various forms of spectral content estimation, statistical condensation, residuals and other derived quantities, are used to extract information useful for classification and analysis of quality in order to develop decision rules from a properly frugal set of derived variables. This step is important because the decision analysis should be applied continuously to all manufactured products, even as changes in the process and product are taking place. If the development of decision rules always requires a very large sample set for training and turning, the system could not be successfully deployed and maintained in real-world manufacturing environments.

13.2.3 Decision analysis or transformation

Decision analysis is the step in which the diagnosis and the quality judgments are made. We may think of it as a 'transformation' to illustrate the decision process as a transformation, or mapping, from the derived variable 'information' space to the classification or fault space. The transformation may be in the form of a decision tree, a look-up table, or, in our case, an implicit transformation realized by an artificial neural system.

While developing the diagnostic system, these steps are often repeated in an iterative fashion at various levels, making necessary adjustments along the way. In this chapter, we focus mainly on the processing and transformation steps, not because the first two steps are less important, but because the technology for those steps is fairly mature and we have focused our work in the latter steps. Specifically, in the processing step, we concentrate on condensation of data without losing information, and in the transformation step, we are interested in developing a quantitative and rigorous method for making the best decisions with limited data available for training and tuning.

Processing methods

The general goals of pre-processing data are to remove noise and condense data to a manageable quantity while retaining as much relevant information as possible. Filtering of waveforms, often useful for the removal of electrical or mechanical noise, was introduced carefully to avoid removing useful information from the signals. We have found median filtering, a form of lowpass filtering in which each sample of a waveform is replaced with the median of a set including itself and neighboring samples, to be expedient, as it suppresses noise while using only genuine samples. Further reduction in the dimensionality of the problem is often accomplished by subsampling the median filtered waveform. Once again, however, the subsampling is not performed if evidence of deterioration in the classification task is found during development. This issue illustrates the iterative nature of the development process in that the results of decision analysis are compared for various parsings of the data, and only those data compression and reduction schemes which maintain or improve classification accuracy are pursued.

We have adopted a special form of principal component analysis (PCA) for data condensation. PCA is fully described in one of the original texts on the subject by I.T. Jolliffe [1]. Specific features of PCA which apply to our diagnostics problems are (1) the ordering of the axes of the principal component (PC) space in terms of decreasing variance represented, and (2) the orthogonality of the axes, allowing them to be treated (i.e. retained or discarded) independently of one another. The basic idea behind PCA is to reduce the dimensionality of a set of interrelated variables while retaining as much information as possible, in the form of variance. PCA accomplishes this by generating a new set of variables, called principal components (PCs), which represent the original data transformed to a new space whose axes are orthogonal and uncorrelated, and are ordered in terms of the variance they represent in the original data set. The quantitative assessment of variance represented provides the means to measure the amount of information retained by the PCA and allows the analyst to make a trade-off between the dimensionality of the representation vs the amount of information retained.

PCA may be performed on any set of vector data, not just analog measurement waveforms. We have used the results of PCA on several synthetic waveforms, such as portions of engine test waveforms representing contributions of specific cylinders, or 'single-cylinder' waveforms composed of the means of such partial waveforms taken over the entire set of cylinders. With respect to diagnostics problems, it should be noted that the principal component projections are simply variables which may or may not contribute to the overall solution of the problem. Their use does not preclude the use of other diagnostic variables, and in fact many successful diagnostics schemes use PCs along with other diagnostic variables. However, we have found PCA methods useful because they minimize the need for engineers to scan large amounts of data to extract useful features upon which to base diagnostic decisions. Evidence from other classification tasks suggest that PCA methods can be applied to achieve classification performance approaching the best human efforts based upon feature extraction in analysis of complex data [2].

There are occasions where straightforward PCA does not prove to be sensitive enough to distinguish certain classes from normal, or 'nearby', faults. These situations usually arise when the set of data from which the PCs are derived contains variations which are not very representative of specific, subtle faults. In such circumstances, we have found it useful to apply a 'sensitization' to PCA which improves the separability of classes and consequently enhances classification performance. In general, we use only a selection of normal or good engines to develop a PC representation, and the PCs that emerge are representative of the variance in the sample of normals. Such a procedure generally produces a set of principal axes within which faulty engines are readily distinguishable from normals. However, for some faults which generate only small differences in signal characteristics from normal, clear identification of the problem is not quite realized with this approach. If we bias the sample of engines we use to compute the principal component axes with a substantial fraction of engines with the given fault, the resulting PCs now better represent that variance in the data due to the presence of the fault. In trials with several subtle faults, which were impossible to identify by conventional methods and which did not appear to form a well-separated cluster apart from the normals without 'sensitization', this procedure increased the separation factor by about 200% for the target cases while maintaining excellent separation for the other classes.

Decision analysis methods

We approached the decision analysis portion of the task of engine diagnostics as a two-step classification problem. The first step, fault detection, may be seen as classifying a system into one of two classes – normal or faulty. The second step, fault identification, would then divide the faulty class into specific faults. We have investigated several different classification methods for both the fault detection and fault identification problems. We have constructed a custom cluster-based classifier which we call the *spherical classifier* since we observed that normal engines tended to cluster near each other in visualizable two- to four-dimensional plots of the PCs. We have successfully employed artificial neural systems, usually as feedforward networks trained to perform fault identification. Additionally, we have developed classifiers based on fuzzy set theory [4] which operate on features in the filtered waveforms, and on interim results of the spherical classifier. We have also employed genetic programming [5,6] to construct a classifier which would make better use of some other interim results of the spherical classifier. Finally, in an attempt to take advantage of the strengths of the different classifiers, we have also composed a hybrid classifier comprised of a voting scheme among the other classifiers.

13.2.4 Spherical classifier

Upon closely examining examples of pre-processed data, we have noted that the most important PCs (those that account for substantially more than 90% of the variance) of good engines consist of 2–4 PCs and appear to cluster together in data plots. Two-dimensional and three-dimensional plots are readily visualized, four dimensions required 3-D and color plotting. We constructed a custom cluster-based classifier called the spherical classifier (SC) in order to exploit this feature. The SC is a simple mechanism to assess an engine's distance from clusters of nominally good engines in several different spaces whose axes correspond to members of groups of related PCs. These groups of PCs correspond to a particular measurement, such as the torque waveform.

The SC is described simply as follows: each element of the PC vector from the engine under test (EUT) is first normalized by subtracting the mean of the corresponding elements of the PC vectors derived from a representative group of normal engines, and then dividing by the standard deviation of those same elements. The EUT's normalized PC vector is then separated into groups of related PCs. A group may consist of the PCs from a particular measurement, such as torque or exhaust pressure, or may consist of PCs related to a portion of the engine, such as the valvetrain or an individual cylinder. For example, the 'Cylinder 1' group may consist of PCs from the portion of the torque waveform influenced primarily by cylinder 1's compression and power strokes, plus the PCs from the portion of the exhaust pressure waveform primarily influenced by cylinder 1's exhaust stroke, plus the PCs from the portion of the intake pressure waveform primarily influenced by cylinder 1's intake stroke. Some PCs may appear in more than one group. The EUT's PCs within each group are plotted in an n-dimensional coordinate system, with n being the number of PCs in that group. The EUT's distance from the center of the cluster of normal engines in that space is then calculated; since the PCs are already normalized by the statistics of the normal engines, this distance is simply the EUT's Euclidean distance from the origin.

Engines are rated according to these distance parameters. Low-valued distance parameters are favorable as they reflect a high degree of similarity to the 'good' engines, while higher-valued distance parameters are unfavorable. An appropriate rating function would therefore be a monotonically decreasing function of the distance parameter. We use a rating function based on two concentric spheres surrounding the cluster of normal engines. If the engine's PCs plot inside the smaller inner sphere (i.e. the engine's distance is smaller than the 'inner radius,' or IR), the engine is considered to be similar to the good engines and receives a score of 1. If the engine's PCs plot outside the outer sphere (i.e. the engine's distance is greater than the 'outer radius,' or OR), the engine is considered to be dissimilar to good engines and receives a score of 0. If the engine's PCs plot between the two spheres, the engine receives a score between 0 and 1, corresponding to its proximity to the inner sphere. Mathematically, the rating function can be expressed as follows:

$$f(d) = \begin{cases} 1 : d < IR \\ \dfrac{(OR-d)}{(OR-IR)} : IR\ d < OR \\ 0 : d > OR \end{cases}$$

We call this rating function the spherical classifier, or SC, which is depicted graphically in Fig. 13.1. A rating function applied to a distance parameter comprises a test, so there is one test for each group of PCs. Each test produces a score between 0 and 1, and these scores are added, producing the total score.

There are many ways to classify the engine based on combined scores;

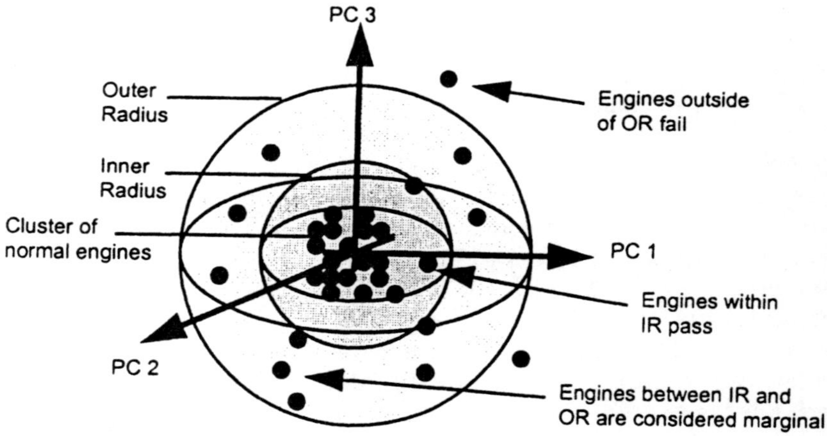

Fig. 13.1 Spherical classifier.

the method we use for fault detection considers both the minimum score and the total score. If the minimum score is greater than zero (i.e. the engine is at least marginal for all tests), and the total score exceeds an adjustable threshold, the engine passes, otherwise it fails.

13.2.5 Neural network

Neural networks (NN) have been used for a wide range of classification problems and offer a simple means to develop highly effective classification performance from small samples of data. We initially used neural nets trained through a modification to the standard backpropagation training algorithm, in which a population of NNs were trained simultaneously, and new populations of NNS were produced periodically by allowing the best network to reproduce itself a number of times with random alterations in weights. However, networks trained by these methods tend to be sensitive to the class population imbalances (normal engines outnumber abnormal engines by a factor of 100 or more) and therefore require that suitable care be exercised in selecting the training set so that proper generalization is realized in the trained systems. We have recently begun using a special form of probabilistic neural networks (PNN), whose training is based on the estimation of probability density functions [8]. These networks appear to have properties better suited to classification training on very asymmetric data sets (substantial population differences between classes) and exhibit more consistent performance as they are retrained and tuned with augmented data sets acquired as production runs yield more data. Developing trainable classifiers which can be trained from parsimonius samples of real data and which properly assimilate new information to generate improved and updated classification systems is still an active research area. Our results using variations of PNNs for these classification problems suggest that these may be the best choice for trainable classifiers on problems of this type, although for stable problems, without continuously augmented training sets, standard backpropagation networks yielded satisfactory results [3].

13.2.6 Fuzzy logic classifier

A fuzzy logic-based classifier (FLC), which is based on fuzzy set theory, may be employed for the transformation. In general, the strategy is to capture the unique behaviors of the engines in fuzzy sets, and compose rules to classify the engines based on this behavior. Rules may be applied to data derived from the PC vector, or to other features of the measured data. The FLC we developed was based upon an analysis of the PC information derived from cold tests.

13.2.7 Genetic program classifier

A method of producing functions suited for a particular purpose is genetic programming (GP), which iteratively optimizes a population of initially arbitrary functions via techniques derived from theories of biological evolution for the purpose of solving a particular task. The basic theory of GP is covered by two texts on the subject [5, 6] and the reader is referred to these for further information.

13.2.8 Hybrid classifier

Although each classification strategy has its strengths and weaknesses, it may be useful to maximize accuracy by constructing a hybrid classifier consisting of a voting scheme among the different classifiers. The basis for the voting scheme would naturally be the set of engines whose conditions are known, and the goal would be to minimize the number of errors made by various hybrid classifiers on this set.

13.3 OPERATIONAL RESULTS

We present two examples of classification performance in this section: (1) an end-of-line (EOL), fully assembled engine test where our goal was the detection of a wide range of faults, including unforeseen faults, and (2) an in-line (IL), partially assembled test, where our goal was the detection and identification of bearing missing and loose connecting rod nut faults in connecting rod assemblies.

We concentrated on the data processing and transformation portions of the diagnostic task in each of these applications, since the instrumentation and data collection were fixed by the systems available at the plants.

13.3.1 Operational result 1: end-of-line test

The goal of the EOL application initially was the detection, but not necessarily the identification, of a wide range of faults. Thus, the diagnostic task was to only provide a 'pass/fail' verdict. The data collected at the EOL test consisted of five waveforms per engine, each having 720 elements corresponding to one engine cycle of information sampled once per degree. Signals measured were torque, intake manifold pressure, exhaust pressure, and oil pressure. Additionally, the DC component was removed from the oil pressure waveform to produce a fifth waveform, dynamic oil pressure. For diagnostic development and testing we had data from 1098 SO (standard output) engines, and 100 HO (high output) engines. The engine was a four-cylinder four-stroke engine with firing order 1–3–4–2.

13.3.2 End-of-line test: data processing

For EOL testing, the pre-processing consisted of three steps:

1. median filtering to remove electrical noise and artifacts of mechanical vibration,
2. subsampling, which allows the data to be condensed while preserving its general shape, and
3. calculation and retention of a subset of the individual waveforms' principal component values.

We performed PCA on the entire dynamic oil pressure waveform and on various partitions and constructs of waveforms which we constructed from the torque, exhaust pressure, and intake pressure waveforms as follows.

The partitions were constructed by separating the torque, intake pressure, and exhaust pressure waveforms into the primary regions of influence of each cylinder. For the torque waveform, the engine cycle (720°) was divided into 180° segments approximately corresponding to each cylinder's most significant torque features, which correspond to the second half of a cylinder's compression stroke (negative torque) and the first half of a cylinder's power stroke (positive torque). For the intake pressure waveform and the exhaust pressure waveform, the engine cycle was divided into similar length segments approximately corresponding to each cylinder's primary 'respiration' features. These partitions permitted easy classification of a wide variety of faults as well as immediate identification of the cylinder associated with the problem.

It was also deemed expedient to construct measures of average cylinder performance and deviation from average performance. Accordingly, two data streams were constructed which captured these features. A synthetic 'mean cylinder event' was produced for the torque, intake pressure, and exhaust pressure waveforms by calculating the point-by-point means of the individual cylinder events. The resulting waveforms were called torque mean cylinder event (TMCE), intake pressure mean cylinder event (IMCE), and exhaust pressure mean cylinder event (EMCE). A synthetic 'standard deviation of cylinder events' was produced for the torque, intake pressure, and exhaust pressure waveforms by calculating the point-by-point standard deviations of the individual cylinder events. The resulting waveforms are called torque standard deviation cylinder event (TSDCE), intake pressure standard deviation cylinder event (ISDCE), and exhaust pressure standard deviation cylinder event (ESDCE).

The result of PCA is a transformation from the original space in which the raw data resides ('data space') to a new space called 'PC space,' whose properties are described earlier in this chapter. We then computed principal component projections (PCs) as follows. We applied the TMCE transformation to each of the individual cylinder events and to the mean cylinder event of the torque waveform to produce four sets of individual cylinder torque PCs plus a set of torque mean cylinder event PCs. We applied the TSDCE transformation to the standard deviation cylinder event of the torque waveform to produce torque standard deviation cylinder event PCs. We handled the exhaust and intake pressure waveforms similarly. We applied the dynamic oil pressure transformation to the dynamic oil pressure waveform to produce dynamic oil pressure PCs. We computed the average value of the static oil pressure waveform to produce a synthetic static oil pressure PC. As is usually the case with PCA, we were able to represent the vast majority of the original information with a relatively small number of PCs. From the sets mentioned above, we assembled a 53-element PC vector which was used for diagnostic analysis.

13.3.3 End-of-line test: decision analysis

Our first requirement for the EOL test was to classify engines into one of two categories: normal or faulty. We analyzed the performance of a variety of classification tools on this problem: neural networks, genetic programming, fuzzy set theory, cluster analysis, and a hybrid technique (voting scheme).

For the neural network, we used a simple three-layer backpropagation network, with 53 nodes in the input layer (one per PC vector element), 12 nodes in the hidden layer, and two nodes in the output layer (one per output class – pass or fail). Bias nodes were used in the input and hidden layers, and a sigmoidal transfer function was used in the hidden and output nodes.

The cluster analysis took the form of the spherical classifier described earlier in this chapter. We used a total of 13 SCs, one for each of the following groups of related PCs.

Cylinder 1 comprised PCs from the torque, exhaust pressure, and intake pressure waveforms' Cylinder 1 partition; Cylinder 2, Cylinder 3, and Cylinder 4 were compiled similarly; Torque comprised PCs from the torque waveform's mean cylinder event; Torque Variation comprised PCs from the torque waveform's standard deviation cylinder event; Exhaust Pressure, Exhaust Pressure Variation, Intake Pressure, and Intake Pressure Variation were compiled similarly; Imbalance comprised PCs from the torque, exhaust pressure, and intake pressure standard deviation cylinder events, Dynamic Oil Pressure comprised PCs from the dynamic oil pressure waveform, and Static Oil Pressure consisted of the average oil pressure pseudo-PC. We combined the individual test scores by adding them with the stipulation that if one of the scores is zero, the total score is zero.

We used the 'distance' results from the spherical classifier as inputs to a fuzzy logic classifier. Recall that an engine's distance parameter is defined as the engine's distance from the center of a cluster of nominally good engines. There are a number of distance parameters, each corresponding to a group of PCs related to a measured parameter. If an engine's distance parameters are large, that engine is quite dissimilar to the good engines and probably has some kind of defect. If an engine's distance parameters are small, that engine has a high degree of similarity to the good engines and probably has no faults present. These ideas are expressed in the following fuzzy rules:

1. If the distance is *small*, the engine is *normal*.
2. If the distance is *medium*, the engine is marginally *normal*.
3. If the distance is *large*, the engine is not *normal*.

The terms in italic above are the names of fuzzy sets. The distance parameters belong, in varying degrees, to the sets *small*, *medium*, and *large*, and upon application of the fuzzy rules, the engine belongs in some degree to the set *normal*. The classification mechanism is to compare the engine's degree of membership in the set *normal* to an adjustable threshold to determine whether the engine passes or fails.

The fuzzy rules are implemented by conducting the fuzzy AND operation (multiplication). That is, for a given distance parameter, rule 1 is scaled by the degree of membership in the *small* set, rule 2 is scaled by the degree of membership in the *medium* set, and rule 3 is scaled by the degree of membership in the *large* set. These operations are conducted for each of the distance parameters. The resulting scaled rules are added vertically to form an 'aggregate' rule.

The FLC assesses the engine's degree of membership in the *normal* set by 'defuzzifying' the aggregate rule. To do this, it first calculates the 'center of mass' of the aggregate rule, which is the horizontal point at which half the area under the graph lies to the left and half lies to the right. It then passes this center of mass through a scaling function which results in a number between 0 and 1 indicating the engine's membership in the set *normal*. A pass/fail verdict is made by comparing this degree of membership to an adjustable threshold.

There are many different scaling functions which could be used. We chose a sigmoidal scaling function, which is a function of the form

$$f(x) = \frac{1}{1+exp(-ax)}$$

in order to counteract the tendency of the center of mass to gravitate toward the middle of the horizontal scale.

Genetic programming was applied to the classification problem, using a pool of operators consisting of the usual mathematical functions (mainly those that appear on the keyboard of a scientific calculator), and a pool of operands comprised of the distance values and test scores from the spherical classifier, plus a number of combinations of these values, that is their sums, averages, products, etc. An objective function was chosen which produced an output that exhibited the greatest 'separation' between the distribution of passing engines and the distribution of failing engines (for engines whose status was known in advance). Separation is defined as the magnitude of the difference between the means of the distributions divided by the sum of their standard deviations. GP produced the following algorithm:

$$\text{Genetic score} = S_2 * Prod * (0.012739 - Ave) * 1.0E + 8$$

where S_2 is the score for spherical test #2, the Cylinder #2 test; *Prod* is the product of the spherical test scores; and *Ave* is the average of the spherical test scores.

We compared the genetic score to a threshold to produce a classification. A threshold of 1.44 was found to give the best results with our data, where engines with a genetic score above this threshold were called 'normal' and engines below this threshold were called 'faulty.' The GP ran for 50 generations with an initial population of 500 random programs.

While each of the classification techniques provided a certain level of accuracy in fault detection, we found that we could improve overall fault detection accuracy on our data set by constructing a hybrid classifier consisting of a voting scheme. We found that the greatest accuracy could be obtained by the following voting scheme:

If the Neural Network and the Spatial Classifier indicate Normal

Then

The engine is called Normal

Else

The engine is called Faulty

13.3.4 End-of-line test: results

With a database consisting of waveforms from over 1000 fully assembled engines, we were able to detect faults with an accuracy of >99% based on evaluations of tagged faulty engines. Fault detection accuracies of the various classifiers used in the EOL test are given in Tables 13.1 and 13.2. Table 13.1 shows results from 1098 pre-production standard output (SO) Zeta engines. Table 13.2 shows results from 100 pre-production high output (HO) Zeta engines.

In the tables, a indicates an alpha error, otherwise known as a type 1 error or false alarm, where a known normal engine is called faulty; b indicates a beta or type 2 error, where a known faulty engine is called normal. A fuzzy logic classifier was not constructed for the HO engines.

The fuzzy logic classifier in the EOL test did not perform very well on the SO engines, and therefore we did invest the time to construct one for the HO engines. The poor results of the genetic program classifier on the HO engines are probably due to the fact that we were testing the generality of the GP solution by applying the SO engines' GPC to the HO

Table 13.1 SO engine classification results

Classifier	*Neural net*	*Fuzzy logic*	*Genetic*	*Spherical*	*Hybrid*
# errors	16	108	55	47	6
a	0	35	8	0	0
b	16	73	47	47	6
Accuracy	98.5%	90.2%	95.0%	95.7	99.5%

Note: The engines used in this study do not represent actual production

Table 13.2 HO engine classification results

Classifier	*Neural net*	*Fuzzy logic*	*Genetic*	*Spherical*	*Hybrid*
# errors	1	N/A	17	8	1
a	0	N/A	10	0	0
b	1	N/A	7	8	1
Accuracy	99.0%	N/A	83.0%	92.0	99.0%

engines. These results indicate that an engine-specific GP classifier should be derived.

The lower accuracy of the HO engines' spherical classifier suggests that 100 engines is not a large enough basis upon which to construct such a classifier. In other evaluations of these methods, we have found that 500–1000 data points provided a reasonable foundation for analysis. Data sets of this size are readily attainable with automated test rigs (the original data was acquired with a manually operated rig with a 15-minute cycle time and only sparse sampling of production) which can test all production engines in plants that produce 1000 to 2000 engines per day. For similar reasons, the HO engines' neural network results may be misleading, as the network was trained with a small database due to our dearth of data for that engine.

13.3.5 End-of-line test: extension to fault identification

Since the data we had available did not represent a full range of faults in the engines, we relied on a series of tests we ran with a single engine in a replica of the production test rig. The engine was dissembled and reassembled with over 20 common faults. In order to mimic the variability we would see in production testing, the individual cylinder data was permuted to vary the signal in a systematic way and random noise was introduced to simulate additional process noise [3]. The data that was used for training consisted of the median filtered, subsampled data reduced by PCA in the manner described above. In these cases, standard backpropagation networks were used, since we could balance the population of samples of normals and each specific fault in the engines to be about equal. The neural network classification performance was virtually error free, in the cases for which the permutations and the noise was introduced. The only situations the network failed to classify properly resulted from introduction of improper oil fill levels. It is believed that marginal levels of lubrication in the engine should not affect engine operation in a detectable manner, and our experience with neural networks confirmed this hypothesis. While these results are extremely encouraging, we must note that we had the luxury of preparing balanced samples of data from different classes for training and therefore the backpropagation methods worked quite well. We would expect similar performance from PNNs, and feel that they will have to be employed when the training samples for the production system will be drawn from the actual production statistical distributions.

13.3.6 Operational result 2: in-line test

The diagnostic task for the in-line (IL) test was to detect and identify specific faults in a 'short block' engine. The short block consists of the engine block, crankshaft, connecting rod and piston assemblies, and camshaft, but no heads or valvetrain components. It may or may not contain the balance shaft. The faults we were to detect and identify were missing connecting rod bearings and loose connecting rod nuts.

The data sets used in developing IL test diagnostics consisted of up to 4096 samples of 'break-away' torque, collected every millisecond until the crankshaft began rotating, 1024 samples per revolution of 'running' torque, collected for four revolutions at 250 rpm, and 1024 samples per revolution of vibrational activity from each accelerometer, collected for eight revolutions at 250 rpm. We had data from 105 engines, including several examples of each of the various faults we were to detect. Figure 13.2 shows the connecting rod assembly.

13.3.7 In-line test: data processing

For IL testing, the pre-processing employed computation of principal components and projection of the data to PC space for the detection and identification of missing bearing faults, and various filtering mechanisms to magnify the effects of loose rod nut faults on the waveforms.

We computed PCs for the accelerometer and running torque waveforms and determined that most of the information present in the waveforms could be represented in the first several PCs from each. We did not use the breakaway torque signal for these tests; this signal is important

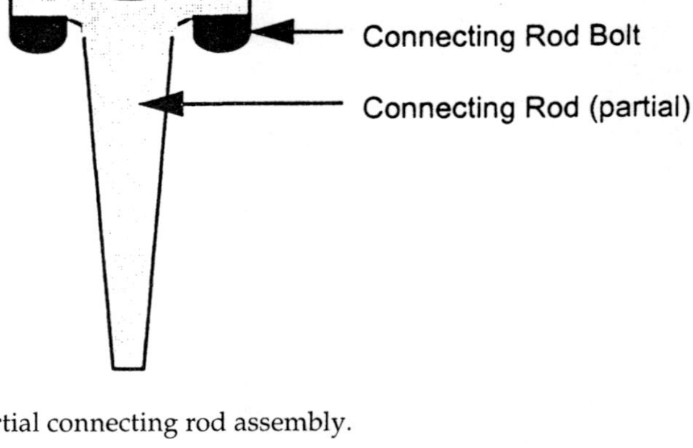

Fig. 13.2 Partial connecting rod assembly.

for detecting other, more severe defects, but was not for the defects we were addressing.

The PCs provided enough information to accurately detect and identify the missing bearing insert faults, but were not sufficiently sensitive to detect most of the loose rod nut faults. We then focused our efforts on pre-processing the data in such a way as to magnify the effects of loose rod nuts, or rather, exaggerating the difference between the loose rod nuts and normal. Specifically, our method of pre-processing consisted of (1) processing the accelerometer data into a magnitude vs time 'vibrational activity' envelope, then (2) filtering random noise and (3) systematic activity from the envelope. We ignored the torque signals as they did not contribute to the diagnosability of the loose rod nut defects.

Vibrational activity envelope

We produced a waveform representing accelerometer activity vs crankshaft angle. This envelope simulates a moving window standard deviation envelope which is calculated by the following equations:

$$e_i = \sqrt{\frac{1}{7-1} \sum_{j=i-3}^{i+3} (a_i - \mu_i)^2}$$

where e_i is point i in envelope waveform, a_i is point i in accelerometer waveform, and μ_i is the estimated mean of seven points in accelerometer waveform, centered about point i. μ_i is calculated by:

$$\mu_i = \frac{1}{7} \sum_{j=i-3}^{i+3} a_i$$

Removal of random noise

Random noise was mitigated by averaging the envelopes of the eight revolutions of data from each accelerometer, to produce one average envelope for each accelerometer. That is

$$ae_i = \frac{1}{8} \sum_{j=0}^{7} e_i + 1024{*}j$$

where ae_i is point i in an average envelope.

Isolation of non-systematic activity

Systematic accelerometer activity is defined here as the activity which is always present in defect-free engines. This can be thought of as the small amount of accelerometer activity that is normally present. Systematic activity is effectively removed by subtracting the average of a set of average envelopes obtained from a group of certified defect-free engines from the average envelopes of the engine under test.

That is, a group of defect-free engines is selected in advance, and steps 1 and 2 are applied to their accelerometer waveforms. The resulting average envelopes are then averaged together to remove variations among engines, and the resulting overall average envelopes (one for each accelerometer) are retained. In step 3, these overall average envelopes are subtracted from the average envelopes from the engine under test.

In cases where the result of subtraction is below zero, the value is clipped at zero, reasoning that there cannot be 'negative' vibrational activity present. The results of step 3 are two envelopes (one per accelerometer) approximately representing the difference over one crankshaft revolution between the engine under test and a defect-free engine. An example set of average envelopes is shown in Fig. 13.3.

13.3.8 In-line test: decision analysis

As with the EOL test, we approached the IL diagnostic as a classification problem; in this case, the engines were to be classified into one of four categories: (1) missing cap bearing insert, (2) missing rod bearing insert, (3) loose connecting rod nuts, and (4) no fault detected. A portion of a connecting rod assembly is shown in Fig. 13.3 to illustrate these parts. We used neural networks to classify the engines.

While neural networks can be trained to perform multiple diagnostics tasks simultaneously [3], we found it expedient to break the problem up

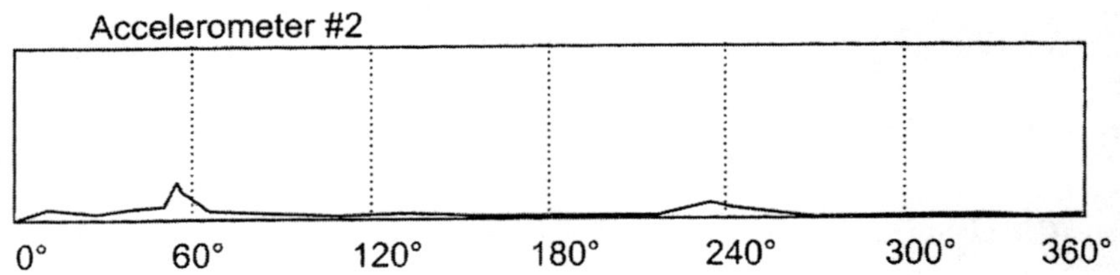

Fig. 13.3 Example of average envelopes.

into smaller discrete diagnostics tasks. Specifically, we trained one network to detect and identify the missing bearing insert faults, one to identify the offending cylinder if the fault was a missing cap insert, one to identify the cylinder if the fault was a missing rod insert, and one to detect loose connecting rod nuts if a missing bearing insert fault was not detected. The proposed diagnostic scheme implementing this strategy is given in Fig. 13.4. The steps employing neural networks are shown in bold.

It may be noted that many other faults could be present at this point in the engine build, however, most of these faults are less subtle in nature and existing diagnostics already in place, which consisted mainly of limit-checking, detected these faults adequately. It is assumed for the purpose of this exercise that the engine under test passed the initial tests. Likewise, as the developer sees fit, further diagnostic tests may be undertaken after the loose connecting rod nuts test.

13.3.9 In-line test: results

With a database of waveforms from 105 partially assembled engines, we were able to detect and identify missing bearing insert faults with 100% accuracy, and loose connecting rod nut faults with 96.5% accuracy. The loose connecting rod nut results are particularly encouraging, as this is a fault that was undetectable by the previous methods. We used three-layer backpropagation networks for each diagnostic. Table 13.3 shows classification results for the 105 partially assembled engines.

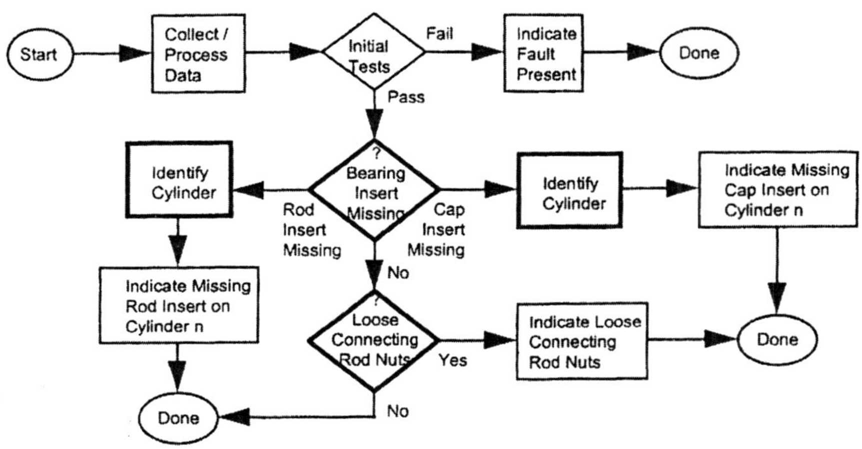

Fig. 13.4 Multiple neural network diagnostic strategy.

Table 13.3 In-line test results

Fault classifier	*Detection Accuracy*
Rod insert missing	100%
Cylinder ID for rod insert missing	100%
Cap insert missing	100%
Cylinder ID for cap insert missing	100%
Loose connecting rod nuts	74.5%

13.4 INFORMATION STORAGE AND RETRIEVAL

In typical production runs, nearly 500 000 engines are built per year at a single plant. In order to maintain quality and recognize deviations from historical data (such as comparison with assuredly high-quality engines which have survived months of actual running), it is important to be able to compare current production data to past data. Furthermore, in the event that some problem arises in a few products, we would like to examine the 'birth records' of those products to determine if any previously unnoticed anomalies can be detected, any similarities exist, or if any other products have similar appearances in the tests. Not only does the PCA representation reduce the data to an easily visualizable set of patterns which can be examined by the best pattern recognition system we have – our own human image processing system – it also compresses the data into an information database for which simple and effective queries about similarities and nearest neighbors can be constructed. Even for on-line computers in the test stands (typically PCs with >16 Mbytes memory, hard disk storage, 32-bit processors), it is possible to have instant access to at least a year's worth of quality records, and display substantial portions of the data for inspection within the single engine test cycle time. Although we expect that commonly available storage devices will be used to store the raw data (a one-Gbyte disk can hold one month's worth of data), access to the raw data and information processing on such large databases may not be practical for real-time, on-line analysis. The data compression methods we applied to the engine test data have already proven an effective tool for recognizing 'novelties' in incoming data for engineers working at the test stands.

13.5 CONCLUSION

We have shown that the problem of evaluating product quality may be approached as a combined process monitoring and diagnostic classification task. By properly processing the raw input data, we can simultaneously reduce the complexity of the classification task and improve the performance of classifiers on the data. The classification task is broken down into two major segments – fault detection and fault identification. We have developed a simple means of data compression which permits us to develop simple rules to perform the first task of fault detection. The same data representation is then used with trainable classifiers to produce a system which learns from experience to perform correctly the fault identification task. Finally, the same representation offers a compact form of information storage so that accurate analysis of product quality trends and effective information retrieval from the engine quality database can be obtained.

REFERENCES

[1] Jolliffe, I.T. (1986) Principal Component Analysis, Springer-Verlag, New York.

[2] Marko, K.A., James, J., Dosdall, J. and Murphy, J. (1989) Automotive control system diagnostics using neural nets for rapid pattern classification of large data sets. *Proceedings of IJCNN*, 11–13.

[3] Bryant, B., (1991) *Neural Network-Based Cold Test Engine Diagnostics*, Master's Thesis, University of Michigan Electrical Engineering and Computer Science Department, Ann Arbor, MI.

[4] Zadeh, L.A. (1965) Fuzzy sets. *Information & Control*, **8**, 338–53.

[5] Koza, John R. (1992) *Genetic Programming: On the Programming of Computers by Means of Natural Selection*, The MIT Press, Massachusetts.

[6] Koza, John R. (1994) *Genetic Programming II: Automatic Discovery of Reusable Programs*, The MIT Press, Massachusetts.

[7] McClelland, James L. and Rumelhart, David E. (1988) *Explorations in Parallel Distributed Processing: A Handbook of Models, Programs, and Exercises*, The MIT Press, Massachusetts.

[8] Chen, C.H. (1996) *Fuzzy Logic and Neural Network Handbook*, McGraw-Hill Inc., New York.

CHAPTER 14

Case Example 3: Measurement of machine performance degradation using a neural network model

Jay Lee

14.1 INTRODUCTION

Machines degrade as a result of aging and wear, which decreases performance reliability and increases the potential for faults and failures. Equipment reliability and maintenance drastically affect the three key elements of competitiveness – quality, cost, and product lead time. Well-maintained machines hold tolerances better, help reduce scrap and rework, and raise consistency and quality of the part. They increase up-time and yields of good parts, thereby cutting total production costs, and also can shorten lead times by reducing down-time and the need for retooling [1]. The recent rush to embrace computer-integrated manufacturing (CIM) has further increased the use of relatively unknown and untested technology. Today, many factories are still performing maintenance on equipment in a reactive, or breakdown, mode. This is due to traditional process monitoring systems that can only detect machine or process faults when they occur. Reactive maintenance is expensive because of extensive unplanned down-time and damage to machinery [2–7]. In high-performance systems one often cannot tolerate significant degradation in performance during normal system operation. This has led many US manufacturers to look to suppliers for smarter equipment that will ease the need for strong technical support for equipment. To avoid machine down-time, some companies have used preventive and predictive maintenance approaches either by using historical maintenance data or by sensing machine condition. However, there are no maintenance data available for many newly developed machines or advanced manufacturing systems. Adding sensors to monitor machine condition will also increase machine complexity and will require more highly trained personnel. Therefore, some manufacturers design a fault detection system that takes system redundancy into account, for example, a system containing several back-up subsystems [8–15].

Currently, no proven technique has been employed to monitor the degradation of a machine and its process in order to adaptively prevent faults or failure [1–15]. To analyze machine behavior quantitatively, a pattern discrimination model (PDM) based on a cerebellar model articulation controller (CMAC) neural network has developed. A stepping motor and a PUMA 560 robot were used to study the feasibility of the developed technique. Experimental results have shown that the developed technique can analyze machine degradation quantitatively. This methodology can help operators set up machines for a given criterion, determine whether the machine is running correctly, and predict problems before they occur. As a result, maintenance hours can be used more effectively and productively. By doing this, maintenance personnel can perform early diagnostics and part replacement during the regular daily maintenance hours. Therefore, the mean-time-between-failure can be extended to an unlimited length. Figure 14.1 shows the concept of the proposed proactive maintenance implementation technique. Figure 14.2 shows the system concept of the proposed methodology, an integrated learning, monitoring, and recognition technique, for a proactive mainte-

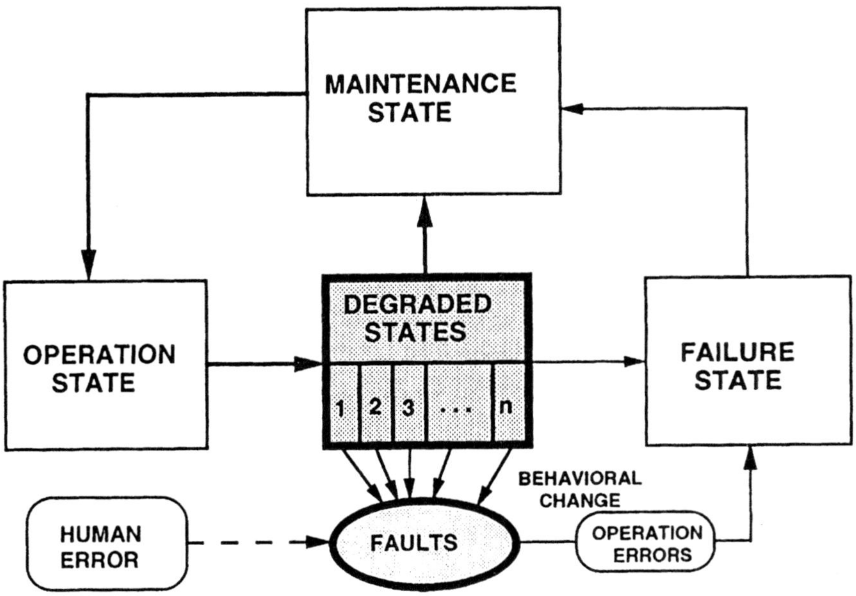

Fig. 14.1 Definition of machine performance states.

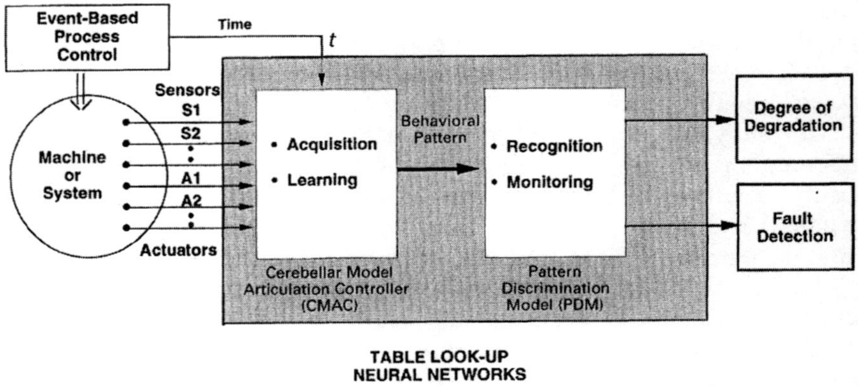

Fig. 14.2 System concept of machine degradation monitoring and fault detection using neural networks approach.

nance system. The cerebellar model articulation controller (CMAC) is used to adaptively learn machine behavior and a pattern discrimination model (PDM) based on the CMAC is used to monitor, recognize, and quantify behavioral changes. The PDM serves as a watchdog to monitor the behavior of the machine by using a confidence value that represents the conditional probability of degradation. The fault then can be detected by comparing the confidence value of the PDM output with a threshold confidence value.

14.2 METHODOLOGY

The author proposes to use the CMAC as a real-time data acquisition and learning tool for the behavior of the actuators and sensors when a system is in operation. The unique part of this approach is the use of CMAC to learn the good or normal patterns under different working conditions rather than to learn the bad or wrong patterns. In many cases, it is very difficult to create bad and wrong conditions for a machine during training.

The Marr-Albus cerebellum model [23–26] was originally a model of the cerebellum, the part of the brain that coordinates fine motor movements. Analytical processing within the cerebellum suggests that, while there are layers of neurons and large numbers of interconnects, there is extensive filtering through the use of negative feedback to confine incoming activity to only a small subset of the most active fibers. This suggests that the most active fibers might be suppressing all less active fibers, rather than distributing signals across all fibers. Further work on this model led to a mathematical formalism, namely, the cerebellar model articulation controller developed by Albus in 1972. The CMAC is essentially a clever adaptive table lookup technique for representing complex, nonlinear functions over multi-dimensional, discrete input spaces. It reduces the size of the lookup table through hashing-coding (many-into-few mapping); provides for response generalization and interpolation through a distributed, topographic representation of the inputs; and learns the appropriate nonlinear function through a supervised learning process that adjusts the content or weight of each address in the lookup table. The authors have developed a PC-based CMAC learning algorithm [30, 31].

14.3 PATTERN DISCRIMINATION MODEL (PDM)

To perform pattern recognition and discrimination tasks, a pattern discrimination model, based on the confidence checking of the CMAC output, is developed. In Figure 14.3, a confidence table is used in parallel with the weight table. During the CMAC training, the data stored in the weight table will activate a '1' value in the confidence table during training. These stored memory locations represent the desired or normal behavior of the machine. During the actual operations, the identical behavior will map to the same memory location in a weight table. A confidence value can be obtained by summing the total '1's and dividing by the number N, which is the size of the address table. It derives a 100% confidence value since the same number of '1's were used for both training and operating conditions. If there is a behavioral change, the change will cause different mapping locations in the weight table. Then, a '0' value will be placed in the confidence table when there is memory location change in the weight table. A confidence value can be obtained by summing the total 1s and 0s and dividing by the size of the address table.

Typically, if the parameters are selected properly, the confidence value is 100% during the normal training and learning process. In case of input pattern changes, the data will be mapped in the locations which are not exactly the same as the ones previously trained or learned. Therefore, a confidence value is calculated to represent the level of change (or level of degradation) of the behavior. A high percent (%) confidence value shows a high possibility of a good pattern. A low confidence value represents the degraded or fault pattern. The confidence value is calculated based on the following formula:

Confidence value = 100 * (Repeatable-Weight-Count)/(Address-Array-Size)

For example, in Fig. 14.3 a good pattern is learned during the training. The data are mapped on four locations in the weight table. The confidence

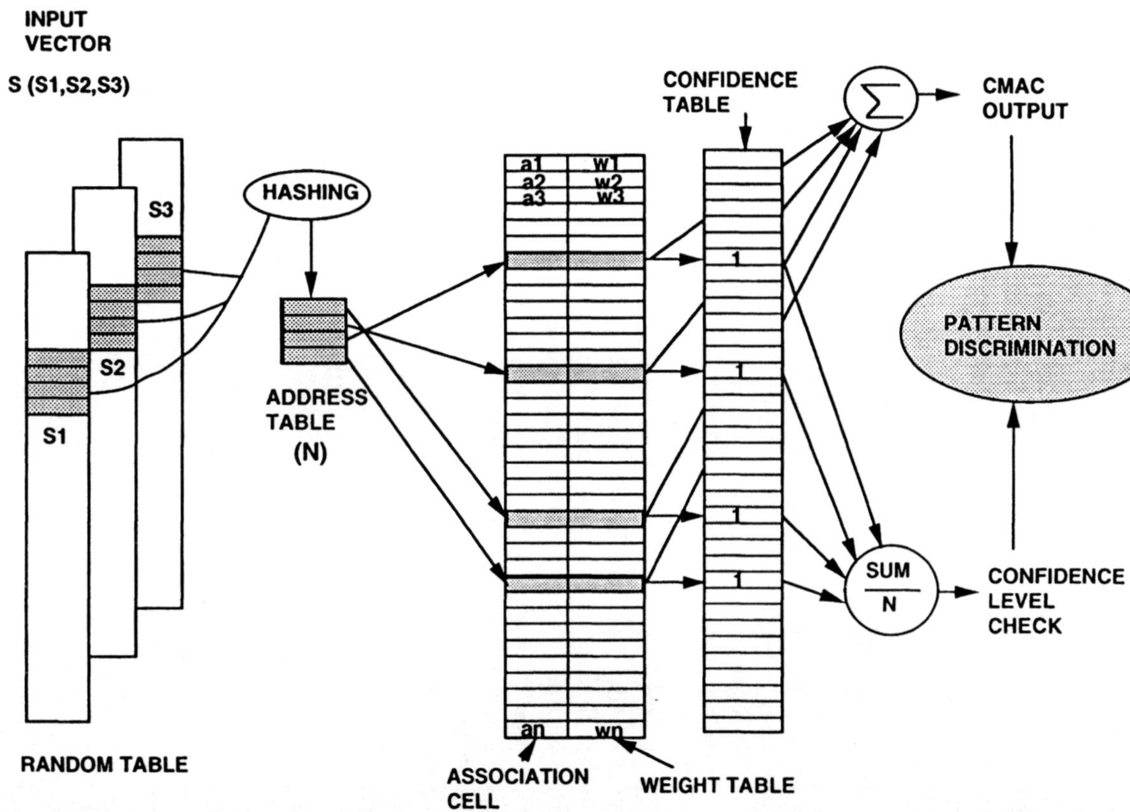

Fig. 14.3 CMAC pattern discrimination model (PDM).

value is 100% [since 100*(1 + 1 + 1 + 1)/4 = 100%]. During operations, a fault pattern is observed by the CMAC, and data are mapped on different locations in the weight table. If two locations are different, two 0s will be placed on two 1s positions. Then, a 50% [100*(1 + 1 + 0+ 0)/4 = 50%] confidence value is obtained.

The confidence value presented here represents a conditional probability of behavioral degradation. If these values can be monitored in an incremental time interval, then the rate of degradation can be calculated. The rate of changes will suggest the urgency of the maintenance.

14.4 EXPERIMENTATION

The new machine behavioral learning, degradation monitoring, and fault detection technique was tested and implemented in monitoring the position of a stepping motor encoder and degradation in the accuracy of a robot. To examine the developed CMAC-PDM and to investigate its capability for monitoring degradation and detecting faults, several test patterns were generated by setting the faults and degradations through different positions of the input encoder value. Seven different degradation test patterns were generated by adjusting the backlash screws of the robot arms and wrist. The experiments were conducted in the Robotics Laboratory at the University of Maryland – College Park. The following sections examine the capability and sensitivity of the developed technique in monitoring degradation and detecting faults.

14.4.1 Experiment 1: Encoder position monitoring and fault detection

The encoder is one of the key elements in a motion control system to provide position feedback for motor or positioning tables. For high-precision motion control systems, such as chip inspection and electronics assembly, the assurance of encoder feedback is critical. Tests were conducted to investigate the feasibility of monitoring degradation and faults by using the developed technique.

Experiment setup

In Fig. 14.4, a stepping motor with an encoder was powered through an HP power amplifier and controlled by a PC. The purpose of this experiment was to examine the developed technique and investigate the relationship among several CMAC parameters, such as address array size, confidence value (performance indicator), and convergence.

There are 13 samples and 14 inputs (one motor speed and 13 position patterns) for CMAC training. Table 14.1 shows that the encoder has the

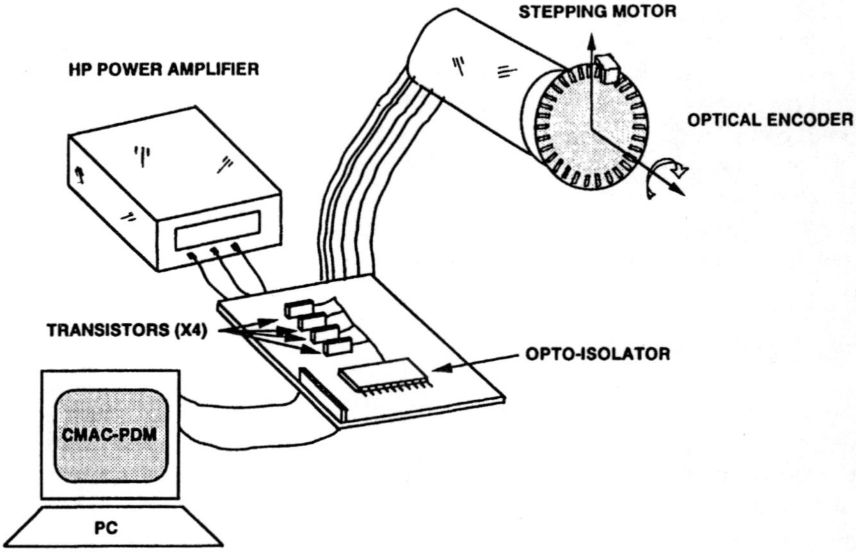

Fig. 14.4 Experimental setup for encoder position monitoring and fault detection.

Table 14.1 Data used for training

Motor speed	*14 Inputs for CMAC training* — *Encoder position signal pattern (13 patterns)*													*CMAC desired output (assigned value)*
3.5165	1.0	0.0	1.0	0.0	1.0	0.0	1.0	0.0	1.0	0.0	1.0	0.0	1.0	70000.000
4.0110	1.0	0.0	1.0	0.0	1.0	0.0	1.0	0.0	1.0	0.0	1.0	0.0	1.0	80000.000
4.5055	1.0	0.0	1.0	0.0	1.0	0.0	1.0	0.0	1.0	0.0	1.0	0.0	1.0	90000.000
5.0000	1.0	0.0	1.0	0.0	1.0	0.0	1.0	0.0	1.0	0.0	1.0	0.0	1.0	100000.000
5.4945	1.0	0.0	1.0	0.0	1.0	0.0	1.0	0.0	1.0	0.0	1.0	0.0	1.0	110000.000
6.0440	1.0	0.0	1.0	0.0	1.0	0.0	1.0	0.0	1.0	0.0	1.0	0.0	1.0	120000.000
6.4835	1.0	0.0	1.0	0.0	1.0	0.0	1.0	0.0	1.0	0.0	1.0	0.0	1.0	130000.000
6.9780	1.0	0.0	1.0	0.0	1.0	0.0	1.0	0.0	1.0	0.0	1.0	0.0	1.0	140000.000
7.5275	1.0	0.0	1.0	0.0	1.0	0.0	1.0	0.0	1.0	0.0	1.0	0.0	1.0	150000.000
8.0220	1.0	0.0	1.0	0.0	1.0	0.0	1.0	0.0	1.0	0.0	1.0	0.0	1.0	160000.000
8.5165	1.0	0.0	1.0	0.0	1.0	0.0	1.0	0.0	1.0	0.0	1.0	0.0	1.0	170000.000
9.0110	1.0	0.0	1.0	0.0	1.0	0.0	1.0	0.0	1.0	0.0	1.0	0.0	1.0	180000.000
9.5055	1.0	0.0	1.0	0.0	1.0	0.0	1.0	0.0	1.0	0.0	1.0	0.0	1.0	190000.000

same input patterns for 13 different rotating speeds. The encoder output pattern should be the same at different rotating speeds. This set of data was used to train the CMAC. Once the behavior of the encoder is trained, the CMAC learns the behavioral patterns and maps the patterns in the weight table. During training, the PDM always has the correct pattern and thus the confidence value is 100 percent. During the testing, however, if there is a fault or degradation, the pattern may be changed. The confidence value would also be changed based on the degree of the degradation or fault conditions. A lower value indicates a higher probability of fault or degradation.

Test results

To examine the developed CMAC-PDM for degradation and fault detection, several fault conditions were created by setting the fault through different positions of the input encoder value. Table 14.2 shows the test fault patterns and degradation patterns. In this table, a 0 represents a fault. Values between 1 and 0 represent degraded conditions. Table 14.3 shows the confidence values for each fault and degradation condition when different address array sizes were used in the CMAC. It shows that address array sizes 20 and 30 can better detect fault and degradation by providing lower confidence values.

14.4.2 Experiment 2: Measurement of robot degradation

The developed model was tested and implemented in monitoring degradation in the accuracy of a robot (Puma 560). Position accuracy, position repeatability, and path straightness are often evaluated and stated in statistical terms. This allows robot manufacturers and machine tool manufacturers to predict the positioning behavior of various moving elements to a stated confidence level, and provides the customer with an assurance of the machine's performance. Typically, poor position accuracy and path straightness are symptoms of the degraded performance which may be due to mechanical wear, backlash of the gear train, vibration, and thermal effects. In this experiment, the position accuracy and path straightness were used as evaluation criteria for measuring robot performance [27–29].

To examine the developed CMAC-PDM and to investigate its capability for analyzing and monitoring the performance degradation of a robot, seven different degradation test patterns were generated by adjusting the backlash screws of the robot arms and wrist. Figure 14.5 shows the experimental setup. The HP 5518A (two-frequency He-Ne laser) interferometer was used to measure the robot performance (position accuracy and path straightness). To cause degradation conditions, the backlash screws were adjusted for the arms and the wrist of the robot. Once backlash was introduced, the robot path straightness and position accuracy were measured by the laser interferometer. Before adjusting the backlash of the robot arm, a set of path straightness and

Table 14.2 Testing data fault detection and degradation recognition

		Fault pattern													
	1	5.0000	1.0	1.0	1.0	0.0	1.0	0.0	1.0	0.0	1.0	0.0	1.0	0.0	1.0
	2	5.0000	1.0	0.0	1.0	1.0	1.0	0.0	1.0	0.0	1.0	0.0	1.0	0.0	1.0
	3	5.0000	1.0	0.0	1.0	0.0	1.0	1.0	1.0	0.0	1.0	0.0	1.0	0.0	1.0
	4	5.0000	1.0	0.0	1.0	0.0	1.0	0.0	1.0	1.0	1.0	0.0	1.0	0.0	1.0
Data set A	5	5.0000	1.0	0.0	1.0	0.0	1.0	0.0	1.0	0.0	1.0	1.0	1.0	0.0	1.0
	6	5.0000	1.0	0.0	1.0	0.0	1.0	0.0	1.0	0.0	1.0	0.0	1.0	1.0	1.0
	7	5.0000	1.0	1.0	1.0	0.0	1.0	1.0	1.0	0.0	1.0	0.0	1.0	0.0	1.0
	8	5.0000	1.0	1.0	1.0	0.0	1.0	1.0	1.0	0.0	1.0	1.0	1.0	0.0	1.0
	9	5.0000	1.0	1.0	1.0	1.0	1.0	1.0	1.0	1.0	1.0	1.0	1.0	1.0	1.0
	10	5.0000	0.0	1.0	0.0	1.0	0.0	1.0	0.0	1.0	0.0	1.0	0.0	1.0	0.0
		Degradation patterns													
	11	5.0000	1.0	0.0	1.0	0.0	1.0	0.0	1.0	0.0	1.0	0.0	1.0	0.0	1.0
	12	5.0000	0.9	0.0	0.9	0.0	0.9	0.0	0.9	0.0	0.9	0.0	0.9	0.0	0.9
	13	5.0000	0.8	0.0	0.8	0.0	0.8	0.0	0.8	0.0	0.8	0.0	0.8	0.0	0.8
	14	5.0000	0.7	0.0	0.7	0.0	0.7	0.0	0.7	0.0	0.7	0.0	0.7	0.0	0.7
Data set B	15	5.0000	0.6	0.0	0.6	0.0	0.6	0.0	0.6	0.0	0.6	0.0	0.6	0.0	0.6
	16	5.0000	0.5	0.0	0.5	0.0	0.5	0.0	0.5	0.0	0.5	0.0	0.5	0.0	0.5
	17	5.0000	0.4	0.0	0.4	0.0	0.4	0.0	0.4	0.0	0.4	0.0	0.4	0.0	0.4
	18	5.0000	0.3	0.0	0.3	0.0	0.3	0.0	0.3	0.0	0.3	0.0	0.3	0.0	0.3
	19	5.0000	0.2	0.0	0.2	0.0	0.2	0.0	0.2	0.0	0.2	0.0	0.2	0.0	0.2
	20	5.0000	0.1	0.0	0.1	0.0	0.1	0.0	0.1	0.0	0.1	0.0	0.1	0.0	0.1
	21	5.0000	0.9	0.0	1.0	0.0	1.0	0.0	1.0	0.0	1.0	0.0	1.0	0.0	1.0
	22	5.0000	0.8	0.0	1.0	0.0	1.0	0.0	1.0	0.0	1.0	0.0	1.0	0.0	1.0
	23	5.0000	0.7	0.0	1.0	0.0	1.0	0.0	1.0	0.0	1.0	0.0	1.0	0.0	1.0
	24	5.0000	0.6	0.0	1.0	0.0	1.0	0.0	1.0	0.0	1.0	0.0	1.0	0.0	1.0
Data set C	25	5.0000	0.5	0.0	1.0	0.0	1.0	0.0	1.0	0.0	1.0	0.0	1.0	0.0	1.0
	26	5.0000	0.4	0.0	1.0	0.0	1.0	0.0	1.0	0.0	1.0	0.0	1.0	0.0	1.0
	27	5.0000	0.3	0.0	1.0	0.0	1.0	0.0	1.0	0.0	1.0	0.0	1.0	0.0	1.0
	28	5.0000	0.2	0.0	1.0	0.0	1.0	0.0	1.0	0.0	1.0	0.0	1.0	0.0	1.0
	29	5.0000	0.1	0.0	1.0	0.0	1.0	0.0	1.0	0.0	1.0	0.0	1.0	0.0	1.0
	30	5.0000	0.0	0.0	1.0	0.0	1.0	0.0	1.0	0.0	1.0	0.0	1.0	0.0	1.0
	31	5.0000	1.0	0.1	1.0	0.1	1.0	0.1	1.0	0.1	1.0	0.1	0.1	0.0	1.0
	32	5.0000	0.9	0.1	0.9	0.1	0.9	0.1	0.9	0.1	0.9	0.1	0.1	0.9	0.9
	33	5.0000	0.8	0.1	0.8	0.1	0.8	0.1	0.8	0.1	0.8	0.1	0.1	0.8	0.8
	34	5.0000	0.7	0.1	0.7	0.1	0.7	0.1	0.7	0.1	0.7	0.1	0.1	0.7	0.7
Data set D	35	5.0000	0.6	0.1	0.6	0.1	0.6	0.1	0.6	0.1	0.6	0.1	0.1	0.6	0.6
	36	5.0000	0.5	0.1	0.5	0.1	0.5	0.1	0.5	0.1	0.5	0.1	0.1	0.5	0.5
	37	5.0000	0.4	0.1	0.4	0.1	0.4	0.1	0.4	0.1	0.4	0.1	0.1	0.4	0.4
	38	5.0000	0.3	0.1	0.3	0.1	0.3	0.1	0.3	0.1	0.3	0.1	0.1	0.3	0.3
	39	5.0000	0.2	0.1	0.2	0.1	0.2	0.1	0.2	0.1	0.2	0.1	0.1	0.2	0.2
	40	5.0000	0.0	0.1	0.1	0.1	0.1	0.1	0.1	0.1	0.1	0.1	0.1	0.1	0.1

Table 14.3 Testing results – confidence value for different address table sizes

		Address table size								
		20	*30*	*40*	*50*	*60*	*70*	*80*	*90*	*100*
	1	0	36.67	55	70	68.33	78.57	77.5	82.22	84
	2	5	33.33	52.5	62	68.33	72.86	75	78.89	83
	3	5	33.33	50	62	73.33	71.43	78.75	77.78	83
	4	5	36.67	50	64	68.33	74.29	78.75	82.22	82
Data set A	5	0	36.67	55	60	68.33	74.29	78.85	78.89	81
	6	5	36.67	52.5	60	71.67	75.71	75	78.89	83
	7	0	40	50	62	70	75.71	76.25	80	80
	8	10	36.67	57.5	60	66.67	71.43	77.5	80	83
	9	0	33.33	50	62	63.33	72;86	76.25	81.11	83
	10	0	36.67	50	62	68.33	75.71	77.5	77.8	81
	11	100	100	100	100	100	100	100	100	100
	12	85	90	92.5	94	95	95.71	96.25	96.67	97
	13	80	86.67	90	92	93.33	94.29	95	95.56	96
	14	65	76.67	82.5	86	88.33	90	91.25	92.22	93
Data set B	15	60	73.33	80	84	88.33	88.57	90	92.22	92
	16	50	66.67	75	80	86.67	85.71	87.5	90	90
	17	40	60	70	76	83.33	82.86	85	87.78	88
	18	30	53.33	72	70	70	72.5	85.56	87.78	86
	19	20	46.67	60	70	76.67	77.14	80	83.33	85
	20	10	40	55	66	73.33	74.29	77.5	81.11	83
	21	85	90	92.5	94	96.67	97.14	97.5	96.67	97
	22	80	86.67	90	92	95	95.71	96.25	95.56	96
	23	65	76.67	82.5	86	90	91.43	92.5	92.22	93
	24	60	73.33	80	84	88.33	90	91.25	91.11	92
Data set C	25	50	70	77.5	80	85	87.14	88.89	88.89	90
	26	40	66.67	72.5	78	81.67	84.29	86.25	87.78	88
	27	30	60	67.5	76	78.33	81.43	85	85.56	86
	28	20	53.33	62.5	72	75	78.57	82.5	83.33	84
	29	10	46.67	57.5	70	73.33	75.71	80	81.11	83
	30	0	40	52.5	66	71.67	72.86	77.5	78.89	81
	31	90	93.33	95	96	98.33	97.14	97.5	98.89	98
	32	75	83.33	87.5	90	93.33	92.86	93.75	95.56	95
	33	70	80	85	88	91.67	91.43	92.5	94.44	94
	34	55	70	77.5	82	86.67	87.14	88.75	91.11	91
Data set D	35	50	66.67	75	80	86.67	85.71	87.5	91.11	90
	36	40	60	70	76	85	82.86	85	88.89	88
	37	30	60	67.5	76	78.33	81.43	85	85.56	86
	38	20	46.67	60	68	78.33	77.14	80	84.44	84
	39	10	40	55	66	75	74.29	77.5	82.22	83
	40	0	33.33	50	62	70	71.43	75	78.89	81

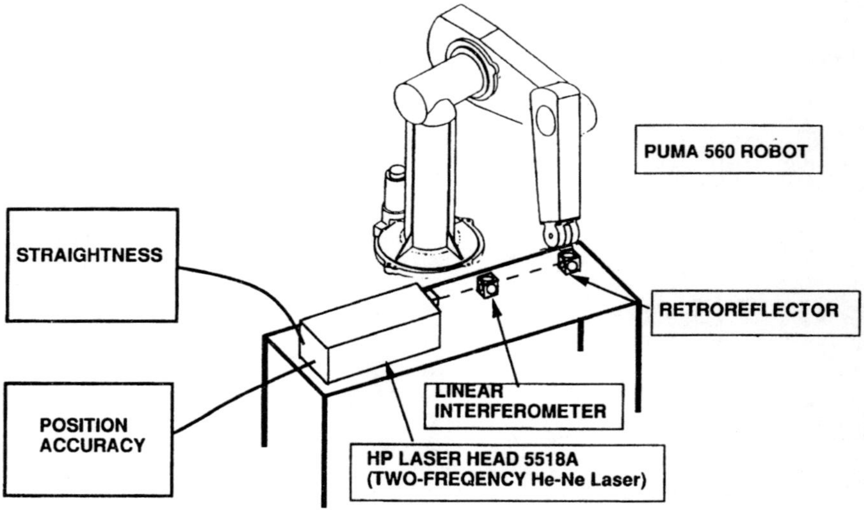

Fig. 14.5 Robot position accuracy degradation measurement using laser interferometer.

position accuracy data was generated to train the CMAC by programming the robot to follow the taught path pattern.

A degraded test pattern was generated by introducing different levels of backlash adjustment. Seven adjustments were made in this experiment, with each level of adjustment representing a different degree of degradation (see Fig. 14.6). These sets of patterns were used to test the CMAC-PDM and to investigate its capability in monitoring degradation. Figure 14.7 shows the confidence plot for each level of adjustment. The graph indicates that an array size of 20 clearly shows the measured variation in the accuracy. For address array size 20, the test was repeated for different weight table sizes. Figure 14.7 shows the confidence plot for these seven levels of degradation. The graph shows that very similar results were achieved for weight tables of size equal to or greater than 128.

14.5 CONCLUSION

This chapter presents a concept for using neural networks to analyze and monitor machine degradation. Since degradation generally occurs before failures, monitoring the trend of machine degradation allows the degraded behavior or faults to be corrected before they cause failure and machine breakdowns. To demonstrate the proposed concept, an

BACKLASH ADJUSTMENT CONDITION	TRAVEL DISTANCE 1				TRAVEL DISTANCE 2				TRAVEL DISTANCE 3			
	POSITION ACCURACY		STRAIGHTNESS		POSITION ACCURACY		STRAIGHTNESS		POSITION ACCURACY		STRAIGHTNESS	
	+	-	+	-	+	-	+	-	+	-	+	-
ADJ. # 1	0.0019	0.07	0.001	0.04	0.0024	0.085	0.0005	0.019	0.001	0.05	0.004	0.04
ADJ. # 2	0.002	0.08	0.001	0.03	0.001	0.06	0.0005	0.0176	0.0018	0.06	0.0001	0.02
ADJ. # 3	0.003	0.12	0.0027	0.09	0.002	0.06	0.026	0.06	0.001	0.04	0.001	0.02
ADJ. # 4	0.0024	0.087	0.004	0.013	0.0016	0.06	0.0002	0.045	0.002	0.07	0.0002	0.0009
ADJ. # 5	0.003	0.11	0.0015	0.05	0.003	0.1	0	0.08	0.0035	0.122	0.0014	0.05
ADJ. # 6	0.002	0.074	0.002	0.068	0.001	0.05	0.001	0.05	0.002	0.008	0.0001	0.005
ADJ. # 7	0.125	4.54	0.126	4.8	0.12	4.1	0.125	4.4	0.199	3.9	0.12	4.3

EXPLANATION OF BACKLASH ADJUSTMENT	ADJ. # 1	ADJ. # 2	ADJ. # 3	ADJ. # 4	ADJ. # 5	ADJ. # 6	ADJ. # 7
	INNER LINK BEVEL GEAR SCREWS 1 TURN	INNER LINK BEVEL GEAR SCREWS 3 TURNS	INNER LINK BEVEL GEAR SCREWS 5 TURNS SHOULDER 5 TURNS	INNER LINK BEVELGEAR SCREWS 5 TURNS	INNER LINK BEVEL GEAR SCREWS 5 TURNS SHOULDER 2 TURNS	INNER LINK BEVEL GEAR SCREWS 1 TURN SHOULDER 2 TURNS	INNER LINK 5 TURNS SHOULDER 5 TURNS WRIST (JOINT 5) 5 TURNS

Fig. 14.6 Testing data sets for different backlash adjustment conditions.

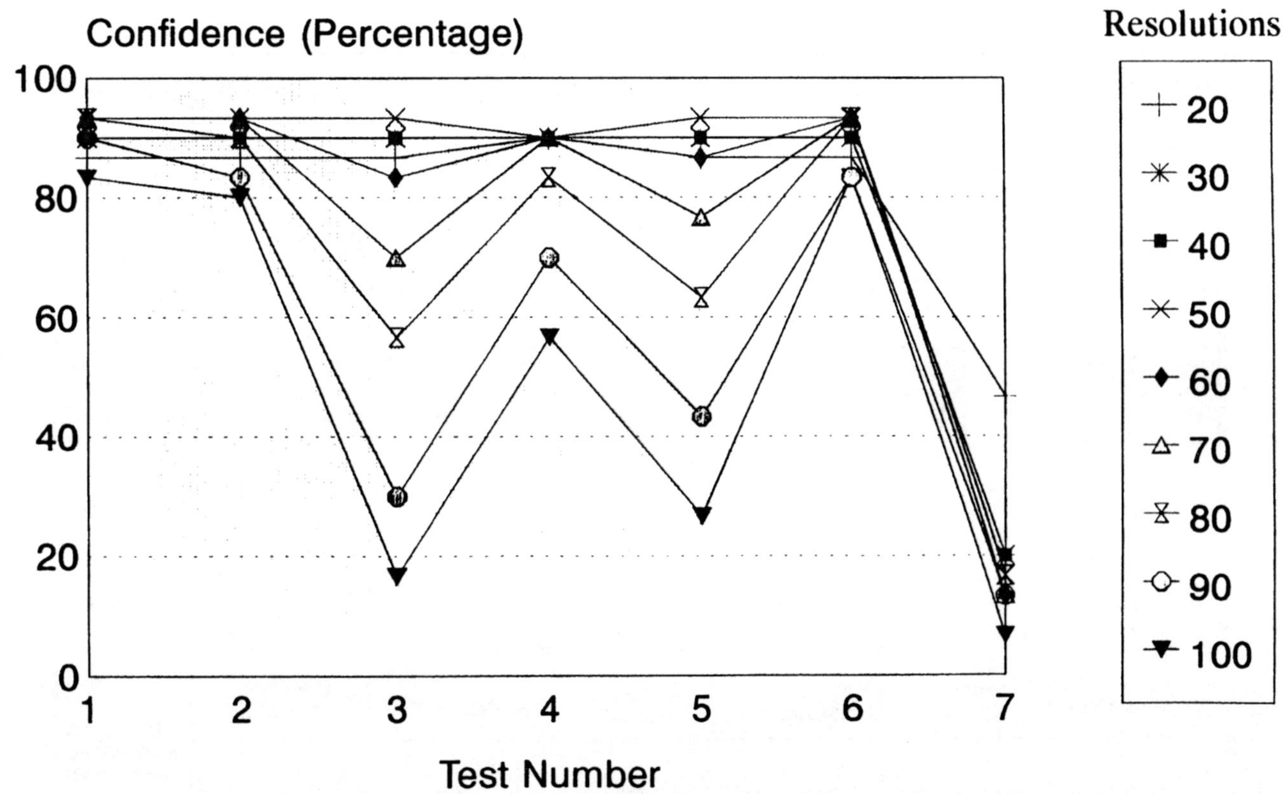

Fig. 14.7 Test number vs confidence (address array size: 30; weight table: 1024).

integrated learning, monitoring, and pattern recognition methodology has been developed. A novel proactive implementation technique, namely, a pattern discrimination model (PDM) based on the cerebellar model articulation controller (CMAC) has been developed and refined. The feasibility of this technique was demonstrated through experimentation.

In conclusion, the developed methodology has demonstrated a learning capability in providing an active and quantitative performance indicator that enables maintenance personnel to perform early fault diagnosis and effective maintenance.

14.6 SUGGESTED FUTURE WORK

The new approach developed in this research has demonstrated a useful technique for achieving proactive maintenance. Future research is needed to broaden the applications. The use of both analog and digital data will provide a better understanding of correlated behavioral patterns of machines. The future work can expand the research focus to a system level in integrating multiple sensors and actuators with the CMAC-PDM. A comparative study of CMAC-PDM, back-propagation neural networks, and probability neural networks (PNN) may yield important information on the functional differences among these different neural network models.

ACKNOWLEDGEMENT

This work is supported by the National Science Foundation and the United States Postal Service.

REFERENCES

[1] National Research Council (1990) *The Competitiveness Edge: Research Priorities for US Manufacturing*, National Academy Press.

[2] Park, H. (1992) Assessing machine performance. *American Machinist*, June, 39–42.

[3] Shainin, D. and Shainin, P. (1990) OK isn't not enough. *Quality Magazine*, February, 19–22.

[4] Papadakis, E.P. (1990) A computer-automated statistical process control method with timely response. *Journal of Engineering Costs and Production Economics*, **18**, 301–10.

[5] Craig, M. (1991) Predicting and optimizing assembly variation. *Quality Magazine*, June, 16–18.

[6] Anon (1991) Equipment should improve through use. *American Machinist Report*, September 81–100.

[7] Mendelbaum, G. and Mizuno, R. (1992) Directions in maintenance. *Maintenance Technology Magazine*, June, 45–8.

[8] Meltzer, R.J. (1992) Sensor reliability of MTBF. *Sensor Magazine*, January.
[9] Netrologic, Inc. (1990) *Report on Space Transportation Analysis and Intelligent Space System*, NASA SBIR NAS9–17995, July.
[10] Marko, K. (1989) Automative control system diagnostics using neural network for rapid pattern classification of large data sets. *Proceedings of International Neural Net Society Meeting*, Washington, DC, June, 13–16.
[11] Franklin, J.A., Sutton, R.S. and Anderson, C.W. (1988) Application of connectionist learning methods to manufacturing process monitoring. *Proceedings of IEEE 3rd International Symposium on Intelligent Control*, August, 709–12.
[12] USPS (1990) *Maintenance in the 90's – A Plant to Support the USPS Goal of Automation and Mechanization*, USPS Report, Washington, DC-HQ, December.
[13] Pardue, E.F., Piety, K.R. and Moore, R. (1992) Element of reliability-based mach maintenance. *Sound and Vibration*, May, 14–20.
[14] Fitch, J.C. (1992) Contaminant monitoring: the overlooked predictive maintenance. *Maintenance Technology Magazine*, June, 41–6.
[15] Lee, J. (1991) *Review of Computer-Aided Predictive Maintenance Technology for Machinery*, USPS Technology Resource Dept. Technical Report, January.
[16] Lee, J. (1986) Adaptive control tool monitoring in machining. SME Technical Paper MR86-131, *Conference Proceedings of Advanced Machining Technology for Cell and FMS*, February.
[17] Fairey, W. (1991) A fault finding hierarchy for PLC-controlled equipment. *Programmable Control Magazine*, November/December, 44–5.
[18] Bitite, U.S. and Ross, A. (1988) PLC based diagnostics system for FMS. *Proceeding of 7th International Conference on FMS*, September, 201–11.
[19] Valette, R., Cardeso, J. and Dubois, D. (1989) Monitoring manufacturing by means of Petri-nets. *Proceedings of IEEE Conference On Intelligent Control.*
[20] Dietz, W.E., Kiech, E.I. and Ali, M. (1989) Jet and rocket engine fault diagnosis in real time. *Journal of Neural Network Computing*, Summer, 5–19.
[21] Stevenson, F. and Greenwood, D. (19??) *Tool Wear Estimation Using Neural Networks*, Netrologic, Inc. Report, San Diego, CA.
[22] Uhrig, R.E. and Guo, Z. (1989) Use of neural networks to identify transient operating conditions in nuclear plants. *Proceedings of SPIE*, **1095**, *Application of Artificial Intelligence VII*, 851–6.
[23] Albus, J.S. (1975) A new approach to manipulator control: the cerebellar model articulation controller (CMAC). *Journal of Dynamic System, Measurement, and Control*, September, 220–28.
[24] Albus, J.S. Data storage in the cerebellar model articulation controller (CMAC). *Journal of Dynamic System, Measurement, and Control*, September, 228–33.
[25] Albus, J.S. (1971) A theory of cerebellar functions. *Mathematical Bioscience*, **10**, 25–61.
[26] Albus, J.S. (1972) *Theoretical and Experimental Aspects of a Cerebellar Model*, PHD Dissertation, Univ. of Maryland.
[27] MTTA (1900) *Standard for the Position Accuracy and Repeatability of Machining Centers and Associated Numerically Controlled Machine Tools*, The Machine Tool Trade Association Report, London, UK.
[28] NMTBA (1972) *Definition and Evaluation of Accuracy and Repeatability for Numerically Controlled Machine Tools*, NMTBA Report, August.
[29] Albertson, P. (1987) Verifying robot performance. *Robotics Today*, October, 33–6.
[30] Lee, J. and Kramer, B.M. (1992) *In-Process Machine Degradation Monitoring*

and Fault Detection Using a Neural Network Approach, doctoral dissertation, Dept of Mechanical Engineering, The George Washington University, September.

[31] Lee, J. and Kramer, B.M. (1993) Analysis of machine degradation using a neural network based pattern discrimination model. *Journal of Manufacturing Systems*, **12**(5), 379–87.

CHAPTER 15

Case Example 4: Detection and isolation of faults in the stamping process using the Haar transform

Christopher K.H. Koh and William J. Williams

15.1 INTRODUCTION

The three-dimensional forming of sheet metal is a process which uses a pair of top and bottom die sets to shape the sheet metal into its final form under considerable pressure. Most deep drawing dies have separate outer binders to regulate the flow of material as the inner punch draws the material into the lower die. Both the forces generated by the outer blankholding binders and the inner punch have significant impact on the quality of the part produced. These forces are also highly sensitive to changes in a number of process variables [2, 6, 7, 14] which may be broadly classified under: material properties, die set variables, press variables and interactive variables. The intricate interaction among these variables affects the forming tonnage which is measured by strain gages mounted on the press. Tonnage signatures are altered either directly by changes in these process variables or indirectly by occasions of splits, wrinkles or slugs in the drawing stage. A typical tonnage signature indicating the different features of the process is shown in Fig. 15.1.

Monitoring systems for peak loads and tonnage variation as a function of time, ram position or crank angle [3, 4] are quite common. They are usually capable of setting upper and lower control limits on the tonnage variation over the entire press cycle. They can also halt the process automatically when these thresholds are exceeded. Although the tonnage signatures are rich in information, there is currently no known fault detection and diagnostic system that is implemented on-line. If the extent of process faults can be quantitatively determined, the traditional $\pm 3\sigma$ tolerance limits can be reduced to yield a higher quality product.

The current practice following a fault alarm is to shut down the press and retrieve a trace of the last few signatures from the recorder. These signatures are then 'eyeballed' for some indications of the faults. This

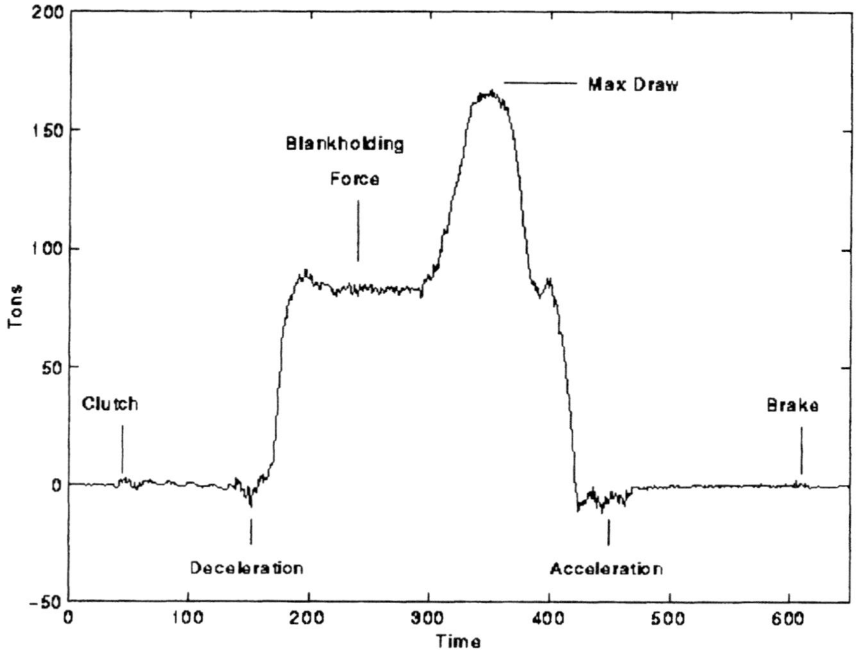

Fig. 15.1 Typical tonnage signature of a double-action, overdrive press.

approach has several disadvantages: (1) it is highly dependent on the experience and skill of the person and is therefore prone to human error, (2) it has a relatively high false alarm rate due to electrical transients or process noise (due to lack of process repeatability in older or poorly maintained presses), (3) frequency or phase shifted fault signals whose amplitudes remain within specification often escape detection, (4) thresholds are usually selected arbitrarily as the exact magnitude of the various fault signals are often unknown.

Some attempts have been made to develop some kind of diagnostic tool for the forming process based on tonnage signatures. Martinez-Heath and Bortfeld [9] used the number of threshold crossings of the control limits on the template signature to discriminate between good and bad processes; and Seem and Knussman [12] used the statistical measurements of peak tonnage to detect worn dies. Both of these methods however are only effective for faults with significant amplitude changes. Faults which alter only the frequency content of the tonnage signal (for example, gib chatter) continue to escape detection by these amplitude-based control limits. These approaches addressed only single fault detection and do not discuss the impact of multiple faults on the detection schemes.

The methodology presented in this paper provides on-line (automatic) fault detection and isolation for the stamping process. It is based on the orthogonal Haar transform and has four major advantages: (1) it is not dependent on *a priori* knowledge of the statistical properties of the fault signatures, (2) it can localize the position of the fault within a press cycle, aiding in its identification, (3) it can detect faults which do not affect the amplitude of the tonnage signal, and (4) it has multiple fault detection capability. Unlike the more familiar orthogonal Fourier transform which uses complex exponentials as its basis functions and is ideal for sinusoidal narrow-band signals, the Haar transform is well suited to represent spectrally wide-band signals [10]. Fault signatures which are frequency or phase shifted and transparent to time domain control limits can now be easily detected. The Haar transform is defined over the entire signal length, and its wavelet structure allows the detection of multiple faults within the same press cycle [6]. As an on-line implementation, the Haar transform is also one of the most efficient algorithms in terms of computational speed and memory usage.

15.2 METHODOLOGY

15.2.1 The Haar transform

The Haar transform is a member of a class of nonsinusoidal orthogonal functions [1]. It consists of rectangular waves distinguished by time scalings and time shifts. The set of continuous Haar functions $\{h(r,m,t)\}$ is periodic, orthonormal and complete, and was proposed by Alfred Haar in 1910. The Haar orthonormal sequence is defined on the open interval [0, 1) and can be generated by the recurring relation (15.1):

$$h(0,0,t) = 1, t \in [0,1)$$

$$h(r,m,t) = \begin{cases} 2^{r/2}, & \frac{m-1}{2^r} \le t < \frac{m-1/2}{2^r} \\ -2^{r/2}, & \frac{m-1/2}{2^r} \le t < \frac{m}{2^r} \\ 0, & \text{elsewhere } \forall\, t \in [0,1) \end{cases} \tag{15.1}$$

where $r \in 0 \le r < log_2 n$ is the scale, n is the number of subintervals in [0, 1) and $m \in 1 \le m \le 2^r$ is the number of functions with scale r.

The first eight continuous Haar functions are shown in Fig. 15.2. Points of discontinuity are defined as the average of the limits approached from both sides of the discontinuity. The corresponding discrete Haar functions are obtained by sampling the continuous Haar

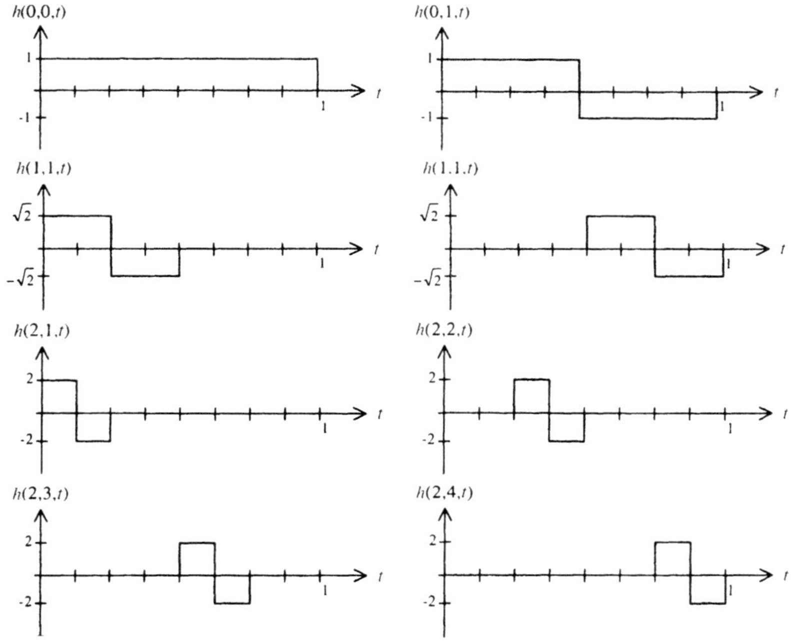

Fig. 15.2 Continuous Haar functions for $n = 8$ subintervals.

functions in Fig. 15.2 at the middle of each subinterval and normalized to produce an array:

$$H_3^T = \frac{1}{\sqrt{8}} \begin{bmatrix} 1 & 1 & 1 & 1 & 1 & 1 & 1 & 1 \\ 1 & 1 & 1 & 1 & -1 & -1 & -1 & -1 \\ \sqrt{2} & \sqrt{2} & -\sqrt{2} & -\sqrt{2} & 0 & 0 & 0 & 0 \\ 0 & 0 & 0 & 0 & \sqrt{2} & \sqrt{2} & -\sqrt{2} & -\sqrt{2} \\ 2 & -2 & 0 & 0 & 0 & 0 & 0 & 0 \\ 0 & 0 & 2 & -2 & 0 & 0 & 0 & 0 \\ 0 & 0 & 0 & 0 & 2 & -2 & 0 & 0 \\ 0 & 0 & 0 & 0 & 0 & 0 & 2 & -2 \end{bmatrix} = \begin{bmatrix} \underline{h}_1^T \\ \underline{h}_2^T \\ \underline{h}_3^T \\ \underline{h}_4^T \\ \underline{h}_5^T \\ \underline{h}_6^T \\ \underline{h}_7^T \\ \underline{h}_8^T \end{bmatrix} \tag{15.2}$$

The discrete Haar transform array is generally denoted by H_N where $n = 2^N$ is the number of discrete data points. Each column of H_N is a discrete Haar function, $\underline{h}_i$, obtained by sampling the corresponding continuous Haar function, $h(r,m,t)$. The matrix H_N is orthonormal by definition.

If $\underline{X} = [x_1\ x_2\ x_3\ ...\ x_n]^T$ is a vector sequence of n discrete points, and $\underline{C} = [c_1\ c_2\ c_3\ ...\ c_n]^T$ the corresponding set of Haar coefficients for the Haar basis functions, $\underline{h}_i$, $i = 1, 2, ...\ n$, then $\underline{X}$ and $\underline{C}$ are related by the transform pair:

$$\underline{X} = H_N \underline{C} = \sum_{i=1}^{n} c_i \underline{h}_i \tag{15.3}$$

and

$$\underline{C} = H_N^T \underline{X} \tag{15.4}$$

15.2.2 Properties of the Haar transform

The Haar transform has several nice properties which make it particularly attractive for this application.

1. *Computational efficiency*. It can be performed in $(2n - 2)$ additions and subtractions and is considerably less than the $n\log_2 n$ multiplications for the fast Fourier transform. Fast algorithms for computing the Haar transform can be found in [5, 11, 13].
2. *Signal bandwidth*. While the more familiar orthogonal Fourier transform is ideal for sinusoidal narrow-band signals, the Haar transform is especially well suited to represent spectrally wide-band signals [10] with step-like discontinuities.
3. *Data reduction*. The uncertainty of process repeatability makes it difficult sometimes to monitor changes in the tonnage at individual crank angles. The lower indexed Haar functions have large regions of support and are proportional to the signal average over these regions. The press is a large inertia system and acts as a low pass filter. The frequency content of the tonnage signals is therefore concentrated in the lower harmonic range and consequently, higher-indexed Haar coefficients represent mostly noise and may be neglected.
4. *Isolation of fault*. Unlike the other orthogonal transforms such as the Fourier, Rademacher, Walsh-Hadamad and the Discrete Cosine transforms, the Haar functions possess a wavelet structure which enables it to isolate localized events in the time domain.

15.2.3 MAPPING BETWEEN THE TIME AND HAAR DOMAINS

The mapping functions between the Haar and time domains determine the regions of support for the Haar functions and consequently the Haar coefficients which represent different segments of the press cycle.

If $\underline{X} = [x_1\ x_2\ \ldots\ x_n]$ is a data vector of length $n = 2^N$ points, and the discrete Haar function $\underline{h}_i$ in (15.1) is indexed by $i = 2^r + m$, the region of support $k_1 \leq k \leq k_2$ is given by

$$(m-1)2^{N-r} \leq k < 2^{N-r}m \tag{15.5}$$

where $k_1 = (m - 1)2^{N-r}$ and $k_2 = 2^{N-r}m - 1$.

Conversely, Haar functions whose regions of support overlap a given data segment can be determined as follows. If the data falls completely between one of the binary boundaries, then i can be determined from

$$r = N - \left[\frac{\log(k_2 - k_1 + 1)}{\log(2)} \right] \tag{15.6}$$

and

$$m = (k_2 + 1)/2^{N-r} \tag{15.7}$$

In general, the data segment of interest, $S_j \in \underline{X}$, may not necessarily span any convenient binary intervals. In this case, the discrete Haar functions $\underline{h}_i$ which completely or partially span S_j must satisfy one of the following conditions

$$\{k_1 < p\} \cap \{p \le k_2 \le q\} \tag{15.8}$$

$$\{p \le k_1 \le q\} \cap \{k_2 > q\} \tag{15.9}$$

$$\{k_1 \ge p\} \cap \{q \ge k_2\} \tag{15.10}$$

$$\{k_1 \le p\} \cap \{q \le k_2\} \tag{15.11}$$

where $k_1 \le k \le k_2$ is the region of support for h_i.

The tonnage signature for a complete press cycle is usually divided into several disjoint segments. Condition (15.10) ensures that the sets of Haar coefficients for disjoint data segments are mutually exclusive.

15.2.4 SENSITIVITY ANALYSIS

The number of Haar coefficients to be monitored can be further reduced by selecting only those coefficients which have high sensitivity to process changes but low sensitivity to process noise. Two sensitivity indices are therefore needed. The first index, ξ_{ij}, is a measure of the sensitivity of coefficient c_i to fault F_j. Assuming homogeneity of variance, ξ_{ij} can be defined as

$$\xi_{ij} = \frac{E[c_i \mid H_j] - E[c_i \mid H_o]}{\sigma_{c_i}} \tag{15.12}$$

where H_o and H_j are the respective hypotheses for no-fault and fault F_j conditions, and σ_{c_i} the standard deviation of the coefficient c_i under H_o (15.7).

If $f_o(t)$ and $f_j(t)$ are the continuous tonnage signals under no-fault and fault conditions, from the definition of c_i

$$E[c_i \mid H_o] = \int E[f_o(t)]h(r,m,t)dt$$

and

$$E[c_i \mid H_j] - E[c_i \mid H_o] = \int E[f_j(t) - f_o(t)]h(r,m,t)dt$$
$$= \int E[e_j(t)]h(r,m,t)dt$$

For a sufficiently large value of scale $\hat{r}$, the support region for $h(\hat{r},m,t)$ is very small compared to the interval [0, 1). If $f(t) = s(t) + n(t)$, the signal of interest $s(t)$ will appear to be slowly changing and $h(\hat{r},m,t)$ will represent mostly the process noise $n(t)$. A second index, ζ_i, can be used to measure the sensitivity of the coefficient c_i to the process noise. This index is merely the standard deviation of the coefficient under normal condition, which is a function of the process noise over the region of support of c_i. Therefore

$$\varsigma_i = \sigma_{c_i} \tag{15.13}$$

For a given region of support, $0 \le p < q < 1$, a subset of $\{h(r,m,t)\}$ exists which span the region $[p : q]$. Of these, an optimal subset of Haar coefficients with scale $\hat{r}$ can be found which maximizes (15.12) and minimizes (15.13).

15.2.5 Moments of the Haar coefficient

A Haar coefficient is a linear combination of a set of consecutive points in the original data. Its statistical properties are therefore dependent on these points. From the analysis in section 15.2.3, $\underline{h}_i$ can be expressed as

$$\underline{h}_i = [h_{i1}\ h_{i2} \ldots h_{in}] \tag{15.14}$$

where $i = 2^r + m$ and h_{ij} is the jth element of $\underline{h}_i$. The Haar coefficient, c_i, which is the projection of the discrete signal $\underline{X}$ on h_i, is given by

$$c_i = <\underline{h}_i, \underline{X}> = \sum_{j=1}^{n} h_{ij} x_j \tag{15.15}$$

The expected value and variance of c_i are thus

$$E[c_i] = E[\sum_{j=1}^{n} h_{ij} x_j] = \sum_{j=1}^{n} h_{ij} E[x_j] \tag{15.16}$$

and

$$\sigma_{c_i} = \text{var}[c_i] = \text{var}[\sum_{j=1}^{n} h_{ij} x_j]$$

$$= \sum_{j=1}^{n} (h_{ij})^2 \sigma_j^2 + \sum_{\substack{j=1 \\ j \ne k}}^{n} \sum_{j=1}^{n} h_{ij} h_{ik} \text{cov}(x_j, x_k) \tag{15.17}$$

Equations (15.16) and (15.17) are used to set the control (or threshold) limits on the respective Haar coefficients. Since c_i is a linear combination of several random variables, x_j, $j = 1, \dots n$, with finite mean and finite variances, the Central Limit Theorem implies normal Gaussian distribution for c_i, $i = 1, \dots n$ if n is sufficiently large. However, if we assume that x_j is Gaussian for all j, then c_j is also Gaussian for all i without the requirement of n being sufficiently large.

15.2.6 Developing the Haar detector

One of the main challenges in the detection of stamping process faults is the lack of *a priori* information on most of the faults. Some faults should not occur at all if the press has been properly used and maintained. Such faults include broken pitmans, cracked structures and loose tie-rods. When faults do occur, the press is often shut down almost immediately by the tonnage monitoring system. As a result, fault signatures are rare to come by for statistical analysis. In this approach, Haar coefficients, which acquire new 'highs' [6] as a result of changes in the tonnage signal, are used as detection signals. The statistical properties of fault signals are therefore unnecessary.

These jumps in the values of c_i can be modeled as a mean shift or a variance shift or both. Assuming that the Haar coefficient c_i has a normal Gaussian distribution under all conditions, the problem can be formulated by the following four hypotheses:

- *Case 1.* No fault condition:
$$H_0\text{:c} \sim \text{N}(m_0, \sigma_0^2)$$
- *Case 2.* Change in mean:
$$H_1\text{:c} \sim \text{N}(m_1, \sigma_0^2)$$
- *Case 3.* Change in variance:
$$H_2\text{:c} \sim \text{N}(m_0, \sigma_1^2)$$
- *Case 4.* Change in mean and variance:
$$H_3\text{:c} \sim N(m_1, \sigma_1^2)$$

where c is the observation of ci arriving at the detector. The mean values of c_i under (H_0, H_2) and (H_1, H_3) are denoted by m_0 and m_1 respectively. The variances of c_i under (H_0, H_1) and (H_2, H_3) are denoted by σ_0^2 and σ_1^2 respectively.

Different algorithms can be derived for each of these hypotheses. In practice, it is easy to estimate the values of m_0 and σ_0^2, but there is often no way of estimating m_1 and σ_1^2 with confidence. It will be assumed in this paper that the process noise remains unchanged throughout so that $\sigma_1^2 = \sigma_0^2$. Hence, the fault condition can be modeled by hypothesis H_1.

The likelihood ratio test

The likelihood ratio test (LRT) for c_i can be expressed as

$$\Lambda(R)=\frac{\dfrac{1}{\sqrt{2\pi\sigma_1^2}}\,e^{-(R-m_1)^2/2\sigma_1^2}}{\dfrac{1}{\sqrt{2\pi\sigma_0^2}}\,e^{-(R-m_0)^2/2\sigma_1^2}}\ \mathop{\gtrless}\limits^{H_1}_{H_0}\ \eta$$

where R is a particular value of c. Taking the natural logarithm of (15.18),

$$l(R)=\frac{1}{2\sigma_0^2}(R-m_0)^2-\frac{1}{2\sigma_1^2}(R-m_1)^2\ \mathop{\gtrless}\limits^{H_1}_{H_0}\ \log(\eta)-\log(\sigma_0)+\log(\sigma_1) \qquad (15.19)$$

Assuming homogeneity of variance, $\sigma_0^2=\sigma_1^2=\sigma^2$ in (15.19)

$$l(R)=\frac{1}{2\sigma^2}(R-m_0)^2-\frac{1}{2\sigma^2}(R-m_1)^2\ \mathop{\gtrless}\limits^{H_1}_{H_0}\ \log(\eta) \qquad (15.20)$$

or

$$R\ \mathop{\gtrless}\limits^{H_1}_{H_0}\ \frac{2\sigma^2\log(\eta)+(m_1^2-m_0^2)}{2(m_1-m_0)}=\gamma>0 \text{ if } m_1>m_0 \qquad (15.21)$$

and

$$R\ \mathop{\lessgtr}\limits^{H_1}_{H_0}\ \frac{2\sigma^2\log(\eta)+(m_1^2-m_0^2)}{2(m_1-m_0)}=\gamma<0 \text{ if } m_1<m_0 \qquad (15.22)$$

where γ is a function of η, m_0 and m_1. The jump in the value of R is the same as choosing a Gaussian density with a larger absolute mean.

It is clear from the structure of the test (15.21)–(15.22) that an ordinary LRT can be designed for a particular m_1. However, m_1 is not a predictable value and therefore cannot be part of the test. The uniformly most powerful (UMP) form of the LRT is proposed to address this issue. A UMP test must be as good as any other test for every m_1 and exists if and only if the LRT for every m_1 can be completely defined (including threshold) without knowledge of m_1 [15].

Proof of existence of the UMP

Using the Neyman–Pearson criterion, the probability of false alarm is denoted by $P_F = \alpha$, where α is the prespecified level of significance. For the case where $m_1 > m_0$

$$\begin{aligned} P_F &= \Pr[\text{choose } H_1 \mid H_0 \text{ true}] \\ &= \int_{\gamma}^{\infty} \frac{1}{\sqrt{2\pi\sigma}} e^{-(R-m_0)^2/2\sigma^2}\, dR \qquad (15.23) \\ &= \operatorname{erfc}\left[\frac{\gamma - m_0}{\sigma}\right] = \alpha \end{aligned}$$

Therefore

$$\gamma = m_0 + \sigma \operatorname{erfc}^{-1}(\alpha) \qquad (15.24)$$

Let $\lambda = (\gamma - m_0)/\sigma$. A value of λ can be found for a given α [8], so that

$$\lambda = m_0 + \lambda\sigma \qquad (15.25)$$

and substituting into (15.21)

$$R - m_0 \underset{H_0}{\overset{H_1}{\gtrless}} \lambda\sigma \qquad (15.26)$$

where $\lambda\alpha \geq 0$ always. The UMP test now reduces to the form of a mean change detector. If $m_1 < m_0$, it can be shown that

$$R - m_0 \underset{H_0}{\overset{H_1}{\lessgtr}} \lambda\sigma \qquad (15.27)$$

For the UMP test to exist, R can take on only values greater than or equal to m_0 (15.26), or R can take on values less than or equal to m_0 (15.27), but not both at the same time for the same fault. From the sensitivity analysis in section 15.2.4, R is known to be highly dependent on the shape of the signal. The same fault would therefore affect the Haar coefficient the same way each time it occurs.

Taking advantage of the symmetry of the two directional tests, (15.26) and (15.27) can be combined as a test of the absolute difference between R and m_0 to give

$$| R - m_0 | \underset{H_0}{\overset{H_1}{\gtrless}} \lambda\sigma \qquad (15.28)$$

The probability of detection, P_D, or power of the test corresponding to (15.23) is

$$P_D = \text{Pr[choose } H_0/H_1 \text{ true]}$$

$$= \int_{\gamma}^{\infty} \frac{1}{\sqrt{2\pi\sigma}} e^{-(R-m_1)^2/2\sigma^2}\, dR \qquad (15.29)$$

$$= \text{erfc}\ [\frac{\gamma - m_1}{\sigma}]$$

Equation (15.29) gives an upper bound on the power of the test for any value of m_1. The threshold γ (15.24) is independent of the value of m_1. The performance of the test, P_D varies with m_1 (15.29), but knowing m_1 will not make it better. This is the uniformly most powerful form of the LRT. The necessity of estimating the magnitude of the fault has been avoided by using the UMP test. The receiver operating characteristic (ROC) curve for the UMP detector is plotted as a function of α in Fig. 15.3 with values of $0 \le \alpha \le 1$, and $m_0 = 0$, $\sigma^2 = 1$. The detector performance is clearly superior to that of a randomized test indicated by the $P_D = P_F$ line.

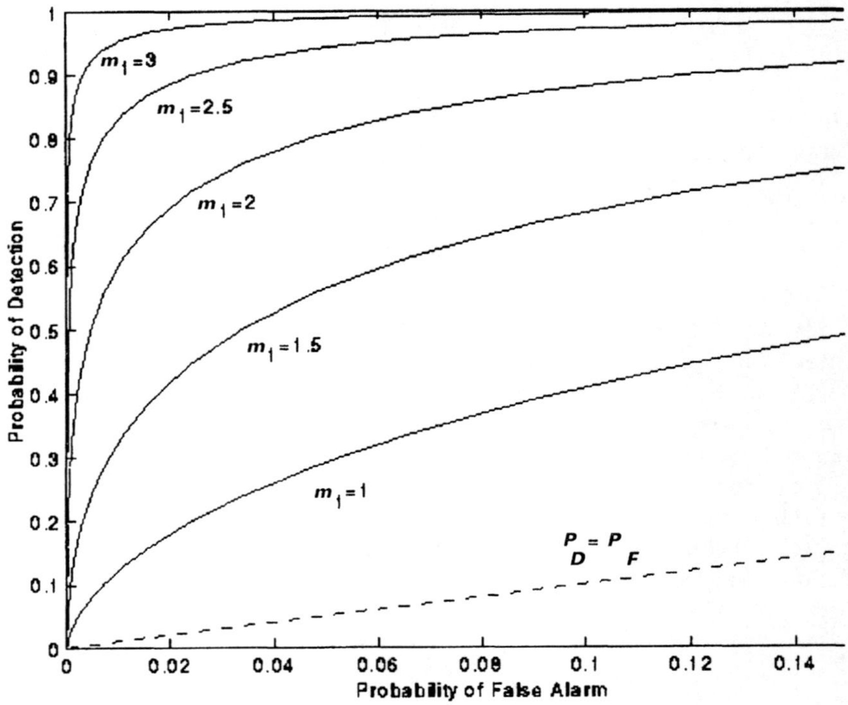

Fig. 15.3 ROC curve for UMP detector for different fault magnitude, m_1, with $m_0 = 0$ and $\sigma^2 = 1$.

15.2.7 Implementation

The Haar detector is implemented in the following steps.

Step 1. Segmentation of signal

Partition the tonnage signal, $\underline{X} = \{x_i\}_{i=1}^{i=n}$, into J disjoint segments, $\{S_j\}_{j=1}^{j=J}$, such that $\underline{X} = \bigcup_j S_j$. Each segment S_j covers a particular stage of the forming process; and a set of faults, F_j, is associated with process stage j. The segments $\{S_j\}_{j=1}^{j=J}$ are partitioned according to the motion curves of the press. These segments will be referenced by the crank angle of the main-shaft. For an N-bit angle decoder, there will be $n = 2^N$ data points for a complete press cycle or 360° of crank rotation.

Step 2. Select sets of Haar coefficients

If $[x_1 : x_2]$ is the width of the data segment and $[k_1 : k_2]$ is the region of support for any Haar function, the Haar function which lies within the segment S_j is bounded by the condition

$$\{k_1 \geq x_1\} \cap \{k_2 \leq x_2\} \tag{15.30}$$

A set of such functions can be found and the corresponding Haar coefficients denoted by C_j. Consequently, J disjoint sets of Haar coefficients $\{C_j\}_{j=1}^{j=J}$ can be found where

$$C_k \cap C_l = 0, \forall_k \neq 1$$

These disjoint sets of Haar coefficients allow detection of multiple faults occurring at different stages of the process during the same press cycle. A further reduction in the dimensions of C_j, $j = 1, \ldots, J$ can be achieved by sensitivity analysis (section 15.2.4) if the fault signals are known in advance, or by simply neglecting higher indexed coefficients which represent the process noise.

Step 3. Establish detection threshold

The UMP threshold for each Haar coefficient is computed from (15.28) using data collected during normal operation of the press. The mean and standard deviation of a Haar coefficient c_i can be determined from (15.16) and (15.17). The value of λ is read off a table [8] when a level of significance, α, has been predetermined.

Step 4. Monitor the sets C_j, $j = 1, 2, \ldots, J$

Apply the UMP to detect jumps in the selected Haar coefficients, that is, all $c_i \in \{C_j\}_{j=1}^{j=J}$. A counter is used to record the number of times a coefficient exceeds its threshold. The UMP threshold for the cumulative *threshold crossing* or *hits* for each of the Haar sets C_j, $j = 1, \ldots, J$, is also determined using (15.28). Let A_j, $j = 1, ..., J$ be the cumulative hits for the Haar sets C_j, $j = 1, ..., J$ after every stroke. The mean cumulative hits under normal conditions is zero (since there should not be any hit under normal conditions). The standard deviation is simply estimated from the number of hits under normal conditions.

Step 5. Identify location of the fault

When A_j crosses its UMP threshold, an alarm is registered at the output of the detector for segment S_j. The use of cumulative threshold crossing of the coefficients in C_j has an averaging effect on the process noise. The number of false alarms due to noisy processes is therefore significantly reduced.

15.3 EXAMPLE

The following test results were obtained from a set of surrogate signals generated from 150 consecutive cycles of a Clearing double-action

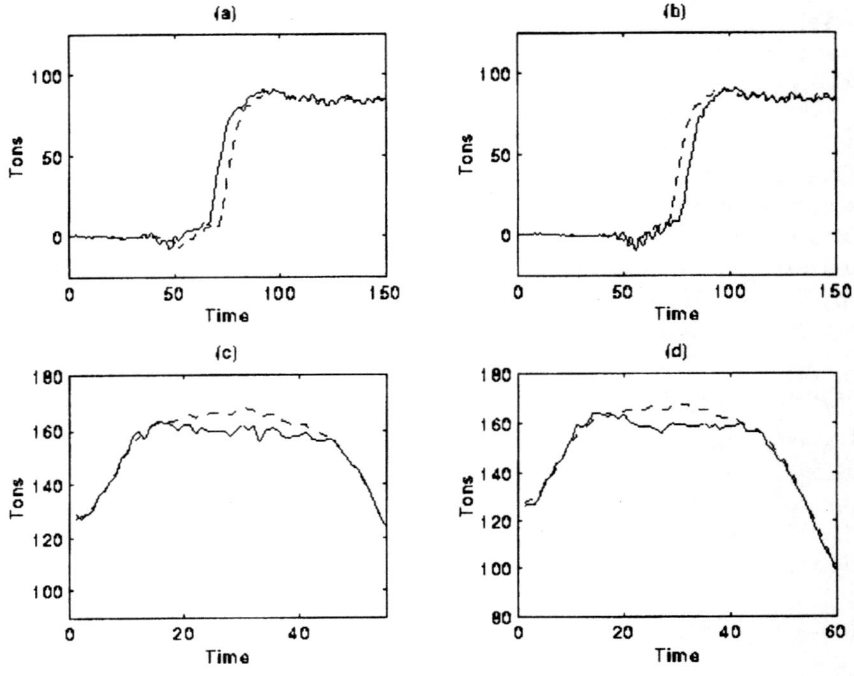

Fig. 15.4 Fault signals used in the test. Template signature (- - -) is shown superimposed on a sample of the faults (a) F_1 – material too thick (b) F_2 – material too thin (c) F_3 – loose tie rods and (d) F_4 – worn bushings. Faults F_1 and F_2 are found in segment S_1 while F_3 and F_4 are found in segment S_2 in Fig. 15.5.

overdrive press. Surrogate signals are normal tonnage signals modified to duplicate the kind of features expected in the fault signatures. These surrogate signals retain the characteristics of the original process noise and are reasonably good surrogates of real fault signals.

Four types of surrogate fault signals were generated from the first 120 observations of 183 consecutive press cycles, with 30 signatures for each type of test faults. Figure 15.4 shows a plot of the four fault signatures. They are (a) F_1 – material too thick, (b) F_2 – material too thin, (c) F_3 – loose tie rods and (d) F_4 – worn bushings. Both (a) and (b) are time-shifted versions of each other. They are difficult to distinguish without alignment. Fault types (c) and (d) have almost the same magnitudes and differ only by an additional low amplitude sinusoidal component in (d). It is nearly impossible to visually differentiate between (c) and (d).

1. *Segmenting the signal*

The size of each realization is $2^N = 512$ data points and segmented according to the motion curve for the press (Fig. 15.5).

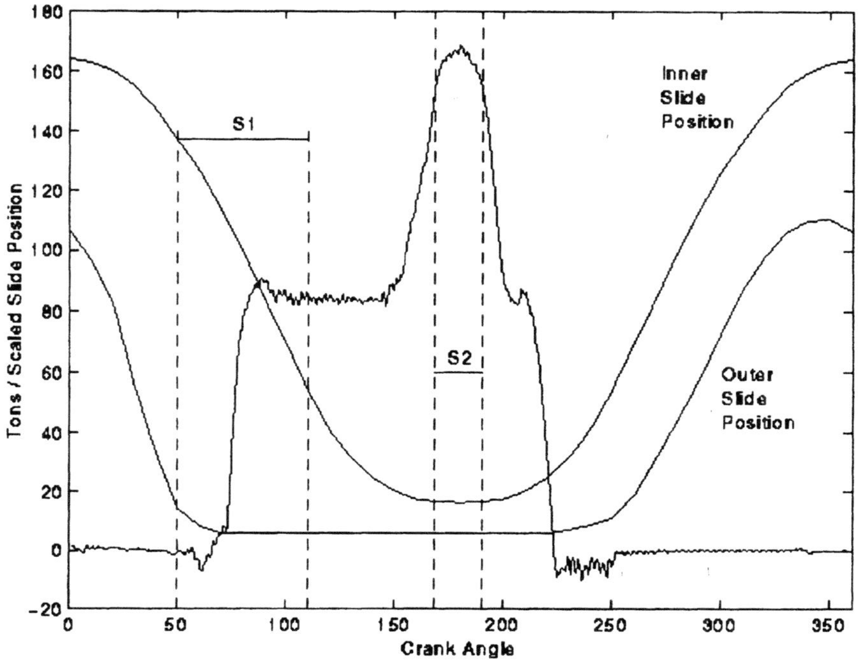

Fig. 15.5 Motion curves and tonnage signature. Faults F_1 and F_2 are found in data segment S_1 and faults F_3 and F_4 are found in data segment S_2.

2. Selecting the sets of Haar coefficients

The disjoint sets of Haar coefficients for data segments S_1 and S_2 were determined from condition (15.30). A total of 94 coefficients for C_1 and 32 for C_2 were obtained.

3. Establishing the detection threshold

The mean ($\bar{c}_i$) and standard deviation (σ_i) for each coefficient in C_1 and C_2 were estimated from the normal data using (15.16) and (15.17). An unbiased estimator is used to determine σ_i. A threshold based on α = 0.01, and hence $\lambda = 2.3$, can be found for each $c_i \in C_1, C_2$.

4. Fault sensitivity

A counter was used to track the number of times a Haar coefficient exceeded its threshold $\lambda\sigma_i$ in (15.28). The 'hits' or number of threshold crossings recorded for each coefficient is shown in Fig. 15.6. The maximum count of 30 is equivalent to 100% hit rate as there are 30 samples of

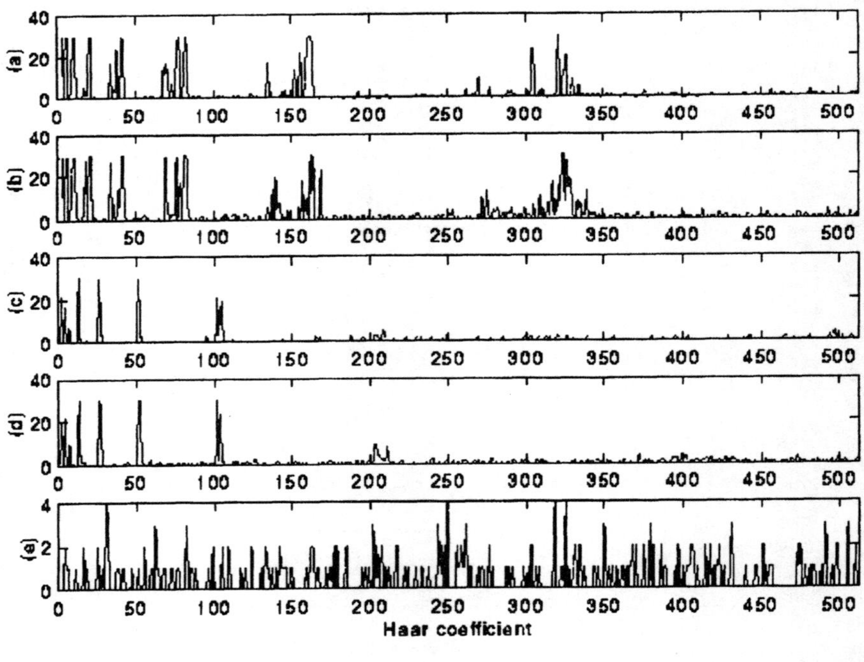

Fig. 15.6 Hits recorded by each of the 512 Haar coefficients for (a) fault F_1, (b) fault F_2, (c) fault F_3, (d) fault F_4, and (e) normal signal, based on a level of significance, $\alpha = 0.01$.

each type of fault. Some coefficients were more sensitive to one type of fault than another. This is consistent with the fact that highly localized faults affect only Haar coefficients in their support region.

Under no-fault condition, the distribution of hits is quite uniform among all the 512 coefficients (Fig. 15.6(e)). There is no bias in the probability of threshold crossing for any coefficient since the threshold is determined by the variance of each coefficient under normal condition.

5. Detector output

The cumulative alarms, A_j for segments S_1 and S_2 were computed for each press cycle. Figure 15.7 shows the trajectories of A_j, j=1, 2 over 150

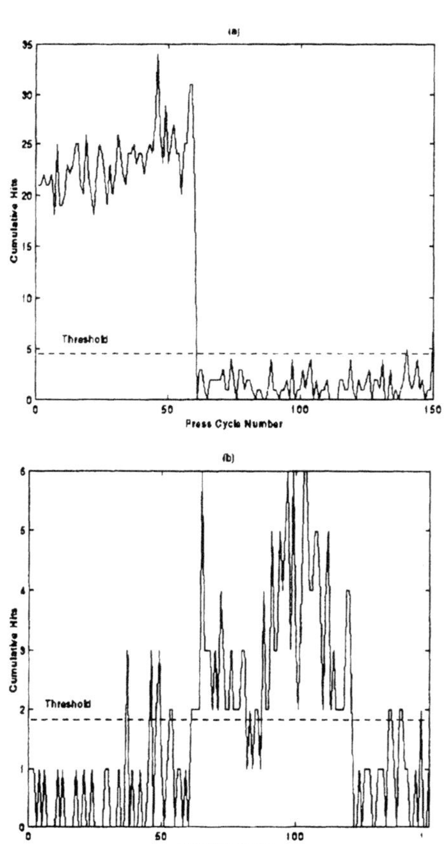

Fig. 15.7 Cumulative alarms for full sets of Haar coefficients for segments (a) S_1 with threshold, $\lambda\sigma_1 = 4.808$, and (b) S_2 with threshold, $\lambda\sigma_2 = 1.868$.

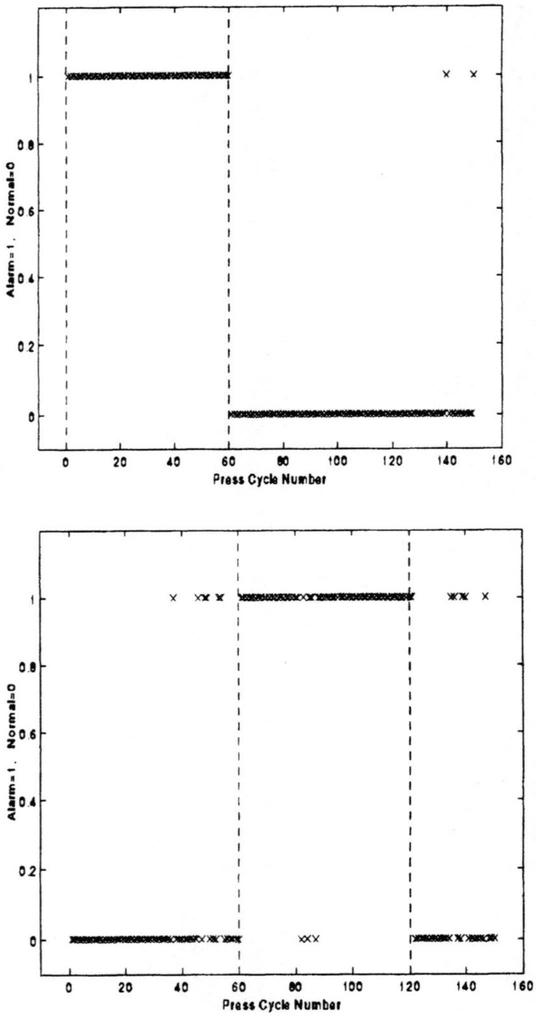

Fig. 15.8 Output of UMP detector for full sets of Haar coefficients for segments (a) S_1 and (b) S_2.

test cycles for the full sets of Haar coefficients C_1 and C_2. The standard deviations for A_1 and A_2 were estimated from the hits registered in Fig. 15.6(e). The thresholds for A_1 and A_2 are respectively 4.808 and 1.868. The detector outputs for S_1 and S_2 are shown in Fig. 15.8. There were 2 false alarms but no missed detection for S_1 (Fig. 15.8(a)) and 11 false alarms and 3 missed detections for S_2 (Fig. 15.8(b)).

6. Detector performance

To complete the discussion, it is necessary to evaluate the performance of the detector. For a Neyman–Pearson test the values of P_F and P_D completely specify the detector performance. The variation of P_F and P_D as a function of the dimension of C_j is shown in Fig. 15.9. For example, when the dimension of C_1 is 20 in Fig. 15.9(a), the false alarm rate is zero.

The detector was able to detect the presence of faults F_1 and F_2 in S_1 with any number of coefficients in C_1 (Fig. 15.9(a)). There is no specific trend in the number of false alarms although zero false alarm was possible using only the first two coefficients of C_1. It may be concluded that the optimal dimension of C_1 is 1 since 100% detection rate is achieved with zero false alarm.

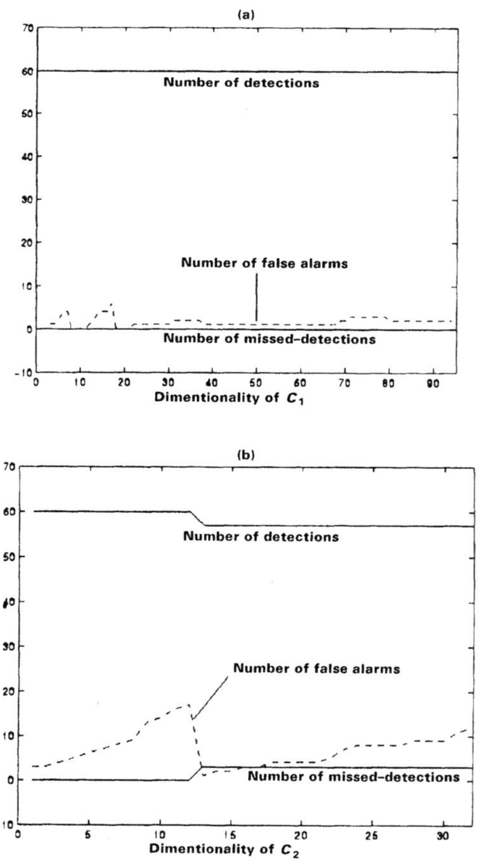

Fig. 15.9 P_F and P_D as a function of dimension of (a) C_1 for S_1 and (b) C_2 for S_2.

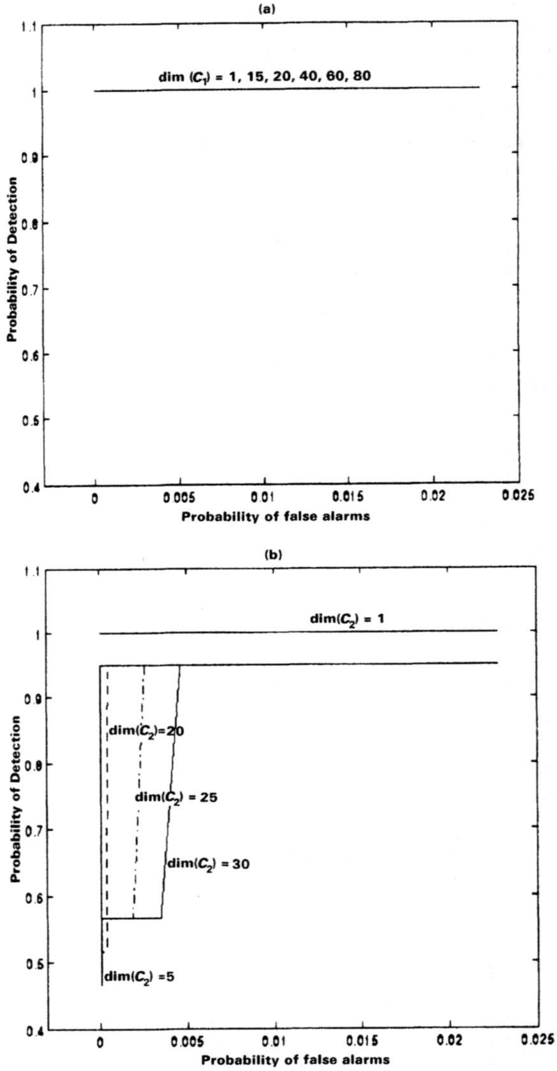

Fig. 15.10 ROC curves for segments (a) S_1 and (b) S_2. The horizontal axes represent the range of values of α corresponding to $2.0 \leq \lambda \leq 4.0$ in (15.28). The vertical axes show the probability of detection , P_D in (15.29).

The detector performance for S_2 dropped slightly when the dimensionality of C_2 exceeded 12 in Fig. 15.9(b). There appears to be two distinct regions of performance. The boundary of the two regions almost coincides with the transition in scale of the coefficients in C_2. The number of false alarms increases monotonically in both regions. This trend is typical of the 'curse of dimensionality' when detector performance declines beyond an optimal number of classification features used.

The detector ROC curves for segments S_1 and S_2 are shown in Fig. 15.10. The level of significance, α, is derived from [8] for $2.0 < \lambda < 4.0$. The probability of detection, P_D, is defined as the ratio of the number of detection to the number of fault cycles, 60. The large gap between the cumulative alarms for no-fault and fault conditions in S_1 (Fig. 15.7(a)) explains why the probability of detection is 1 for all dimensionality of C_1; that is, for all $\{c_i\} \in C_1$, $i = 1, 2, ..., 94$, $P_D = 1$. The situation is different, however, for S_2 where the cumulative alarms for both conditions are not so well separated (Fig. 15.7(b)). As the dimensionality of C_2 is increased, the ROC curves indicate a slight deterioration in the detection rate due to the 'curse of dimensionality'.

15.4 CONCLUSION

The complexity of the stamping process and the number of variables that affect the process pose a very difficult challenge for any kind of on-line fault detection system. A solution has been proposed which exploits the information in the tonnage signature using the orthogonal Haar transform. The resulting coefficients, if properly selected, have been shown to perform remarkably well with a UMP detector based on the Neyman–Pearson criteria. The Haar coefficients are ideal detection signals for a wide range of different faults. The location of these faults can be determined by partitioning the original data into disjoint segments, and associating each segment with an exclusive set of detection signals. The UMP algorithm takes care of fault signals whose statistical properties are not known in advance. The performance of the detector was investigated with respect to threshold, dimensionality and false alarm rate. This approach can be extended to other applications which face similar constraints.

REFERENCES

[1] Ahmed, N. and Rao, K.R. (1975) *Orthogonal Transforms for Digital Signal Processing*, Springer-Verlag, Berlin.

[2] Ahmetoglu, M.A., Altan, T. and Kinzel, G.L. (1992) Improvement of part quality in stamping by controlling blank holder force and pressure. *Materials Processing Technology*, **33**, 195–214.

[3] Carabbio, R. (1993) Optimizing process control through signature verification. *Proceedings of IPC 93 Technical Symposium.*

[4] Clark, A.V., Izworski, N. *et al.* (1991) Progress towards non-destructive, online measurement of sheet metal formability. *Autobody Stamping Technology Progress*, SAE SP-865, February.

[5] Fakruddin, D.B. and Parthasarathy, K. (1985) Simplified algorithms based on Haar transforms for signal recognition in protective relays. *Proceedings of the IEEE*, **73**(5), 940–2.
[6] Koh, C.K.-H, Shi, J. and Williams, W.J. (1995) Tonnage signature analysis using the othogonal Haar transform. *Transactions of the NAMRC*, May, 229–34.
[7] Koh, C.K.-H. (1995) *Tonnage Signature Analysis for the Stamping Process*, PhD Thesis, University of Michigan, August.
[8] Leon-Garcia, A. (1989) *Probability and Random Processes*, Addison-Wesley, Table 3.3, p. 127.
[9] Martinez-Heath, M.R. and Bortfeld, D.P. (1987) Signature analysis for machine control. *Sensors for Manufacturing*, PED-vol 26, ASME Winter Annual Meeting.
[10] Mikhael, W.B. and Ramaswamy, A. (1992) Residual error formulation and adaptive minimization for representing nonstationary signals using mixed transforms. *IEEE Transactions on Circuits and Systems – II: Analog and Digital Signal Processing*, **39**, July, 489–92.
[11] Roeser, P.R. and Jernigan, M.E. (1982) Fast Haar transform algorithms. *IEEE Transactions on Computers*, February, 175–7.
[12] Seem, J.E. and Knussmann, K.D. (1994) Statistical methods for on-line fault detection in press-working applications. To be published, ASME Winter Annual Meeting.
[13] Shore, J.E. (1973) On the application of Haar functions. *IEEE Transactions in Communications*, March, 209–216.
[14] Siekirt, J.F. (1986) Process variable effects on sheet metal quality. *Journal of Applied Metalworking*, **4**(3), 262–9.
[15] Van Trees, H.L. (1968) *Detection, Estimation, and Modulation Theory*. John Wiley, New York.

CHAPTER 16

Case Example 5: Fault monitoring in manufacturing systems using template models

Lawrence E. Holloway

16.1 INTRODUCTION

Faults in manufacturing systems can be an expensive problem. Manufacturing faults may result in equipment downtime or unsatisfactory product. System downtime can cost thousands of dollars per minute in lost production [1], whereas poor quality can lead to dissatisfied customers and loss of market share.

Ideally, a manufacturing system would be designed to be robust enough that faults would not occur. Realistically, however, there will always be some likelihood of failure, perhaps as a consequence of normal wear, and at best the designer must make trade-offs between increasing reliability, performance, and cost. Realizing that faults will occur, we then have the issues of detecting their occurrence, diagnosing the particular problem, and recovering to normal operation. In this chapter, we focus on the first problem, namely, detecting faults in manufacturing systems.

The type of manufacturing systems considered in this chapter are low-level manufacturing systems with discrete sensors and actuators. Such sensors and actuators have only a limited set of states, such as 'on' and 'off', 'high' and 'low', etc. These sensors and actuators are very common. Programmable logic controllers (PLCs) are in widespread use, and are designed especially for using such discrete signals [2].

An example of the type of such a manufacturing system we consider is the automated bottling line shown in Fig. 16.1. As bottles enter the conveyor, they pass a proximity switch which generates an event $PS \uparrow$. A limit switch is tripped ($LS \uparrow$) when the bottle enters the fill station. A blocking device is activated ($BD \uparrow$) to stop the bottle. The fill valve is then opened ($FV \uparrow$). When the fluid in the bottle reaches a given level, a

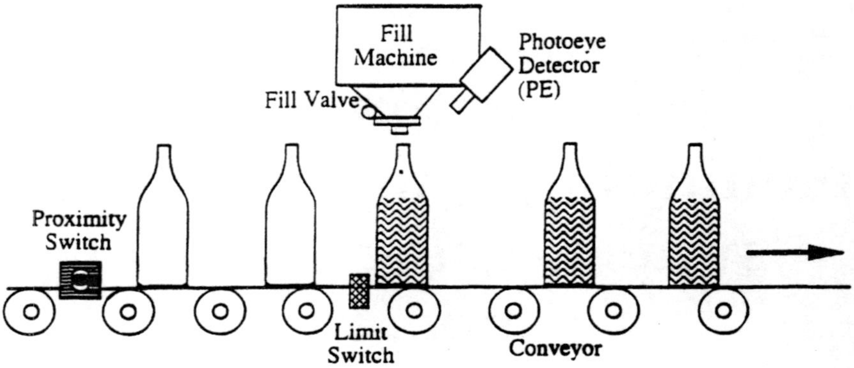

Fig. 16.1 An automated bottling line.

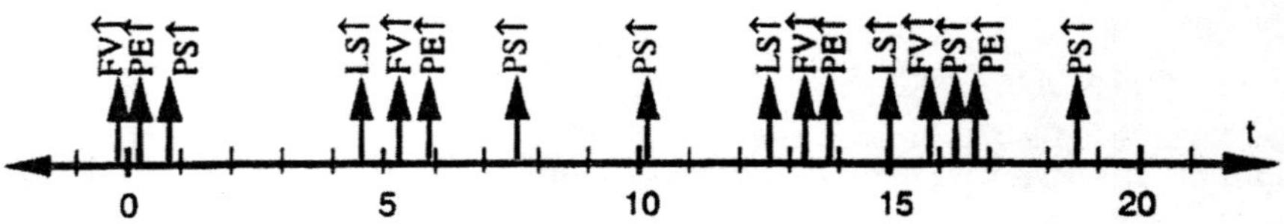

Fig. 16.2 Example observation trajectory from the bottling line with multiple bottles (selected events only).

photoeye detector is triggered ($PE \uparrow$) and fill valve is closed ($FV \downarrow$). The blocking device is released and the bottle is released from the station.

A simplified example of a timed observation for the bottling line is shown in Fig. 16.2 (some events have been omitted for ease of illustration). A fault monitoring system must analyze this observation information to determine if the system is operating correctly. Several characteristics of this example should be noted. First, information regarding the system operation must be determined from the sequencing and timing of events. Thus an initial system state may not always be known. Secondly, if multiple bottles are permitted on the line simultaneously, then events which are sequential or close in time may not be related due to the multiple entities in the system that are producing the events. For example, a bottle triggering the limit switch $LS \uparrow$ at a given time may not trigger the proximity switch until much later, after several other unrelated events $PS \uparrow$ may have occurred in the meantime due to the entry of other bottles on the conveyor.

We define a *fault* to be a deviation of a system from its specified acceptable operations. The purpose of a *fault detection* method is to examine available on-line observations from the system and determine if the observations are consistent with a specification of acceptable operations.

In the manufacturing systems we consider, this means the fault detection system must look at the sensing and actuation signals available to the controller.

Three important points should be made about fault detection as defined above:

- Fault detection is limited to faults which are apparent in the observations. If a fault does not in any way change current or future sensing signals, then it cannot be detected by any method that relies only on those signals.
- Fault detection uses on-line observations from the operating manufacturing system, and does not intervene in the system operation for the purpose of detection. This is in contrast to fault testing techniques, where the system control inputs are used to probe the system and confirm correct operation.
- Fault detection requires a specification of correct versus incorrect system operation. This specification defines what is to be considered a fault.

In the subsection below, we present a framework for comparing fault detection techniques and overview several techniques for fault monitoring for manufacturing systems using discrete signals. In section 16.2, we more closely examine the template monitoring approach, and in section 16.3, we present an example of the method. Section 16.4 concludes the chapter and discusses several extensions to the technique.

16.1.1 Fault detection methods: comparison and contrast

This section provides an overview of the various approaches of fault detection as applied to manufacturing systems and suggests methods for comparing and contrasting them. We explicitly separate the tasks of detection and diagnosis and focus exclusively on fault detection. The distinction between detection and diagnosis is not always made clear in the literature because these tasks can be closely intertwined. In general, however, *fault detection* is the recognition of an unacceptable behavior, and *fault diagnosis* is the identification of a component or set of components in the system which caused the fault.

The task of fault detection is basically that of taking observations of the system, filtering these observations to extract features of interest, and then comparing these features to a set of criteria distinguishing correct or incorrect operation. Based on the result of the comparison, the observation is classified as a fault (of possibly varying degrees) or not a fault.

This paradigm of detection suggests several methods of comparison of techniques. First, detection techniques differ in the system information they presume are available. In this chapter, we consider only techniques which use discrete sensing and actuation signals. Detecting and diagnosing faults through a discrete interface offers many challenges. The observations provide indications of system operation only at certain points in time. For the example of Fig. 16.1, the location of a bottle is sensed only when the bottle is in front of the proximity sensor or the limit switch. Because these indications of system operation are infrequent, there may be delays in being able to detect a fault. Furthermore, because of the sequential nature of many systems with these interfaces, simple sensor checks provide very little information about correct or incorrect operation. Rather, it is more often the context of the sensor reading in terms of past observations which determines whether the current observation represents correct operation.

Detection techniques also differ in how the distinguishing criteria between satisfactory and unsatisfactory operations are defined. Correct or incorrect operation may be specified through a model of the system, or through a set of patterns that the system observations must satisfy. Many methods use a model to track an estimate of the system state, declaring a fault if the observations become inconsistent with the state estimate.

Finally, detection techniques may be classified in terms of the features of the observed signals they consider. One of the simplest forms of monitoring manufacturing systems is verifying instantaneous conditions among sensor signals. For the bottling line, an example of this is the condition that whenever the limit switch (LS) does not indicate the presence of a bottle, the fill valve (FV) should not be in state 'open'. Such condition checking is commonly done in PLC programs. Although such condition evaluation is simple to do, it is very limited in the types of behaviors that can be confirmed.

Simple condition evaluation is rarely adequate for monitoring systems with discrete observations. Many characteristics of correct operation depend on timing between events. Monitoring of timing relationships between signal changes can be easily implemented on most PLCs, and so timing relationships are often tested within the control code. In PLC programs, timers to verify these relationships are commonly called *watchdog timers*. Often, such timers are used only to confirm that the time between two events did not exceed some maximum, rather than also confirming that the time between events was not less than some minimum.

In many systems, events are expected to follow a specific sequence. To be able to confirm such a sequence for fault monitoring, it is necessary for the fault monitoring method to maintain some form of 'state', to indicate what the last event heard was and what the next expected event in the sequence is. For this reason, fault monitoring methods that use sequencing features to distinguish between correct and faulty behavior typically are based on a model of the system. One method of fault detection in manufacturing systems is based on *state followers*. A state follower uses a finite state machine or Petri net or similar discrete state model to track the sequence of observations and then alarms upon detecting a deviation from the sequence (see [3, 4] for introductory surveys of Petri nets). A state follower is implemented on a PLC in [5].

It should be mentioned that fault monitoring methods may not be able to track the state of the system as defined in the state follower. One reason is that the discrete sensing does not allow the monitoring system to know the location of all entities in the system, since some state changes in the system (successful movement of parts, etc.) may not be immediately observed through the sensing. In such cases, tracking the state in the state follower may not be straightforward. The state may be uncertain, and an observer scheme must be used to try to maintain the state estimate [6, 7]. Such a state observer may become quite complicated. Similar methods must be used when the system is started up without knowledge of an initial state, or to try to determine the state after an unexpected (and thus faulty) event.

An obvious extension of the above methods is to use both timing and sequencing information for characterizing the correct operation of the system. A state follower system can be extended to contain timing information within the model. As above, such state following methods may have difficulty with determining the state after start-up under unknown conditions, or trying to determine the state following a fault. Dersin and Florine describe a state follower based on state automaton for use in high-speed printing, packaging, and welding applications [8]. Viswanadham and Johnson [9] consider timed Petri net models for state following, and Srinivasan and Jafari [10] use a back-tracking in timed Petri nets for confirming system behavior. Valette, Cardoso, and Dubois consider state following using timed Petri net representations with uncertain markings [6, 7]. The *template monitoring method* [11, 12] by Holloway and Chand uses sets of interrelated simple timing and sequencing relationships to monitor the system, rather than explicitly maintaining a state estimate. This technique is described in more detail in section 16.2.

Another feature one may use for fault detection is the statistical features of the timing or sequencing. If it is known that the timing between two events has a given probability distribution, then if the timing deviates from this distribution a fault can be alarmed. Fault detection using stochastic features typically use techniques similar to statistical process control. With statistical process control, control limits are established which represent the maximum deviation of sample averages from the expected mean. If the average of a set of times is beyond these limits, then the process is considered to be 'out of control', and a fault is alarmed, indicating that the process needs to be examined more closely. Statistical process control techniques have found widespread use in industry for measuring quality attributes of a product, as well as for measuring process variables to verify consistent equipment operation. Statistical process control can be very effective for detecting subtle drifts of the process operation, unlike the previously discussed techniques. However, if SPC techniques are used by themselves, they may not be as prompt at detecting anomolous timings because of their reliance on sample averages. In processes where immediate shutdown may be desired after a fault, one of the other techniques may be necessary to supplement SPC techniques.

16.2 METHODOLOGY

As noted above, fault monitoring with sequencing and timing characteristics is often done with a model specifying the correct characteristics where the model is run in synchrony with the system to track the state and to determine which events are expected next. Such a state follower scheme works well if an initial state is known and all events occur as expected. However, state following is more difficult if an initial state is not known or if the state estimate is corrupted following a fault (see [13] for one method). State following also does not work well if there are multiple concurrent sequences that must be tracked, since an event from one sequence may confuse the state follower for another sequence.

The template monitoring method was developed to address these problems. The method uses simple sequencing and timing relationships, called templates, to specify correct operation of the system. These relationships are confirmed against on-line observations in order to detect when a fault occurs. The templates can be generated from state-machine representations of manufacturing systems for monitoring of either single-instance processes or multiple-instance processes (e.g. it can be used to monitor for the case when only one bottle may travel the conveyor line, or the case when there may be multiple bottles simultaneously on the conveyor line, each generating events) [11, 14]. Timing relationships between selected events can also be learned from past observations such that the potential for false alarms and missed detections is minimized [15]. The method is simple enough that it can be implemented with very low processor requirements.

The remainder of this section will describe the template monitoring method, as well as techniques for generating the templates from state machine process models for both single-instance monitoring and multiple-instance monitoring.

16.2.1 On-line monitoring with templates

The template fault monitoring method was introduced by Holloway and Chand in [11, 12], but was generalized as *condition templates* by Pandalai in [14]. Readers are referred to those sources for a more analytical presentation.

The template monitoring method uses a set of templates (defined below) to define characteristics of observations from a correctly operating system. These templates are used online to create a set of *expectations* regarding future event occurrences. Each expectation is of the form (t, e_{trig}, C), where t is a time, e is an event (the *trigger event*), and C is a set of *consequences*, each of which is a pair of the form (e, [$\tau_{\min}$, τ_{max}]), where e is also an event and $\tau_{\min}$ and τ_{max} are delay times. An expectation (t, e_{trig}, C) means that

1. The event e_{trig} occurred at time t,
2. We expect that for some (e, [$\tau_{\min}$, τ_{max}]) in C, event e will occur at some time between $t+\tau_{\min}$ and $t+\tau_{\max}$. (The delays may be positive or negative, as is seen in section 16.2.2.)

An expectation is illustrated in Fig. 16.3, where e_1 is the trigger event that occurred at time 10. The consequences (e_2, [3, 4]) and (e_3, [10, 12]) each define a time window over which events e_2 or e_3 are expected to occur, respectively.

The template monitoring method is a method of creating and removing expectations according to on-line observations. The method is presented in Fig. 16.4. Let $E(t)$ represent the set of expectations maintained at time t, where $E(t)$ is a function of time because of the removal and addition of expectations over time. Expectations are removed when they are either satisfied or violated. An expectation is *satisfied* if the event of at least one of its consequences occurred within its time window. If none of its consequences is satisfied, and if the time windows of each consequence lies within the time over which we have observed the system (i.e. does not require knowledge about future events not yet observed or events in the past prior to our observations), then the expectation is said to be *violated*. This will indicate a fault detection, as shown in line 12. As given in line 14, expectations will continue to be in the set $E(t)$ over time until they become violated or satisfied.

Templates define what expectations are created in response to events. A template is of the form (e_{trig}, P), where P is a set of pairs (S,

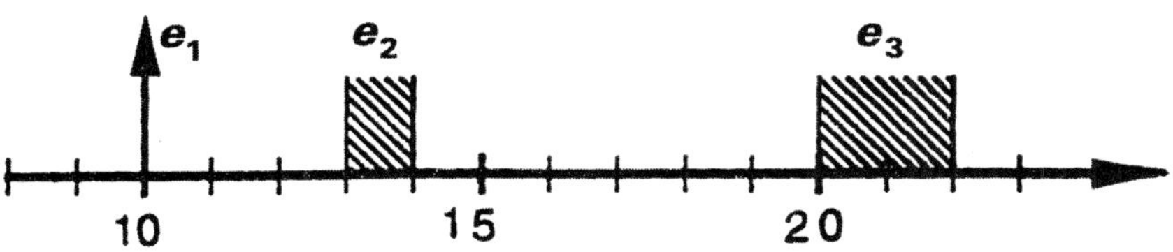

Fig. 16.3 An expectation (10, e_1, {(e_2, [3,4]), (e_2, [10, 12])}).

```
1   Given time t, template set M, and previous expectation set E(t⁻) {
2   For each event e received at time t {
3       For each template (e, P) ∈ M with matching trigger event {
4           For each (S, C) in P with S satisfied by E(t⁻)
5               Add the expectation (t, e, C) to the set E(t)
6           If no (S, C) in P has S satisfied by E(t⁻)
7               Declare Fault
8       }
9   }
10  For each expectation (t', e', C') ∈ E(t⁻) {
11      If (t', e', C') is violated for observations up to time t
12          Declare Fault
13      Else if (t', e', C') not satisfied at time t, add (t', e', C') to E(t)
14  }
```

Fig. 16.4 Procedure to update monitor for time *t*.

C), where *S* is a set of expectation conditions and *C* is a consequence set. An expectation condition is a pair (*e*, *C*) which is said to be satisfied at time *t* if the expectation set $E(t^-)$ (where t^- indicates the time just before time *t*) contains some expectation with trigger event *e* and consequence set *C*. As shown in lines 3–5, if an event *e* occurs at current time *t*, for each template (e_{trig}, *P*) with corresponding $e_{trig} = e$, for each (*S*, *C*) in *P* with all conditions in *S* satisfied by $E(t^-)$, an expectation (*t*, e_{trig}, *C*) is created and added to *E*(*t*). If no expectation condition *S* is satisfied for the template, then a fault is declared (lines 6–7). If *S* is the empty set, then all of its conditions are trivially satisfied, resulting always in an expectation with its corresponding consequence set.

As noted earlier, the time delay interval $[\tau_{min}, \tau_{max}]$ associated with a consequence may represent a positive or negative delay. A template with all positive delays among its consequences is referred to as a *forward template*. A forward template thus indicates a set of events, one of which should occur (following its given time delay) in the future after the occurrence of the triggering event. A template with all negative delays among its consequences is referred to as a *backward template*. The consequences in a backward template indicate a set of events, one of which should have occurred (within its given negative time delay) prior to the triggering event. Forward templates thus establish predictions following the trigger event, whereas backward templates indicate an expected context in which the trigger event occurs. Note that a template can have both negative and positive consequence delays, and thus be neither a forward nor a backward template.

16.2.2 Template models for multi-instance monitoring

The timing and sequencing relationships in a manufacturing system can be represented in many ways. In this section we consider systems in which the event timing and sequencing relationships are represented by a state machine model, but where the observed event occurrences correspond to the concurrent operation of some unknown number of these state machines. An example of this is the bottling line. The travel and filling of each bottle generates a timed sequence of events that can be represented by a state machine. Since the bottling line may have multiple bottles on it at the same time, the observations from the system will correspond to the concurrent interleaving of several traces from the state machine. We refer to this situation as the multi-instance monitoring problem. In this section, we summarize how a timed finite state machine specification can be converted into a template specification or multi-instance monitoring. Details of the method are given in [11, 12].

We consider a timed finite state machine to be an automaton where each state q has an associated delay time interval $d(q)$. When the state machine enters a given state q at a time t, it will remain in that state for at least t_{min} time units and for at most t_{max} time units, where $d(q) = [t_{min}, t_{max}]$. Figure 16.5 shows a simple example of a timed finite state machine, where the graph nodes are states, and the nearby intervals represent the time delay interval $d(q)$ of the state. We will assume that no delay interval is zero, and that for each state and any given event label, there exists at most one incoming arc with that event label to that state, and at most one outgoing arc with that event label from that state. For a given state q, we let in(q) denote the set of events associated with arcs leading into state q, and out(q) as the set of events associated with arcs leaving state q.

The generation of templates from a timed finite state machine model is done in two parts, corresponding to generating a set of forward templates and then a set of backward templates. First, a set of forward templates $M_f(G)$ is generated from the timed state machine model G. Then, a reversed timed finite state machine G^{rev} is constructed from G by

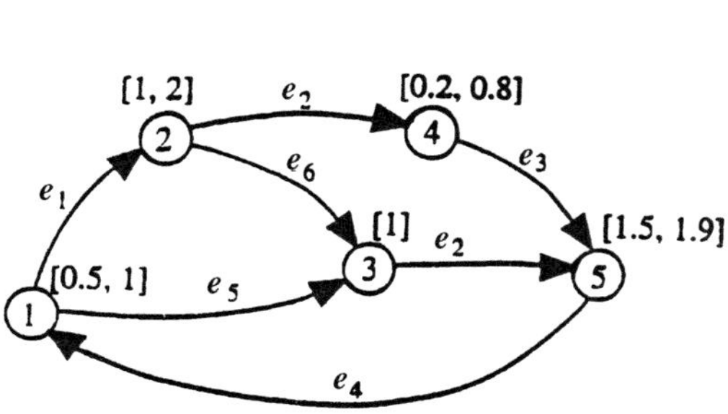

Fig. 16.5 An example of a timed finite state machine.

reversing all arcs and negating all state delay times. The set of forward templates from this reversed state machine then becomes the set of backward templates for G, i.e. $M_b(G) := M_f(G^{rev})$. The technique described in this section will only generate empty expectation conditions, i.e. $S = \theta$, which will thus always be evaluated as true when monitoring.

The set of forward templates is generated in two parts. For each event e in the state machine, a template is generated with a set of consequences defined as follows: for each state q for which e is an input, and for each output event of that state, there will be a consequence in the template corresponding to that output event with a consequence time delay of $d(q)$. This is a rather obvious set of consequences, indicating that whenever an event occurs, that identifies a set of possible states that the system must be in, so we expect the system to generate a future event corresponding to leaving one of those states.

If events occur in multiple places in the state machine, then hearing an event alone is not enough to keep track of the state of the state machine that the system is in. It is necessary to establish templates that extend around the shared event (or a shared sequence of events) in order to maintain context to track which part of the state machine the system is in. This is done with the notion of *exit consequences*. Given any state sequence σ representing a path in the state machine, define $C_{exit}(\sigma)$ as the set of all events and time delays within the model which represent a deviation or extension of that path in the automata. In other words, a consequence in $C_{exit}(\sigma)$ represents an event which leaves the state sequence or goes beyond the state sequence σ, and the consequence time is the sum of the time delays in the sequence before the consequence event occurs. To illustrate the concept of exit consequences, consider the state sequence $\sigma = (q_1q_2q_4q_5)$ for the state machine of Fig. 16.5. There are three possible deviations or extensions of the sequence: (q_1q_3), $(q_1q_2q_3)$, and $(q_1q_2q_4q_5q_1)$. The set of exit consequences is then

$$C_{exit}(\sigma) = \{(e_5, [0.5, 1]), (e_6, [1.5, 3]), (e_4, [3.2, 5.7])\}.$$

Note that when a state sequence consists of only one state, then the exit consequences are the outgoing events for that state along with the state's delay. Thus, for $\sigma = (q_1)$, then $C_{exit}(\sigma) = \{(e_1, [0.5, 1]), (e_5, [0.5, 1])\}$.

The second part of generating forward templates $M_f(G)$ corresponds to repeated state sequences (the first part above can be considered as a special case of this in [11]). Consider any event sequence which could be generated by two or more state sequences in the state machine. For example, in the state machine of Fig. 16.5, the event sequence (e_2) could result from state sequence (q_2, q_4) or (q_3, q_5). For each such state sequence σ, and for each event e that leads into the state sequence σ (i.e. is an entry event to the first state in the sequence σ), a template (e, C) is created. The set C of consequences for this template consists of the union of the set $C_{exit}(\sigma)$ of exit consequences for σ and any set $C_{exit}(q)$ of exit consequences for any other state q for which e is an incoming event. The template thus will maintain context over a repeated event sequence; it is triggered by an event preceding the repeated sequence, and its consequence represents events that follow the repeated sequence. The set $M_f(G)$ of forward templates consists of all templates generated in this way. The set $M_f(G)$ will be finite as long as there are no infinite sequences of events which could be generated by two or more infinite state sequences. This condition is typically easily verified by ensuring that there are no two distinct cycles of G which generate identical event sequences, The set of such duplicated event sequences can efficiently be determined since a longer duplicated sequence must necessarily be an extension of a shorter duplicated sequence.

The generation of $M_f(G)$ can be illustrated using the state machine of Fig. 16.5 again. As already mentioned, the event sequence e_2 could result from state sequence (q_2q_4) or (q_3q_5). For $\sigma = (q_2q_4)$, this results in template $(e_1, \{(e_6, [1, 2]), (e_3, [1.2, 2.8])\})$. Note that we may also consider the empty event string ε. The state sequences corresponding to the empty event string are all of length one and correspond to each state in G. Thus, for state sequence (q_1), we have template $(e_4, \{(e_1, [0.5, 1]), (e_5, [0.5, 1])\})$. For state sequence (q_4), we have template $(e_2, \{(e_3, [0.2, 0.8]), (e_4, [1.5, 1.9])\})$. The consequence with event e_4 results from e_2 being an incoming event for state q_5 as well as q_4. It can be seen that the consequences of the templates derived from these single state sequences represent the simple relationships of subsequent expected events following a trigger event. The templates generated by considering longer state sequences are used to maintain context of future events when there may be intermediate events which could be ambiguous. These ambiguous intermediate events could be generated by several different state sequences in the model. The templates that represent relationships between the events preceding and the events following these ambiguous intermediate event sequences ensure that the state machine event language can be correctly monitored.

In [11], two results are given on the effectiveness of using a template monitor to determine if system observations are consistent with the state machine as a specification of model timing and sequencing. The first result says that if the observations correspond to some number of concurrently operating state machines G, then the template monitoring method using the templates generated from G will not declare a fault. This means that no false alarms will be declared by the template monitoring method if G is a correct specification of the system operation.

The second result states that if the finite state machine G has no repeated events in it (i.e. no two arcs have the same event label), then an observation is declared as a fault if and only if the observations do not correspond to the concurrent operation of some number of the state machines. If the state machine is an accurate specification of correct and incorrect behavior, then this implies that there are no false alarms and no missed observable faults.

16.2.3 Template models for single-instance monitoring

Sometimes correct operation of a system is best represented by a single event trace from a state machine, in contrast to the multiple interleaved traces considered in the previous section. For example, maybe the bottling line is only supposed to allow a single bottle on the conveyor at a time, so a monitoring method should in this case declare a fault when observations correspond to multiple bottles. Monitoring to confirm that observations correspond to the trace of a single instance of a state machine (i.e. no concurrent traces) is referred to as single-instance monitoring.

A method to determine a template model for single-instance monitoring of a specified finite state machine is given in [14]. The approach of the method is to produce a set of templates such that for any time t, the expectation set $E(t)$ will uniquely identify the state of the system at that time. In contrast to the template generation method for multi-instance monitoring, the method for the single-instance case assumes an initial state of the state machine to create the initial set of expectations, and also assumes that no two events from the state machine will occur simultaneously in the observations.

Due to space limitations, we only describe a very special case of the method presented in [14]: each event label is associated with exactly one arc in the state machine. In this case single instance monitoring is particularly straightforward since each event uniquely identifies which state the system is in. Consider any event e in the automata, and let q_1 be the state of the system which has e as an output event, and let q_2 be the state of the system which has e as an input. The event e will have two templates, each of the form $(e, \{(S, C)\})$, each with exactly one condition S and one consequence C. The first template is a forward template, with $S = \{(e, C_{exit}(q_1)) \mid e \in in(q_1)\}$ and $C = C_{exit}(q_2)$. This template says that there must be an expectation corresponding to the previous state q_1 before hearing event e and creating an expectation corresponding to q_2. The second template associated with e is a backward template, with $S = \{(e, C_{exit}(q_1)) \mid e \in in(q_1)\}$ and $C = \{(e, \tau) \mid e \in in(q_1), \tau = d(q_1)\}$. This backward template ensures that the event e followed the input event to q_1 with the appropriate delay.

Monitoring using such templates tracks the language of the state machine exactly. In other words, an observation trace is accepted by the monitoring method without declaring a fault if and only if there exists an execution of the state machine which generates such a timed trace (not considering event labels not used in the state machine). For any time t, the set $E(t)$ will always contain exactly one expectation. The trigger event of this expectation uniquely identifies the state at time t, since the event is an input to only one state. If an output event of the state does not occur before the end of the appropriate time period, then the expectation is violated and a fault will be declared (Fig. 16.4 line 13). If the output event occurs too soon, then its backward template will be violated and a fault will be declared (line 13 again). Finally, if an event occurs out of sequence, then it will not have a satisfied condition, leading to a fault declaration (Fig. 16.4 line 7).

In the next section, we illustrate single-instance monitoring and multi-instance monitoring, and show how the templates are determined from state machine models.

16.3 EXAMPLE

The operation of the bottling line described in section 16.1 is useful for illustrating features of the template monitoring method. Figure 16.6 shows two state machines used for specifying the correct operation of the bottling line. State machine G1 describes the sequence of events generated by a bottle moving along the line, as discussed in section 16.1. Since there may be several (or even no) bottles on the line, we will permit interleavings of sets of traces from G1, which means we will use the multi-instance monitoring method of section 16.2.2 for generating templates from G1. The templates for this state machine are given in Table 16.1.

Other state machines can be used to indicate timing relationships which may not be representable in the bottle state machine. For example, the state machine G2 in Fig. 16.1 indicates that the photoeye should reset

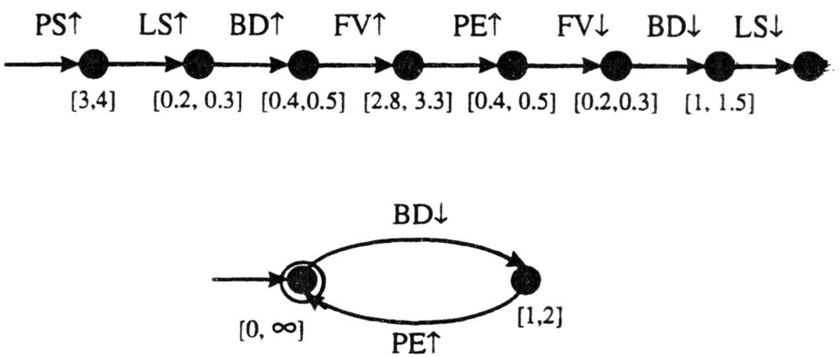

Fig. 16.6 Two state machine models for specifying the behavior of the bottling line.

Table 16.1 The set of templates derived from the state machine models

Forward templates	*Backward templates*
$(PS\uparrow, \{\varnothing, \{(LS\uparrow, [3,4])\})\}$	$(LS\uparrow, \{\varnothing, \{(PS\uparrow, [-4,-3])\})\}$
$(LS\uparrow, \{\varnothing, \{(BD\uparrow, [0.2,0.3])\})\}$	$(BD\uparrow, \{\varnothing, \{(LS\uparrow, [-0.3,-0.2])\})\}$
$(BD\uparrow, \{\varnothing, \{(FV\uparrow, [0.4,0.5])\})\}$	$(FV\uparrow, \{\varnothing, \{(BD\uparrow, [-0.5,-0.4])\})\}$
$(FV\uparrow, \{\varnothing, \{(PE\uparrow, [2.8,3.3])\})\}$	$(PE\uparrow, \{\varnothing, \{(FV\uparrow, [-3.3,-2.8])\})\}$
$(PE\uparrow, \{\varnothing, \{(FV\downarrow, [0.4,0.5])\})\}$	$(FV\downarrow, \{\varnothing, \{(PE\uparrow, [-0.4,0.5])\})\}$
$(FV\downarrow, \{\varnothing, \{(BD\downarrow, [0.2,0.3])\})\}$	$(BD\downarrow, \{\varnothing, \{(FV\downarrow, [-0.3,-0.2])\})\}$
$(BD\downarrow, \{\varnothing, \{(LS\downarrow, [1,1.5])\})\}$	$(LS\downarrow, \{\varnothing, \{(BD\uparrow, [-1.5,-1])\})\}$
$(BD\downarrow, \{(\{(PE\uparrow, \{(BD\downarrow, [0,\infty])\})\}, (PE\uparrow, [1,2])\})\})$	$(PE\uparrow, \{(\{(PE\uparrow, \{(BD\downarrow, [0,\infty])\})\}, (BD\downarrow, [-2,-1])\})\})$
$(PE\uparrow, \{(\{(BD\downarrow, \{(PE\uparrow, [1,2])\})\}, (BD\downarrow, [0,\infty])\})\})$	$(BD\downarrow, \{(\{(BD\downarrow, \{(PE\uparrow, [1,2])\})\}, (PE\uparrow, [-\infty,0])\})\})$

(*PE* ↓) after the bottle leaves the station (*BD* ↓). (The double circle in the diagram indicates the specified initial state.) The additional state machine allows us to indicate that both the *PE* ↓ and the *LS* ↓ (shown in G1) are related to signal *BD* ↓, but may not be sequentially related to each other. Since there will only be one bottle in the filling station at a time, it is not appropriate to have interleavings of multiple traces from the state machine, so we monitor with templates determined from the single-instance method from section 16.2.3.

Note that the templates from both state machines can be unioned together to define the template set for the combined system. This is a feature of the template monitoring method, that the method can simultaneously monitor multiple state machine models, some of which are used for multi-instance and some of which are used for single-instance monitoring.

16.4 CONCLUSIONS AND EXTENSIONS

This chapter has overviewed the template monitoring method as a method for monitoring manufacturing systems with discrete signals. The method uses a template model to specify expected sequencing and timing relationships between signal changes (events) in control system I/O, and then confirms these relationships on-line as the signal changes occur.

One advantage of template models is that they can model event sequences corresponding to a finite state machine (single-instance monitoring), or corresponding to interleavings of unknown but finite numbers of sequences from such a state machine (multi-instance monitoring). The multi-instance monitoring is especially useful when modeling systems with conveyors, where the number of products on the conveyor may vary. This multi-instance monitoring operates even under unknown initial state. The single-instance monitoring is useful for modeling single devices, where interleavings of multiple traces would not represent correct operation. Note that the templates for multi-instance monitoring of one system and for single instance monitoring of other systems can be monitored together as shown in section 16.3.

Template monitoring has been extended in several ways. *Time-scaled templates* were introduced in [16] for modeling systems where interevent timings may change depending on other events. For example, changes in the conveyor speed would result in changes in the time delays for a product to travel along the conveyor. Das presents a method for determining templates, such that the time delays between particular events are selected to minimize the weighted cost of false alarms and missed detections [15]. A method of using single-instance templates to automatically generate executable PLC ladder-logic code for monitoring is given in [14]. (The implementation considers only single-instance monitoring because of limitations of the ladder-logic language.) The multi-instance monitoring method has been implemented in C++ on a Sun Sparcstation5.

ACKNOWLEDGEMENTS

This work is supported in part by NASA grant NGT-40049, Rockwell International, NSF grant ECS-9308737, and the Center for Robotics and Manufacturing Systems at the University of Kentucky.

REFERENCES

[1] Ball, K.E. (1989) Diagnostics: The frontal assault on downtime. *Programmable Controls*, April, 32–4.

[2] Pessen, D.W. (1989) *Industrial Automation: Circuit Design and Components*. John Wiley and Sons, New York.

[3] Murata, T. (1989) Petri nets: Properties, analysis and applications. *Proceedings of the IEEE*, **77**(4), 541–80.

[4] David, R. and Alla, H. (1994) Petri nets for modeling of dynamic systems – a survey. *Automatica*, **30**(2), 175–202.

[5] Wyss, J.G. Peer processing diagnostics for plant automation. *In Proceedings of IPC'91*, Detroit, Michigan, April, 163–170.

[6] Valette, R., Cardoso, J. and Dubois, D. (1989) Monitoring manufacturing systems by means of Petri nets with imprecise markings. In *Proceedings of IEEE Conference on Intelligent Control*, Albany, New York, September, 233–8.

[7] Cardoso, J., Valette, R. and Dubois, D. (1989) Petri nets with uncertain markings. In *10th International Conference on Application and Theory of Petri nets*, Bonn, Germany, June.

[8] Dersin, P. and Florine, J. (1986) Firmware transitional logic for on-line monitoring and control. In *Proceedings of IEEE Conference on Decision and Control*, Athens, Greece, December, 1102–7.

[9] Viswanadham, N. and Johnson, T.L. (1990) Fault diagnosis of automated manufacturing systems. In K.F. Martin, J.H. Williams and D.T. Pham (eds), *Intelligent Diagnostic Systems*. IFS/Springer-Verlag, Berlin, AI in Industry Series.

[10] Srinivasan, V.S. and Jafari, M.A. (1993) Fault detection/monitoring using time Petri nets. *Systems, Man, and Cybernetics*, **23**(4), July/August.

[11] Holloway, L.E. and Chand, S. (1994) Time templates for discrete event fault monitoring in manufacturing systems. In *Proceedings of the 1994 American Control Conference*, Baltimore, June, 701–6.

[12] Holloway, L.E. and Chand, S. (1996) Distributed fault monitoring in manufacturing systems using concurrent discrete-event observations. *Integrated Computer-Aided Engineering*, **3**(4), October, Special Issue on Detecting and Resolving Errors in Manufacturing Systems.

[13] Ozveren, C.M. and Willsky, A.S. (1990) Observability of discrete event dynamic systems. *IEEE Transactions on Automatic Control*, **35**(7), 797–806.

[14] Pandalai, D. (1996) *Condition templates: Improved template model for fault monitoring in discrete event manufacturing systems*. Master's thesis, Dept of Electrical Engineering, University of Kentucky, August.

[15] Das, S.R. and Holloway, L.E. (1996) Characterizing discrete event timing relationships for fault monitoring of manufacturing systems. In *Proceedings of 1996 IEEE International Conference on Control Applications*, Dearborn, Michigan, September.

[16] Holloway, L.E. (1996) On-line fault monitoring of a class of hybrid systems using templates with dynamic time scaling. In *Proceedings of DIMACS Workshop on Verification and Control of Hybrid Systems*, Lecture Notes in Computer Science, Springer-Verlag, to appear.

CHAPTER 17

Case Example 6: In-process diagnosis of tool failures in milling

Kazuo Mori

17.1 INTRODUCTION

One of the most formidable challenges in intelligent automation and numerical control of machining processes is the presence of phenomena that cannot be modeled *a priori*, but which leads to unpredictable changes in machining conditions, and usually deterioration in the product quality and the productivity. In milling operations, due to the absence of a practical model to efficiently characterize undesirable process conditions, such as tool failures, we must rely heavily on sensor-based in-process techniques for the diagnosis of current machining conditions.

In this chapter, two different monitoring methods for in-process diagnosis of tool failures for milling are introduced: the application of the Wavelet Transform as an indirect monitoring method and a tool geometry measurement system using a laser sensor as a direct monitoring method.

17.2 METHODOLOGY

17.2.1 Application of the Wavelet Transform as an indirect monitoring method

During real time machining processes, when a cutting tool or a workpiece rotates at high speed, chips and coolant drops are scattered near the machining point, and direct observation of machining processes is almost impossible. In detecting tool failures, for instance, we must usually rely on the indirect monitoring methods based on the analysis of signals generated by the interaction between the tool and workpiece. Such techniques include the analysis of the vibrations on different parts of the machine tool, acoustic noise generated by stressed materials of the workpiece and tool, cutting sound, and cutting forces. An impressive array of tool sensing systems have been developed, most of which are described in Cook (1980), Dan and Matthew (1990), Micheletti *et al.* (1976), and Tlusty and Andrews (1983). For the indirect monitoring methods, signal processing techniques are usually required to find the correlation between the tool geometry condition and the sensory signals. The following section describes the technique of the application of the Wavelet Transform for signal analysis to detect tool failures in milling operations.

Theoretical background of the Wavelet Transform

The Wavelet Transform (WT) is a time-frequency decomposition of a signal onto a set of 'wavelet' basis functions. The basic formula of the Continuous Wavelet Transform (CWT) of a time function $f(t)$ is given by

$$T(a,b) = \int_{-\infty}^{\infty} f(t)\ \psi_{(a,b)}(t)\ dt \qquad (17.1)$$

where * denotes complex conjugation and

$$\psi_{(a,b)}(t) = \frac{1}{\sqrt{a}}\ \psi\left(\frac{t-b}{a}\right),\ (a,\ b \in R,\ a \neq 0) \qquad (17.2)$$

is a member of the wavelet basis, derived from the basic analysis wavelet $\psi(t)$ through translation and dilation.

As seen in Equation (17.1), the transformed signal $T(a, b)$ is defined on the a–b plane, where a and b are used to adjust the frequency and time location of the wavelet in Equation (17.2). A small a produces a high-frequency (contracted) wavelet when high time resolution is required, while a large a produces a low-frequency (dilated) wavelet when high frequency resolution is needed. The WT's superior time-localization properties stem from the finite support of the analysis wavelet: as b increases, the analysis wavelet traverses the length of the input signal, and a increases or decreases in response to changes in the signal's local time and frequency content. Finite support implies that the effect of each term in the wavelet representation is purely localized. This sets the WT apart from the Fourier transform, where the effects of adding higher frequency sine waves are spread throughout the frequency axis.

Discrete methods are required for computerized implementation of the WT. The Discrete Wavelet Transform (DWT) is derived from the CWT through discretization of the wavelet $\psi_{(a,b)}(t)$. The most common discretization of the wavelet is the dyadic discretization, given by

$$\psi_{(j,k)}(t) = \frac{1}{\sqrt{2^j}}\ \psi\left(\frac{t-2^j k}{2^j}\right) \qquad (17.3)$$

where a has been replaced by $2j$ and b by 2^jk (Daubechies 1988; Mallet 1989). Under suitable conditions Equation 17.3 is an orthonormal basis of $L^2 R$ (Daubechies 1988), and the original time function can be expressed as

$$f(t) = \sum_{j=-\infty}^{\infty} \sum_{k=-\infty}^{\infty} Cj,k\psi_{(J,k)}(t) \tag{17.4}$$

$$C_{j,k} = \int_{-\infty}^{\infty} f(t)\ \psi^*_{(j,k)}(t)\ dt \tag{17.5}$$

where $C_{j,k}$ are referred to as wavelet coefficients. A second set of basis functions $\theta(t)$ called scaling function is then obtained by applying multi-resolution approximations to obtain the orthonormal basis of $\psi(t)$:

$$\theta_{(j,k)}(t) = \frac{1}{\sqrt{2^j}}\ \theta\left(\frac{t-2^jk}{2^j}\right) \tag{17.6}$$

The original time function can now be written as

$$d_{j,k} = \int_{-\infty}^{\infty} f(t)\ \theta^*_{(j,k)}(t)\ dt \tag{17.7}$$

Here, the $d_{j,k}$, which are called the scaling coefficients, $d_{0,k}$ being the sampled version of $f(t)$, represent a jth order resolution discretization of f(t). The scaling coefficients and the wavelet coefficients for resolutions of order greater than j can be obtained iteratively by

$$d_{j+1,k} = \sum_{i=-\infty}^{\infty} h(i-2k)\, d_{j,k} \tag{17.8}$$

$$C_{j+1,k} = \sum_{i=-\infty}^{\infty} g(i-2k)\, d_{j,k} \tag{17.9}$$

The sequences h and g are low-pass and high-pass filters derived from the original analyzing wavelet $\psi(t)$. The scaling coefficients $d_{j,k}$ represent the lower frequency approximations of the original signal, and the wavelet coefficients $C_{j,k}$ represent the distribution of successively higher frequencies. The inverse DWT yields a difference series representation for the input signal $d_{0,k}$ in terms of the filters h and g and the wavelet coefficients $C_{j,k}$:

$$d_{j+k} = \sum_{i=-\infty}^{\infty} h(k-2i)\, d_{j+1,i} + \sum_{i=-\infty}^{\infty} g\,(k-2i) C_{j+1,i} \tag{17.10}$$

The wavelet filters adopted determine the quality of the wavelet analysis. Here the Daubechies system of wavelets (Daubechies, 1988) is used, because it has the advantages of orthogonality, compact support in the time domain and computational simplicity. The Daubechies wavelets of length 4 chosen by the authors are

$$h(0)=(1+\sqrt{3})/4\sqrt{2}, \quad h(1)=(3+\sqrt{3})/4\sqrt{2}, \quad h(2)=(3+\sqrt{3})/4\sqrt{2}, \quad h(0)=(1+\sqrt{3})/4\sqrt{2}, \quad g(n)=(-1)^n h(3-n) \tag{17.11}$$

Since the input signal $f(t)$ is discretized into N samples, Equations (17.8) and (17.9) can be written in the form of matrices

$$\begin{bmatrix} d_{1,1} \\ C_{1,1} \\ d_{1,2} \\ C_{1,2} \\ \cdot \\ \cdot \\ d_{1,N/2} \\ C_{1,N/2} \end{bmatrix} = T_N \begin{bmatrix} d_{0,1} \\ d_{0,2} \\ d_{0,3} \\ d_{0,4} \\ \cdot \\ \cdot \\ d_{0,N-1} \\ d_{0,N} \end{bmatrix} \tag{17.12}$$

where

$$T_N = \begin{bmatrix} h(0)h(1)h(2)h(3) & & & 0 \\ g(0)g(1)g(2)g(3) & & & \\ & h(0)h(1)h(2)h(3) & & \\ & g(0)g(1)g(2)g(3) & & \\ & & \cdots \quad \cdots & \\ & & \cdots \quad \cdots & \\ & 0 & & h(0)h(1)h(2)h(3) \\ & & & g(0)g(1)g(2)g(3) \\ h(2)h(3) & & & h(0)h(1) \\ g(2)g(3) & & & g(0)g(1) \end{bmatrix} \tag{17.13}$$

The scaling coefficients $d_{j+1,k}(k=1\sim N/2^{j+1})$ and the wavelet coefficients C_{j+1k} of the $(j+1)$th order resolution can be obtained by applying the $N/2^j \times N/2^j$ matrix $T_{N/2j}$ to the scaling coefficients of the jth order $d_{j,k}(k=1 \times N/2_j)$. When the number of data points is $N = 2^i$, all of the wavelet coefficients are obtained after $i-1$ iterations of Equation (17.12). The inverse DWT is performed in a similar manner by straightforward inversion of the orthogonal matrix T_N.

Tool failure detection using simulated cutting force signals

This section illustrates the use of the DWT in on-line detection of small tooth chipping and eccentricity of the tool center in face milling using simulated cutting force signals. Pattern recognition techniques are used to determine the tool state (normal, chipping, or eccentricity) from the DWT coefficients obtained from the force signal.

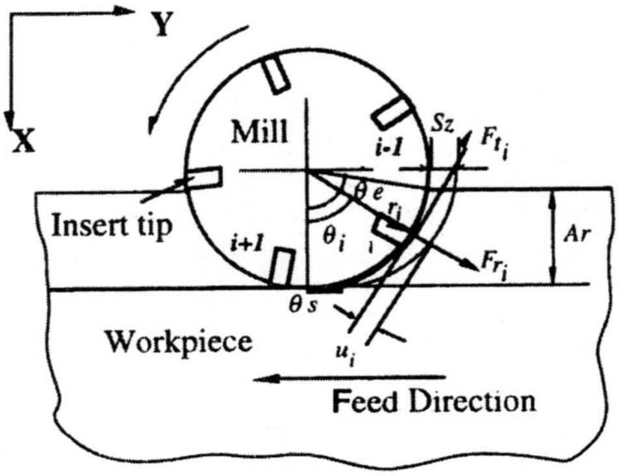

Fig. 17.1 Simulation model.

A computer is used to simulate the cutting force exerted by an M-tooth face milling cutter, based on the model shown in Fig. 17.1. The thickness of the cutting chip in the ith tooth is given by

$$u_i = \begin{cases} S_z \sin\theta_i + (r_i - r_{i-1}) + S_i \\ \quad : S_z \sin\theta_i + r_i > r_{i-1} \\ 0 \quad : S_z \sin\theta_i + r_i \leq r_{i-1} \end{cases} \tag{17.14}$$

where

$$s_i = \begin{cases} S_z \sin\theta_i + (r_{i-1} - r_{i-2}) \\ \quad : S_z \sin\theta_i + r_{i-1} > r_{i-2} \\ 0 \quad : S_z \sin\theta_i + r_{i-1} \leq r_{i-2} \end{cases} \tag{17.15}$$

where θ_i is the cutting edge rotational angle, S_z is the feed rate per tooth and r_i is the radius of the cutting edge. It is assumed that the cutting force is proportional to the volume of workpiece material removed by each tooth, such that the tangential and radial cutting forces F_{t_i}, F_{r_i} are related to the chip thickness u_i and the axial depth of cut k_w by

$$F_{t_i} = P_t k_w u_i \tag{17.6}$$

$$F_{r_i} = P_t k_w u_i \tag{17.7}$$

where the specific cutting forces P_t and P_r are given by the Technical Research Institute (1978):

$$P_t = 2.6*10^8 u_i^{-0.206} \tag{17.8}$$

$$P_r = 0.6 P_t \tag{17.9}$$

In the simulation model, tool failure occurs when one of the cutting teeth becomes chipped, leading to a diminished value of r_i. Tooth chipping can be recognized by the changes in the cutting force signal: there is a decrease in the force exerted by the damaged tooth and an increase in the force exerted by the next tooth in the cutting path. As shown in Fig. 17.2, the effects of eccentricity can be simulated using the same model. Let us begin by replacing the radius r_i of the ith cutting edge in Equations (17.14) and (17.15) with

$$r'_i = \sqrt{\varepsilon^2 + r_i^2 - 2r_i\varepsilon\cos(\theta_i - \alpha)} \tag{17.20}$$

where ε is the distance of eccentricity from the tool center and α is the angle of eccentricity. The edge angle θ_i depends upon the eccentricity, but because the change is on the order of ε / r_i, where $\varepsilon << r_i$ in practical cases, this dependence can be neglected. It follows that the $\underline{X}$ and $\underline{Y}$ components of the cutting force are

$$F_X = \sum_{i=1}^{M} \delta(\theta_i)(F_{ni}\cos\theta_i - F_{ti}\sin\theta_i) \tag{17.21}$$

$$F_Y = \sum_{i=1}^{M} \delta(\theta_i)(F_{ti}\cos\theta_i + F_{ni}\sin\theta_i) \tag{17.22}$$

where

$$\delta(\theta_i) = \begin{cases} 0 : \theta_s \leq \theta_i \leq \theta_e \\ 1 : \text{otherwise} \end{cases} \tag{17.23}$$

θ_s is the entry cutting angle and θ_e is the exit angle. The resultant cutting force F_r is then

$$F_r = (F_X^2 + F_Y^2)^{1/2} \tag{17.24}$$

We perform the simulation for two radial depths of cut $A_r = 25$ mm and $A_r = 50$ mm, and two cutting tools $M = 4$ and $M = 8$, where M is the number of inserts. For each of these four cases, four conditions are simulated: normal (no chipping or eccentricity), 0.05 mm chipping, a fully broken tooth and an eccentricity of $\varepsilon = 0.02$ mm. The tool diameter $d = 50$

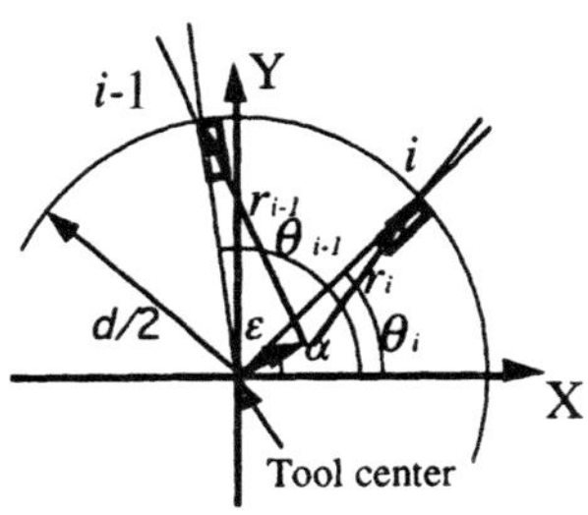

Fig. 17.2 Effect of the eccentricity.

mm, the feed rate $S_z = 0.2$ mm/tooth and the sampling rate is 256 points per revolution. Figs 17.3 and 17.4 show the simulated resultant force for $M = 4$ and $M = 8$, respectively. For $M = 4$, the damage is simulated in the second tooth and for $M = 8$, damage is simulated in the third tooth.

Figs 17.5 and 17.6 show the DWT coefficients for the $M = 4$ and $M = 8$ simulated cutting force signals, where the vertical axis represents coefficient magnitude and the horizontal axis represents time over an interval corresponding to a single insert. The coefficients are shown at resolutions $j = 5$ and 6, where the effects of the tool condition are most clearly evident. In the case of normal tool condition, the coefficients are approximately equal. Tooth damage is indicated by a 'zigzag' pattern which is observable during chipping but especially pronounced during breakage. Eccentricity is indicated by a continuously changing wave pattern which clearly distinguishes it from the normal and tooth failure conditions.

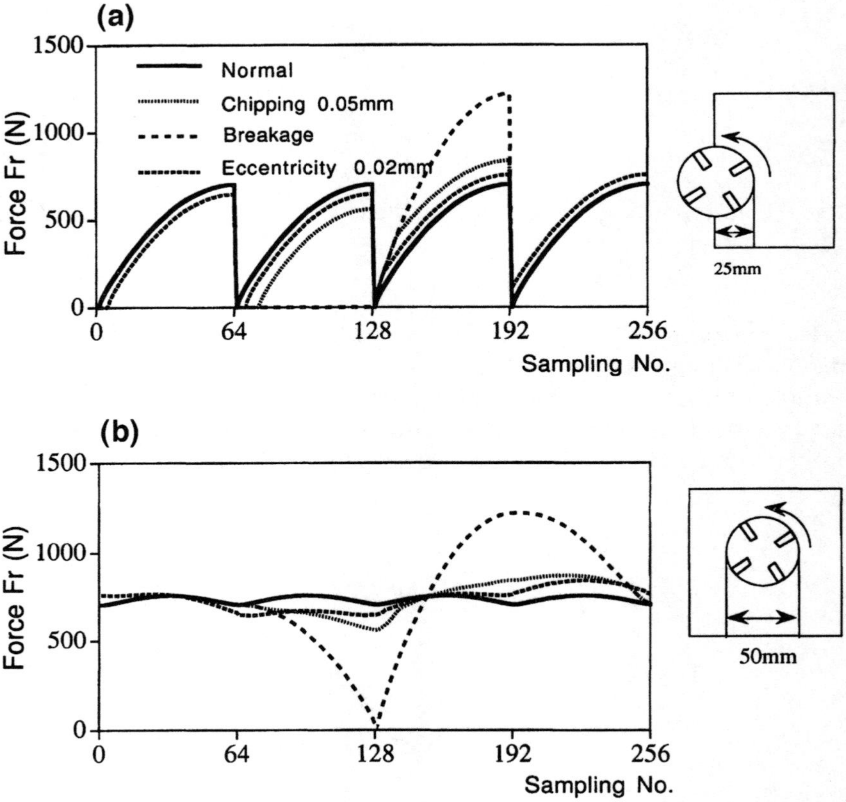

Fig. 17.3 Simulated cutting force of four inserts (M=4). Radial depth of cut: (a) 25 mm, (b) 50 mm.

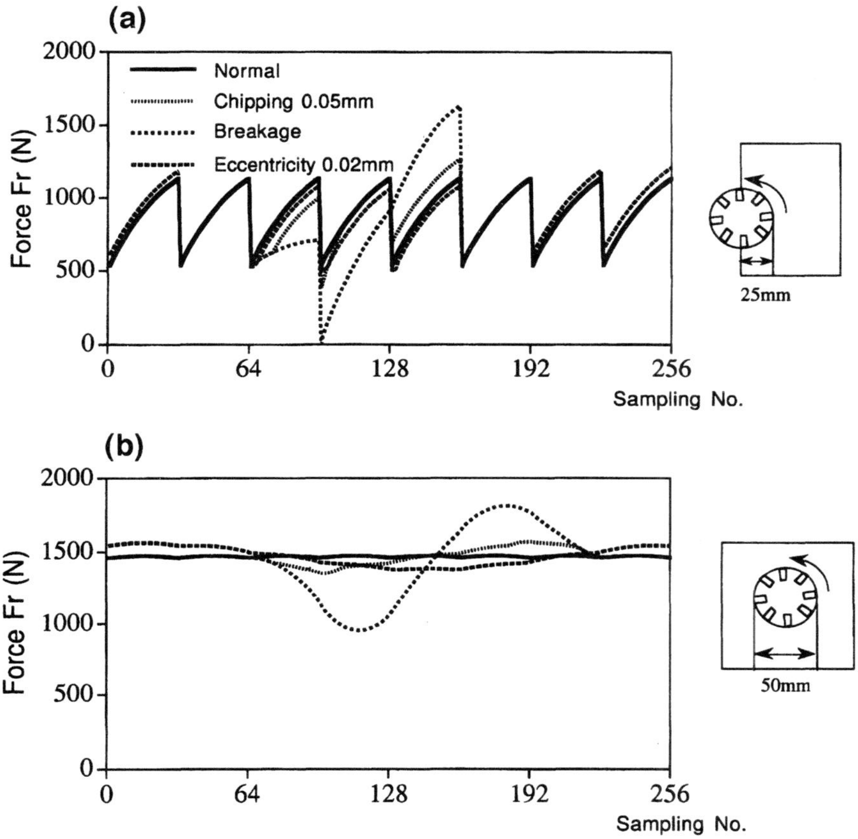

Fig. 17.4 Simulated cutting force of eight inserts (M=8). Radial depth of cut: (a) 25 mm, (b) 50 mm.

For purposes of comparison, the FFT spectrum of the simulated cutting force signals is shown in Fig. 17.7 for $M = 4$. Though small differences between the normal and chipping conditions can be observed in the low frequency region of the spectrum, the peak values and peak positions are nearly identical, making small chipping difficult to detect. The FFT is relatively poor at detecting tool failures because it spreads the effects of time-localized high frequency events throughout the length of the input signal (one revolution). This problem can be overcome by adjusting the sampling time width to equal the cutting time required for one tooth. Thus, for the DWT one transformation per revolution is required, whereas with the FFT the number of transformations is M, the number of inserts. Clearly, it is faster and more straightforward to use the DWT.

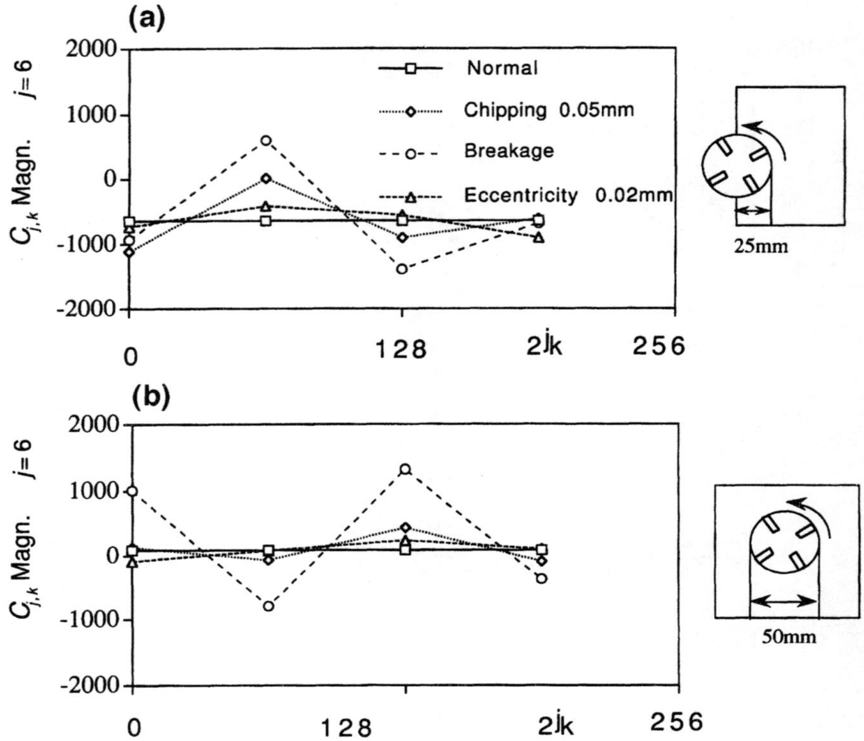

Fig. 17.5 Wavelot coefficient $C_{6,k}$ of simulated cutting force of four inserts (M=4). Radial depth of cut: (a) 25 mm, (b) 50 mm.

17.2.2 Application of a laser sensor as a direct monitoring method

The potential uses of a laser displacement meter in direct on-line measurement of milling tool condition is presented. A laser displacement meter measures simultaneously the distance from the laser source to a reflecting surface and intensity of reflected laser light. From the general point of view, we can reconstruct the tool cross-sectional geometry by focusing the laser beam on the tool shape and measuring variations of the distance and the intensity during tool rotation. The main drawbacks of the direct monitoring systems are the difficulty of collecting direct signals under actual cutting conditions and the limited accuracy of direct measurements, which makes the systems unsuitable for most industrial applications. This section presents an on-line tool geometry measurement system based on a laser displacement sensor, and the technique to obtain 3-D geometrical tool image to evaluate tool chipping and wear.

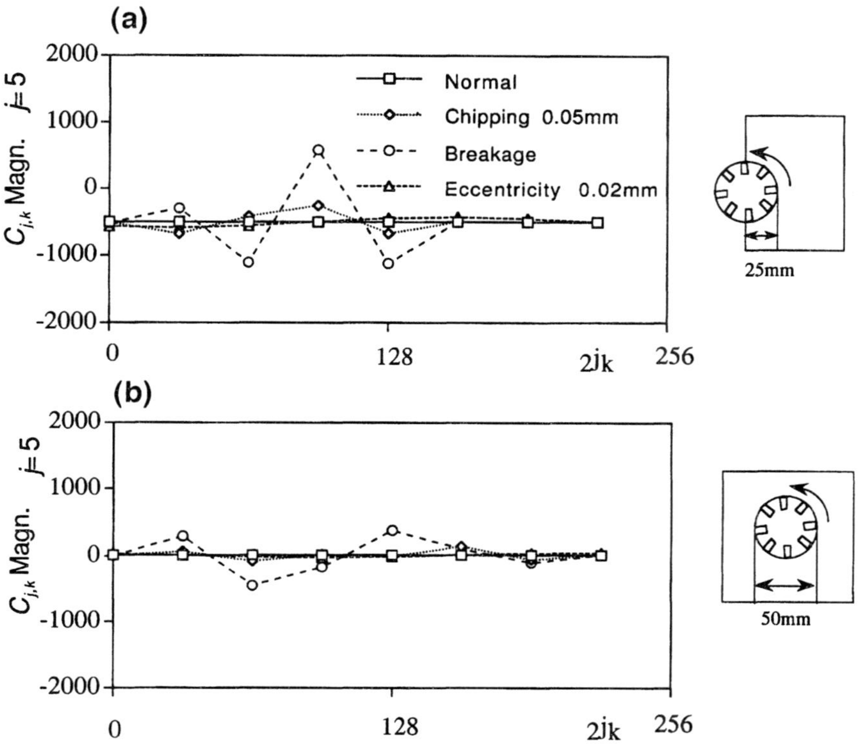

Fig. 17.6 Wavelot coefficient $C_{5,k}$ of simulated cutting force of eight inserts (M=8). Depth of cut: (a) 25 mm, (b) 50 mm.

Overview of in-process tool geometry measurement system

Fig. 17.8 shows the in-process tool geometry measurement system developed by the Mechanical Engineering Laboratory (Ryabov *et al.* 1996). It consists of a support and location mechanism mounted on the main spindle, two laser sensors used to scan the tool shape and synchronize the signals to the tool rotation, a servo motor that moves the main sensor along the tool's axis, laser and servo motor controllers, and a PC for the acquisition and processing of data. The position adjuster is used to immobilize the main laser head at a precise position in the XY-plane; the Z position of the main laser head is determined by the servo motor which controls the vertical motion of the support and location mechanism. The auxiliary laser synchronizes the main laser with the tool rotation by detecting a marker on the tool which denotes the rotational origin of the tool. The main laser is equipped with a protective acrylic cover, as shown in Fig. 17.8. The cover has a trapezoidal cross-section, and it has openings at the bottom and top. The walls of the cover contain

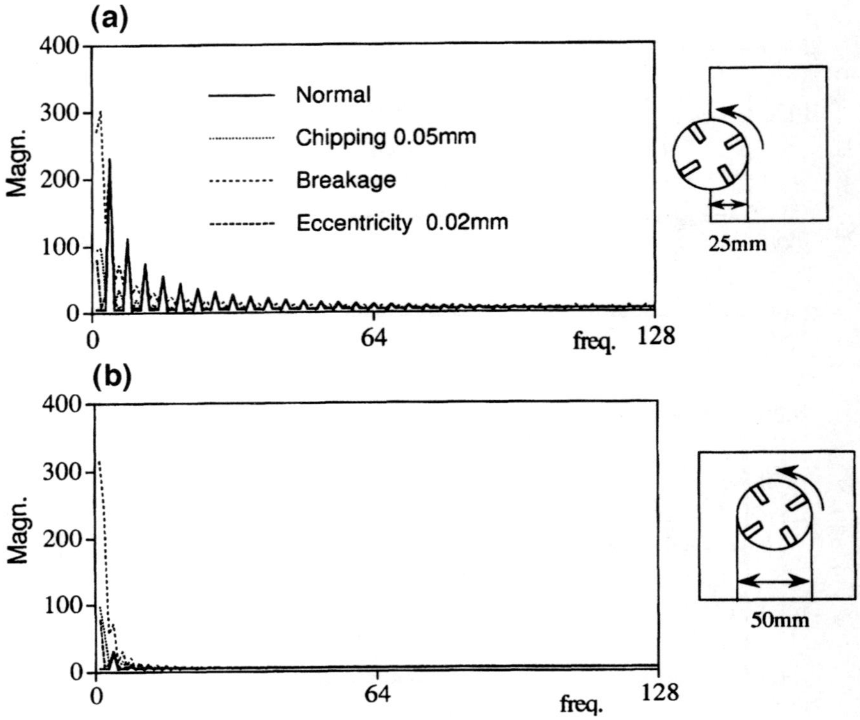

Fig. 17.7 FFT analysis of simulated cutting force of four inserts (M=4). Radical depth of cut: (a) 25 mm, (b) 50 mm.

double-walled nozzles which are also open at the tops. The main laser sensor sits under the cover and the laser beam passes through the top opening of the cover. Compressed air, which is blown into the nozzles and jetted through the top nozzle openings, prevents obstacles from getting inside the cover and hitting the sensor. The laser sensors are 670 μm semiconductor commercially available displacement meters which detect the angle and intensity of reflected light by means of photo detectors. The laser's spot diameter is 40 × 20 μm, its maximum distance resolution is 0.5 μm, and its maximum sampling rate is 50 kHz.

Hybrid measuring method to detect tool failure

The laser displacement sensor generates two simultaneous outputs – a distance measurement and a light intensity measurement – which the system processes in different ways to deal with normal and small tool

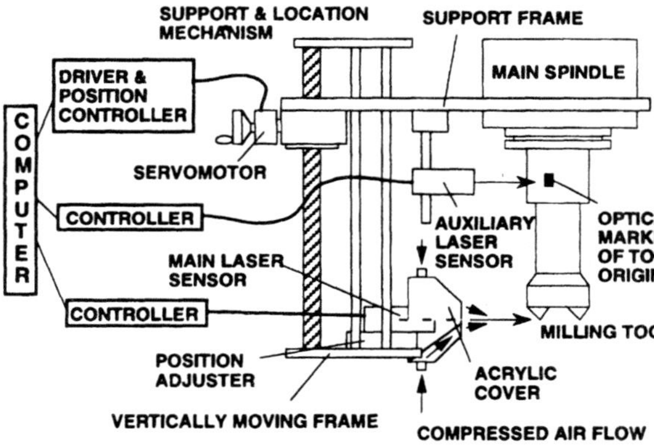

Fig. 17.8 System configuration.

wear. In the case of normal tool wear, the distance measurements are used to generate estimates of the tool geometry, which are sufficient for detection and measurement purposes. However, in the case of small tool wear, displacement output is not sufficient, and we use a hybrid approach involving tool geometry and the distribution of light intensity on the tool.

Fig. 17.9(a) shows an example of the laser sensor displacement output. As the tool rotates, the measurement system continuously measures the distance between the sensor and the tool. From the sensor output for a complete tool rotation, the system generates a cross-section of the tool shape at the Z level of the scanning laser beam. By using the support and location mechanism to change the Z level of the beam, the system is able to construct a 3-D image of the tool geometry.

The resolution of the tool geometry measurement depends upon many parameters, such as the tool's rotational speed, the laser scanning rate, the tool diameter, and the Z-intervals at which the tool's cross-sections are measured. Since any variations of the scanned shape geometry within the laser spot are unresolved by the sensor, its diameter (40 μm, in this system) limits the shape resolution along the tool circumference. The radial accuracy of the shape reconstruction depends on the sensor distance resolution and striking angle of the beam. In the system, it ranged from 5–10 μm.

Fig. 17.9(b) shows characteristic cross-sectional output for three types of tool failure: new tool, chipped tool, and worn tool (more than 400 μm of flank wear length). The figures illustrate the system's ability to detect chipping and normal flank wear, and to estimate the scale of the tool deterioration.

As described above, the system can measure medium and large-scale

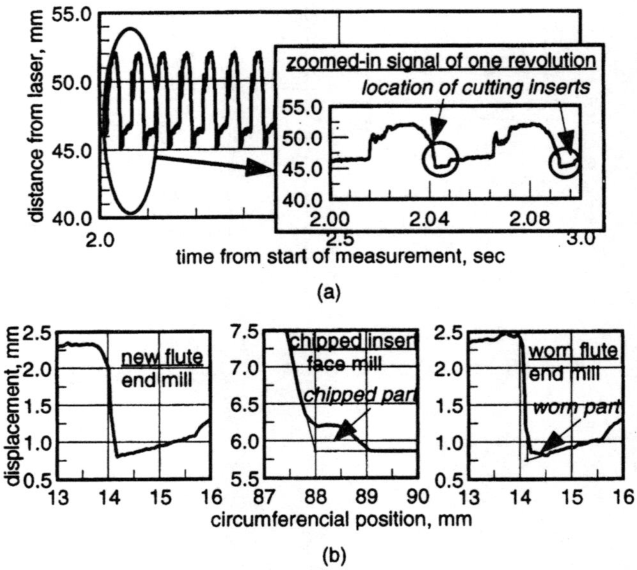

Fig. 17.9 Examples of displacement output under air-cutting conditions. (a) Signal output of face mill sensing with two inserts; (b) detailed geometric images of cutters with different failures.

tool wear using the distance measurements generated by the laser displacement sensor. However, since the depth of the flank wear is usually less than ten times its length, the geometric resolution of the system is not sufficient for small tool wear (flank wear less than 200 μm in length). To detect small tool failure, another approach based on the analysis of the distribution of the reflected light intensity along the tool flank was developed. The characteristics of a surface reflection depend on the surface's orientation relative to a falling light beam and its light dispersion properties. It is found experimentally that though normal and work flank surfaces do not differ significantly in their dispersion properties, they can be clearly distinguished through different light intensity patterns arising from differences in surface orientation. This is the main reason why the presence of tool wear is visible to the naked eye, though the geometry of the tool wear is impossible to discern

Fig. 17.10 shows how the simultaneous distance and intensity outputs signal the presence of tool wear. A spike in the intensity signal (Fig. 17.10(a)) indicates flank wear, even in cases where flank wear is not clearly evident in the displacement signal (Fig. 17.10(b)). But there are no intensity spikes at the cutting edge of an unworn tool (Fig. 17.10(c)). The length of the worn area can be determined by the width of the intensity spike.

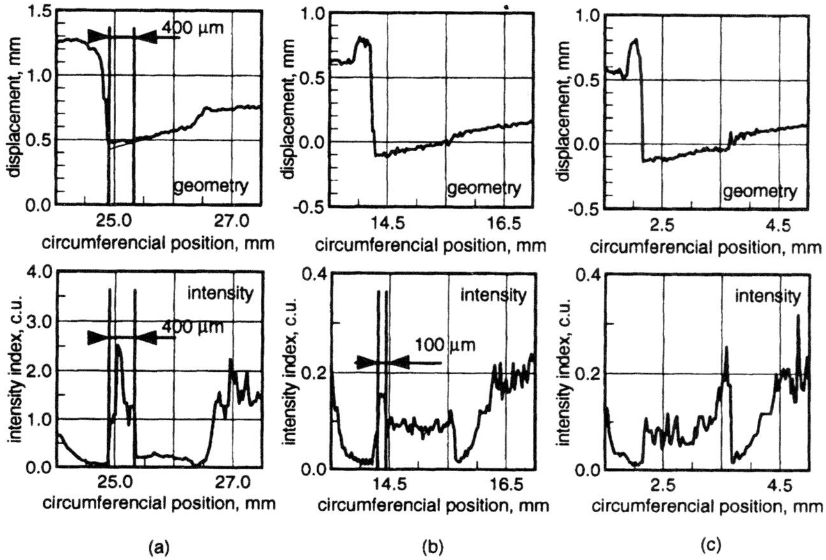

Fig. 17.10 Geometry and reflected light intensity distribution along flank faces under air-cutting conditions. Spindle speed, 1200 rpm; sampling rate, 20 kHz. (a) Actual flank wear, 400–500 μm, (b) 80–100 μm, (c) new tool.

Thus, a hybrid approach based on analysis of intensity and distance measurements is more effective for the detection and measurement of small tool wear than the geometric method.

Histogram approach to eliminate noisy signals under real cutting conditions

When no obstacles complicate the measurement environment, laser sensor outputs can be used directly for analysis without any preprocessing. However, under real cutting conditions, chips and drops of coolant near the cutting point interrupt the path of the laser beam and create noise in the signals, as shown in Fig. 17.11. To filter the useful part of the signal from the noise, a signal processing method based on histogram techniques will be described here.

Signal noise caused by interruption of the laser path by cutting chips and coolant drops is random in the time and space domains: both the timing of the interruptions and their locations relative to the sensor are random. This contrasts with the distance measurements to the tool, which are predictable in the space domain because of the basic tool shape and in the time domain because of the periodicity of the tool rotation.

The histogram technique is effective in separating the predictable components of a signal from the random components. It works as

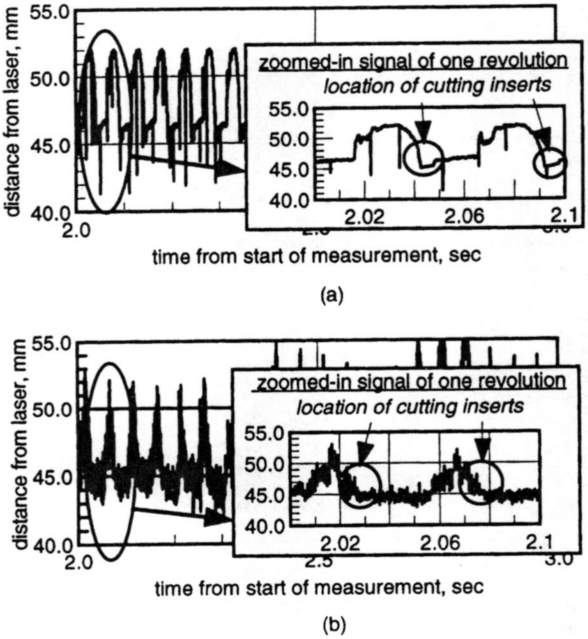

Fig. 17.11 Displacement output under different sensing conditions. Two-insert face mill: diameter, 24 mm; spindle speed, 600 rpm; sampling rate, 5 kHz. (a) Dry cutting: feed rate, 100; axial depth of cut, 3 mm; radial depth of cut, 5 mm. (b) Air-cutting with coolant: coolant supplying rate, 6 liter/min.

follows. We begin by collecting displacement signals for several tool revolutions, and then separate the measurements into groups depending on the point on the tool to which they correspond. Each resulting group of measurements consists of correct distance measurements – that is, actual distances between the sensor and the given reflection point – and distance measurements that have been corrupted by obstacles in the laser path. For each group of measurements, we can construct a histogram of the distance between the sensor and the tool by plotting the number of occurrences of each value of the distance measurements. When enough data has been collected, the maximum value of the histogram should indicate the correct distance to the cutting tool at the given point on the tool.

The laser displacement sensor generates displacement and intensity outputs simultaneously. Correct intensity values are filtered from noise-corrupted intensity values after displacement output processing when we can indicate the moments of correct measuring in time domain and pull out the correct values from the original intensity signal.

17.3 EXAMPLE

In this section, experiments are conducted to enlarge upon the results of the simulated results of tool failure detection by applying the DWT and also to demonstrate 3-D tool image reconstruction using the in-process tool geometry measurement system.

17.3.1 Experimental tool failure detection based on the Wavelet Transform

Fig. 17.12 shows the schematic of the experimental setup. The cutting force components F_X and F_Y are measured independently by a dynamometer mounted on the table of a vertical machining center. The force data are sent via an A/D converter to a personal computer, where 256 samples per revolution are extracted for each component, and the resultant force F_r is calculated using Equation (17.24). The resultant force is input to the DWT algorithm described in section 17.2.1. Table 17.1 shows the tool parameters and cutting conditions.

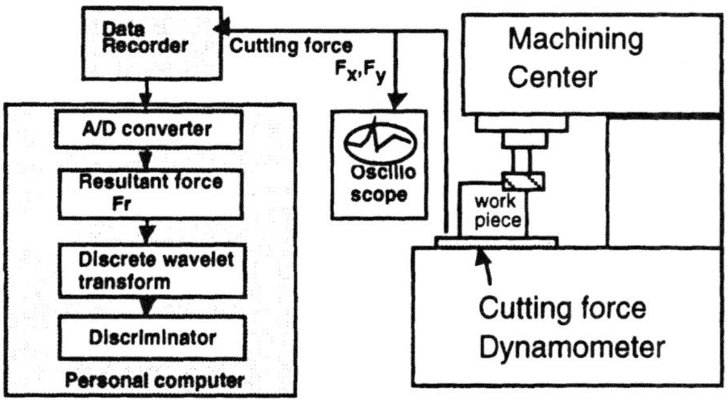

Fig. 17.12 Experimental setup.

Table 17.1 Experimental conditions

Tool	Throwaway face milling cutter	
	Inserts	4
	Approach angle	45°
	Diameter	50 mm
Cutting conditions	Spindle speed	637 rpm
	Feed rate	510 mm/min
	Axial depth of cut	2 mm
Workpiece	Radial depth of cut	25 mm, 50 mm
	Stainless steel (SUS304)	

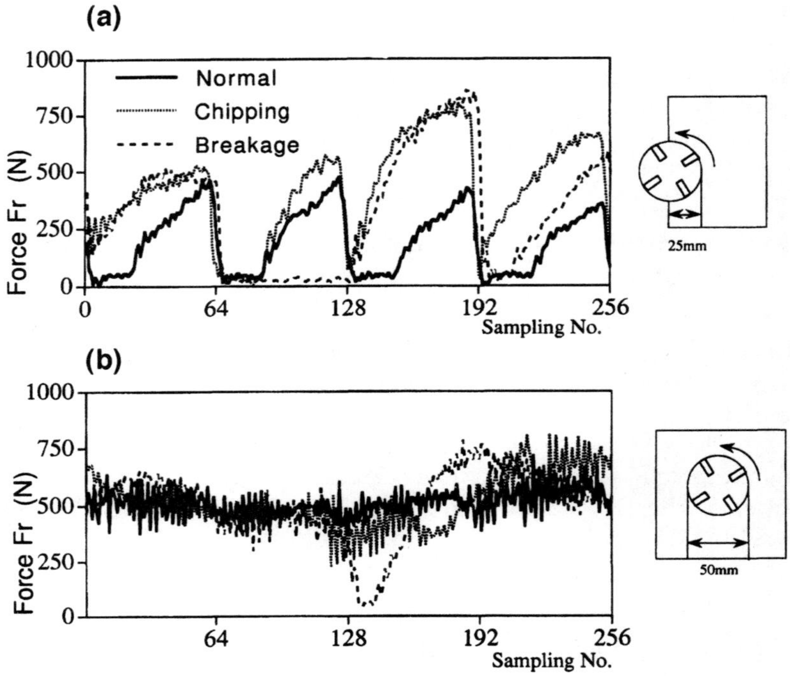

Fig. 17.13 Real cutting force data. Radial depth of cut: (a) 25 mm, (b) 50 mm.

Fig. 17.13 shows the live cutting force data. The damage is in the second insert as the simulated case of $M = 4$.

Fig. 17.14 shows the results of DWT analysis at resolution $j = 6$. The effects of chipping are evident, just as in the simulation, though the experimental cutting force signals are considerably contaminated by noise signals. Tooth damage is indicated by 'zigzag' between wavelet coefficients, whereas eccentricity is signaled by an arc-like pattern. Thus it has been confirmed that the DWT can clearly diagnose the tool failures and the tool eccentricity.

The 'zigzag' number of the coefficient line Z is counted up when $\text{sign}(C_{j,k} - C_{j,k-1})$ changes for automatic classification. To measure the extent of the zigzag at resolution j, the following expression is introduced:

$$V_j = \frac{\sum_{k=0}^{M-1} | C_{j,k} - C_{j,k-1} |}{F_m} \tag{17.25}$$

where F_m is the average resultant cutting force over one tool revolution. Fig. 17.5 shows the value of V_j for the simulated and live milling experiments, plus a 'zigzag' index Z. Normal condition is readily distinguished

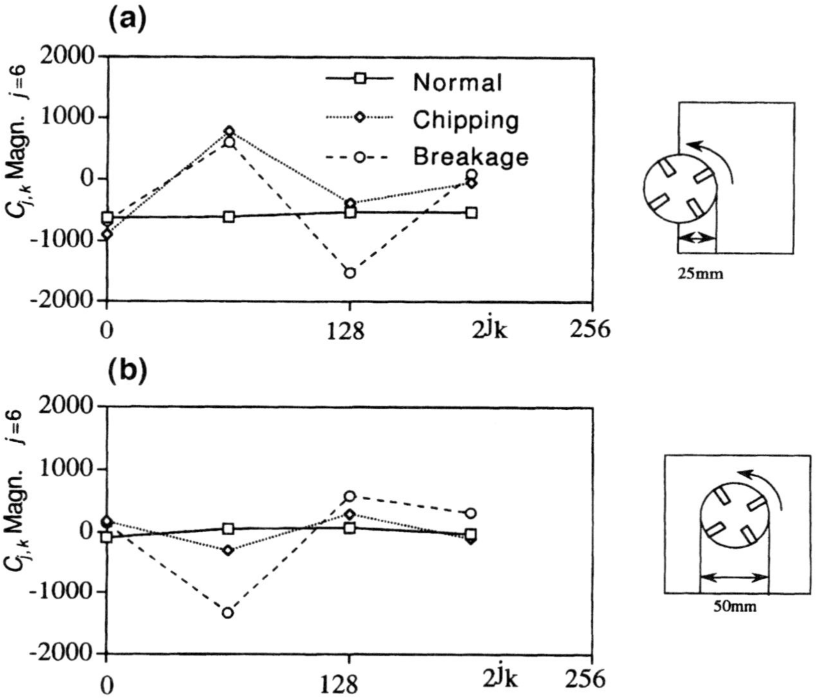

Fig. 17.14 Wavelet coefficient $C_{6,k}$ obtained from physical experiment. Radial depth of cut: (a) 25 mm, (b) 50 mm.

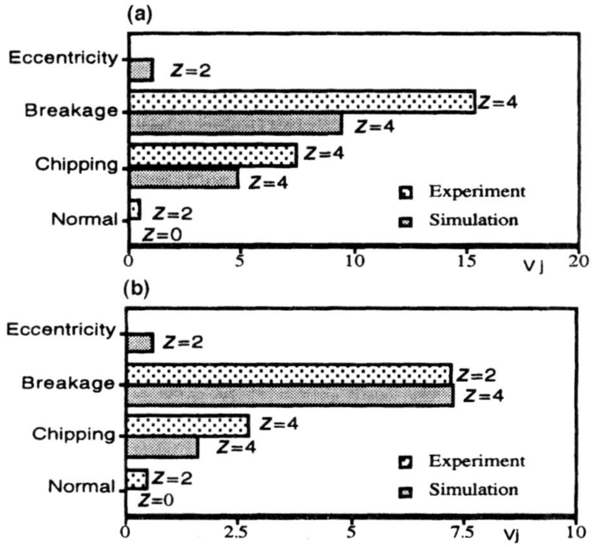

Fig. 17.15 Diagnosis of tool conditions. Radial depth of cut: (a) 25 mm, (b) 50 mm.

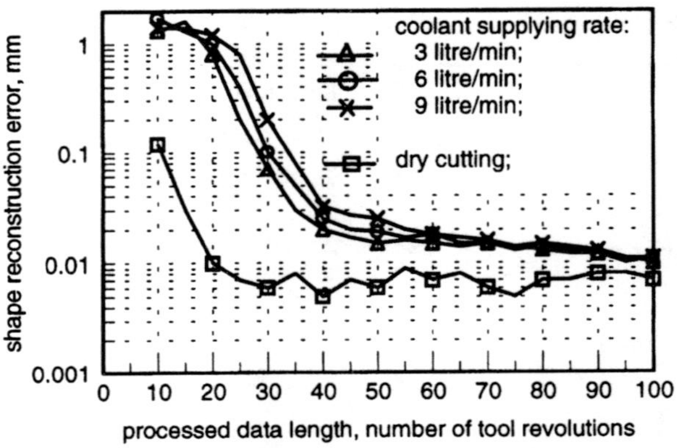

Fig. 17.16 Accuracy of histogram process. Two-insert face mill: diameter, 24 mm; spindle speed, 1200 rpm; feed rate, 100 mm/min; axial depth of cut, 3 mm; radial depth of cut, 5 mm; sampling rate, 10 kHz.

from tool failure by a small V_j. Eccentricity is set apart from other types of tool failure by the 'zigzag' number Z; that is $Z = 4$ in the case of chipping and $Z = 2$ in the case of eccentricity. A small eccentricity is observed under normal conditions in the live experiment.

17.3.2 3-D tool shape reconstruction

Before showing 3-D tool image reconstruction experiments, the signal processing error based on the histogram is evaluated. Fig. 17.6 shows the processing error for the histogram method versus the processed data length, where the processing error is defined as the absolute value of the difference in radial direction between the tool image reconstructed under cutting conditions and the tool image reconstructed in a clean environment, averaged over all measurement points on the tool flutes. The processing error is in inverse proportion to the length of the signal. It may be determined that at 40 rotations per tool image, we can obtain an acceptable level of resolution while meeting the speed requirements necessary for on-line tool monitoring. Under these conditions, the error does not exceed 40 μm.

To obtain a 3-D tool image on the display, the system shown in Fig. 17.8 performs signal acquisition, processing and analysis in the following six steps:

- *Step 1*: Fix the Z-level of the tool scanning, and simultaneously receive the geometry/intensity measurements from the main sensor and the synchronizing signal from the auxiliary sensor.
- *Step 2*: Prepare the output signals for histogram analysis by extracting from each signal revolution the values corresponding to specific

points on the tool. The number of points extracted depends upon the desired resolution of the tool shape.

- *Step 3*: Perform histogram analysis of the distance signal, reconstruct the tool cross-section for the current Z-level from the distance measurements, and reconstruct the tool intensity distribution.
- *Step 4*: Perform tool shape analysis and tool failure detection.
- *Step 5*: Shift the main sensor vertically by a Z-increment chosen in accordance with the desired vertical resolution for the tool shape, and repeat steps 1–4.
- *Step 6*: Reconstruct the 3-D tool image from the tool cross-sections.

An image of a scanned end mill is shown in Fig. 17.17. In this case one cutter is chipped, and it is possible to detect the chipping and to estimate scales of lost parts from the picture.

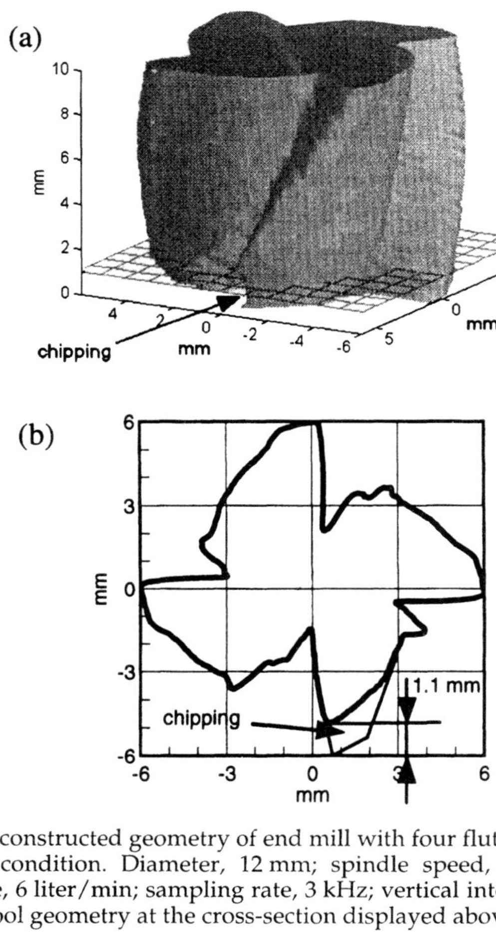

Fig. 17.17 Reconstructed geometry of end mill with four flutes under air-cutting with coolant condition. Diameter, 12 mm; spindle speed, 1200 rpm; coolant supplying rate, 6 liter/min; sampling rate, 3 kHz; vertical interval, 0.5 mm. (a) 3-D image, (b) tool geometry at the cross-section displayed above.

An image of a worn tool is presented in Fig. 17.8. Actual flank wear length is 100–200 μm. In this case of small wear the geometrical resolution of the tool image is not enough to detect the failure, as the depth of flank wear is usually ten times less than its length. Results of simultaneous outputs of intensity and geometry are presented in Figs 17.19 and 17.20. Spike of intensity (Fig. 17.19) indicates flank wear in the case when it is not so clean in geometry signal (Fig. 17.18), and width of the spike corresponds to the wear length that is the most important parameter of the flank wear. There is no intensity spike in the case of normal tool sensing (Fig. 17.20). Thus, from the image shown on the computer display, we can clearly discern the type of the damage and its extent.

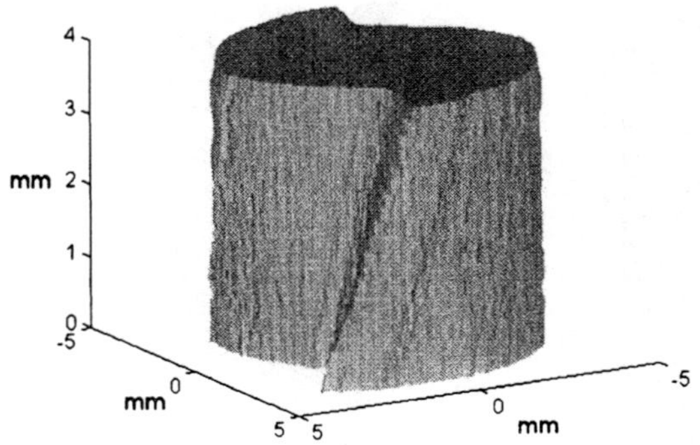

Fig. 17.18 3-D image of worn end mill with flank wear length 100–200 μm.

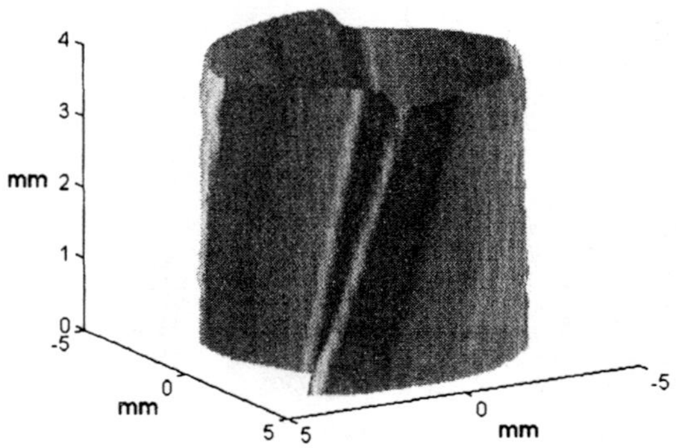

Fig. 17.19 Hybrid output of intensity and geometry of worn end mill with flank wear length 100–200 μm.

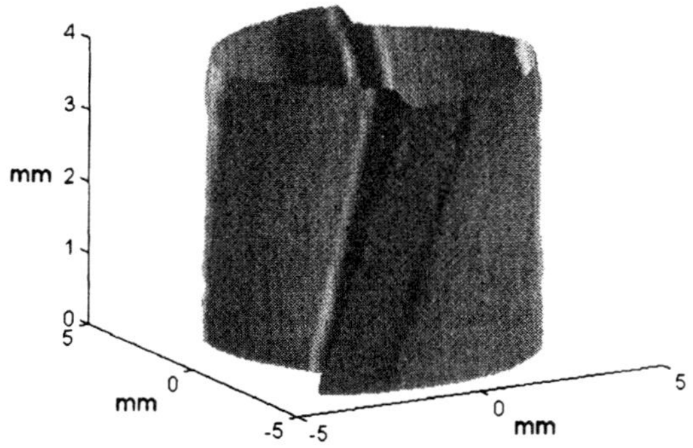

Fig. 17.20 Hybrid output of intensity and geometry of new tool.

17.4 CONCLUSION

In this chapter, two different approaches have been presented for in-process tool failure diagnosis for milling operations. Since direct observation of tool failures is generally impossible under actual machining conditions, indirect monitoring methods based on analysis of sensory signals which are indirectly correlated to the phenomena of interest are applied. The method using the Discrete Wavelet Transform has been presented as the indirect method. It has been confirmed that the method based on the DWT can clearly classify the types of tool failures including tool eccentricity.

The tool geometry measurement system using a laser sensor, which can be referred to as the direct monitoring method, allows qualitative and quantitative observation of tool chipping and wear with high resolution directly from the 3-D tool image displayed on a computer. The hybrid measurement using displacement and intensity techniques of the laser sensor simultaneously and the histogram for pre-processing of noisy signals enables us to diagnose the current tool geometry status under actual machining conditions.

The above methods can easily be implemented into individual in-process systems.

REFERENCES

Cook, N.H. (1980) Tool wear sensors. *Wear*, **62**, 49–57.

Dan, L. and Mathew, J. (1990) Tool wear and failure monitoring techniques for turning a review. *International Journal of Machine Tools Manufacturing*, **30**(4), 579–98.

Daubechies, I. (1988) Orthonormal bases of compactly supported wavelets. *Communications on Pure Appl. Math.* **XLI**, 909–96.

Mallat, S.G. (1989) A theory for multiresolution signal decomposition: the Wavelet representation, *IEEE Transactions on Pattern Analysis and Machine Intelligence*, **11**(7), 674–93.

Micheletti, G.F., Koenig, W. and Victor, H.R. (1976) In Process tool wear sensors for cutting operations. *Annals of CIRP*, **25**(2), 483–96.

Technical Research Institute (1978) *Machining Data Files*, 704F00, Japan Society for the Promotion of Machine Industry.

Tlusty, J. and Andrews, G.C. (1983) A critical review of sensors for unmanned machining. *Annals of CIRP*, **32**(2), 563–72.

Ryabov, O., Mori, K. and Kasashima, N. (1996) An in-process direct monitoring method for milling tool failures using a laser sensor. *Annals of CIRP*, **45**(1), 97–100.

CHAPTER 18

Case Example 7: Monitoring and predicting surface roughness and bore tolerance in end-milling

A. Chukwujekwu Okafor

18.1 INTRODUCTION

The work described in this chapter is a result of a series of research conducted in the Laboratory for Industrial Automation and Flexible Manufacturing at the Department of Mechanical and Aerospace Engineering and Engineering Mechanics, University of Missouri-Rolla. The main objectives of the research effort, which is still in progress, are to develop and implement an on-line (real-time) intelligent machining monitoring and diagnostic system for predicting and assuring the quality characteristics of machined parts, in the form of surface roughness and bore tolerance, based on the integration of multi-sensors using neural networks and expert systems.

This chapter (case study) focuses on monitoring and integrating multi-sensor data (cutting force components, acoustic emission, spindle vibration (acceleration) and machining time) acquired during a series of circular end-milling experiments, and predicting surface roughness and bore tolerance of end-milled parts using a neural network. It describes the machining experiments, multi-sensor data acquisition, signal processing and feature selection, and presents a systematic evaluation of the individual effects of training parameters: learning rate, momentum rate, number of hidden layer nodes, transfer function and learning rule on performance of backpropagation networks used for predicting the quality characteristics of end-milled parts based on multi-sensor integration. The individual effects of varying each of these parameters was investigated when the other parameters were kept at (or near) their optimum values (Okafor and Adetona 1995). A detailed investigation of the interaction effects of varying more than one of these parameters at a time was also carried out and is reported in Okafor and Adetona (1994). The multi-sensor signatures acquired during circular end-milling of 4140 steel and 6061-T6 aluminum and their corresponding measured quality characteristics (surface roughness and bore tolerance) were used to train the network. The network is part of a proposed intelligent machining monitoring and diagnostic system for quality assurance of machined parts. The results obtained will be useful for the successful implementation of the proposed intelligent machining monitoring and diagnostic system for quality assurance of machined parts proposed by Okafor (1993) and Okafor *et al.* (1991). After reading this chapter, the reader should be able to select optimum neural network training parameters more systematically rather than the trial and error approach currently being used. Before proceeding further, it is necessary to discuss the motivation behind the development of the proposed intelligent monitoring and diagnostic system.

The current intense global and domestic market competition has focused the attention of manufacturers and users of machine tools on the automation of manufacturing systems as a means of increasing the productivity and quality of manufactured products. This increasing trend towards automation has created a need for the development of a comprehensive approach to the problem of assuring the quality of manufactured parts (Wright and Bourne 1988). To successfully realize unmanned/lean manufacturing systems, a real-time machining process monitoring and diagnostic system is needed to integrate quality assurance into manufacturing processes. Such a monitoring system must also replicate the overseeing function and qualitative judgment of the expert machinist.

For a manned machining system, the expert machinist can make use of the four senses of hearing, touch, vision and smell to detect or predict deteriorating conditions and take appropriate corrective actions. Many of these decisions may be arrived at by intuitive or 'gut feelings'. For example, changes in chip color (temperature), vibration of the machine tool, and the surface finish obtained on the machined component are the first sign that the tool is becoming worn out and should be changed at the end of the machining pass. This implies that there is a trend analysis active in the mind of the expert machinist.

Due to the random and complex nature of machining, and its many variables, it is very difficult to formulate a realistic and accurate mathematical model of the process. But the properly recorded sensor data of the process signatures contain adequate information about the tool, process and part quality status. These characteristics of machining processes and noise in the sensor signals make it necessary to have a reliable and robust process monitoring and diagnostic system. Such a monitoring and diagnostic system would be a part of an intelligent controller which will effectively control and optimize machining operations.

Present quality assurance methodology based on post-process monitoring is a post-mortem process that offers detection after the fact. The problem with this post-process monitoring technique is that a number of non-conforming parts may have been produced between manufacturing and inspection and rework of any rejected feature increases manufacturing costs. An in-process real-time monitoring technique is, therefore, needed to overcome these problems. In the event of tool breakage, the diagnostic method employed must act in a short time period to retract the cutting tool to prevent costly damage to the workpiece and machine tool. Computational speed is therefore an important consideration in any real-time diagnostic system. In-process monitoring techniques are classified as direct and indirect techniques. The direct in-process monitoring techniques slow down the manufacturing cycle. Various indirect in-process monitoring techniques have been proposed by many researchers, but most of them have been limited to tool wear monitoring and tool breakage detection using various process variables such as cutting force (Filippi and Ippolito 1969; Altintas *et al.* 1985; Lan and Yngve 1986; Lee *et al.* 1989; Okafor and Chyou 1988), vibration (Mehta *et al.* 1983; Yee and Evans 1985; Kim and Klamecki 1990), temperature (Colwell 1975), and acoustic emission (Moriwaki 1980; Lan and Dornfeld 1982; Liu and Liang 1990; Dornfeld 1991). Excellent reviews of these techniques have been made by Micheletti *et al.* (1976) and Tlusty and Andrew (1983). Most of the proposed indirect monitoring techniques only characterized the variation of the process variables with tool wear and provide little information about the quality characteristics such as surface roughness and/or dimensional accuracy of the machined parts. Since it is the part quality that matters, an intelligent real-time machining monitoring and diagnostic system that provides information on product quality is needed.

A recent and very useful development in the indirect in-process monitoring is the application of artificial neural networks (ANN) to integrate multi-sensor information. Rangwala and Dornfeld (1987) applied multi-layer, feedforward backpropagation neural networks to classify tool wear stages in turning. They used power levels in selected frequency hands of the cutting force spectrum and AE spectrum, as recommended by Weck (1986), as input to the neural network and predetermined flank wear as the desired output. Chryssolouris and Domroese (1988) evaluated three approaches: neural networks; group method of data handling (GMDH); multiple least-square regression for integrating multiple sensors and estimating tool wear rates in machining. Their simulation results identified neural networks to be more effective at learning the relationship and providing the best parameter estimates. The use of neural networks to predict tool wear and surface roughness in turning has been reported by Elernayas *et al.* (1991), and for estimating only tool wear in turning by Masory (1991).

For the backpropagation network, the choice of the training parameters and training data are the most important criteria that determine the

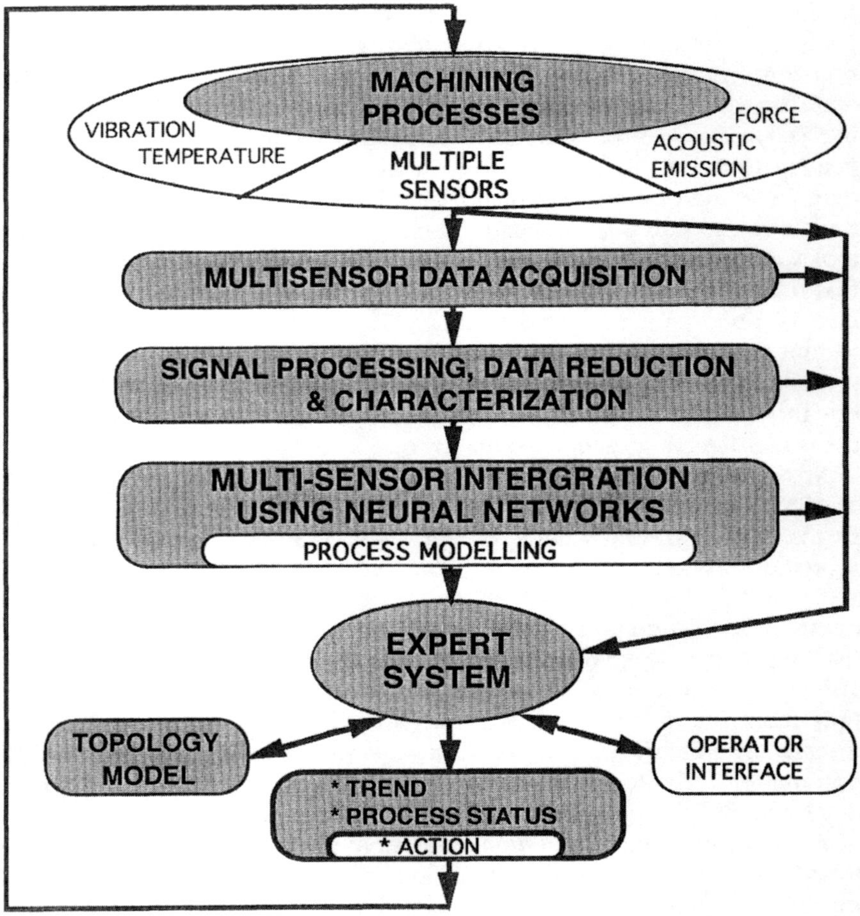

Fig. 18.1 Proposed intelligent machining monitoring and diagnostic system.

degree of success of a network used to perform a specified task. Even if a set of training parameters and a corresponding architecture have been selected and successfully implemented, the question of whether or not the selected parameters are the optimum for that task will still remain to be answered. This situation is worsened by the fact that published literature recommends different combinations of training parameters.

18.1.1 The proposed intelligent machining monitoring and diagnostic system

A modular structure of an intelligent machining monitoring and diagnostic system for quality assurance of machined parts has been proposed by Okafor *et al.* (1991; 1993) and Okafor (1993). The conceptual diagram of a modified version of the proposed intelligent machining monitoring and diagnostic system is shown in Fig. 18.1. The system consists of five modular structures: multiple sensors and a multi-sensor data acquisition module; signal processing, data reduction and characterization module; multi-sensor integration using neural network modules for process modeling (prediction of surface roughness and bore tolerance). Trend classification suggested an expert system. The multi-sensor integration module is implemented using a feedforward fully interconnected by layers, backpropagation neural network. The networks developed were used to predict surface roughness and bore tolerance of circular end-milled parts made from 6061-T6 aluminum, 7075-T6 aluminum and 4140 steel. The results obtained for 4140 steel are reported in this chapter and in Okafor and Adetona (1995), while the results for 6061-T6 aluminum are published in Okafor and Adetona (1996).

18.2 METHODOLOGY

18.2.1 Backpropagation Artificial Neural Network models

Neural networks backgrounds

Artificial Neural Networks (ANN) are systems with inputs and outputs composed of many simple interconnected parallel processing elements. These systems are inspired by the structure of the brain. Some of the key features of ANN are their processing speed through massive parallelism, their proven ability to be trained, to produce instantaneous and correct responses from corrupted inputs once trained, and their ability to generalize information over a wide range. For our application, the backpropagation network is generally considered to be highly suitable. The strength of an interconnection between processing elements or nodes of a neural network is denoted by a numerical quantity, W_{ji}, which are referred to as interconnection weights and are adjustable (Lippman 1987; Rumelhart and McClelland 1989). Changing these weights will change the input-output relationship of the network. The systematic procedures for adjusting the weights are called 'training' or 'learning' algorithms. A neural network is characterized by the processing elements, the network topology, and the learning algorithm. These characteristics are described in the following sections.

The network topology (architecture)

The network topology depends on the way the processing elements (nodes) are connected to each other and to the input and output nodes.

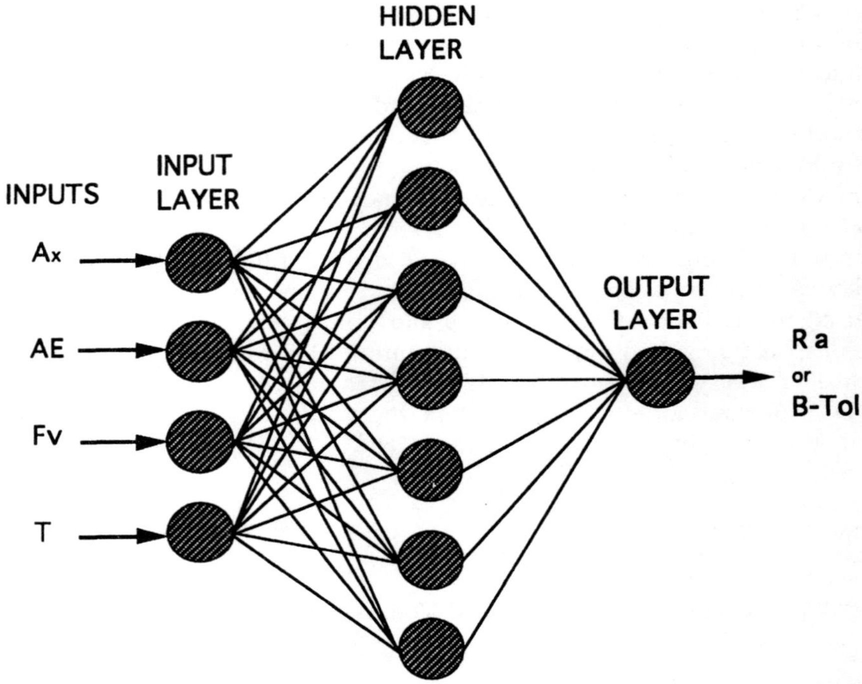

Fig. 18.2 The implemented backpropagation network.

According to their topology, neural networks are classified as single or multi-layer networks. The illustration in Fig. 18.2 is an example of a three-layer feedforward fully interconnected by layers network. The first (left) layer of processing elements (nodes) is the input layer (buffer), where data is presented to the network. The last (right) layer is the output layer (buffer) which holds the network response. The layers between the input and output layers (buffers) are called hidden layers. However, in Fig. 18.2, the network shown has only one hidden layer. In Fig. 18.3, the attributes of individual nodes or interconnections are represented by lower case letters with subscripts. The letters x, y, w and n denote input, output, interconnection weight, and the number of nodes in a layer. The subscripts i, j, and l refer to the input, hidden and output layers respectively.

The processing elements (nodes)

A basic backpropagation processing element is illustrated in Fig. 18.3. The processing element may have any number of input paths but only one output. Inputs X_i can originate from the output paths of other

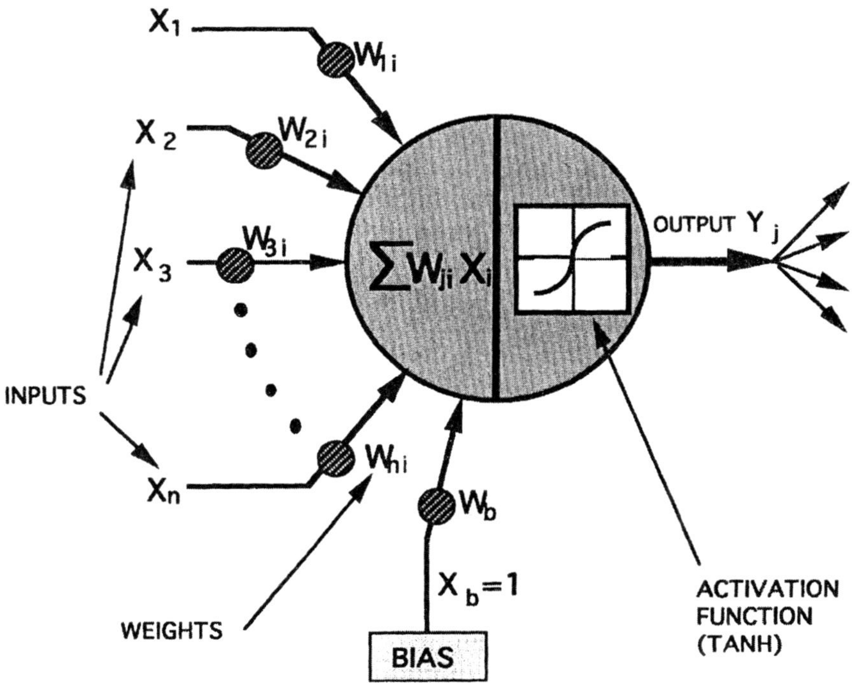

Fig. 18.3 A backpropagation processing element.

processing elements or from themselves, in the case of a feedback. The input signals are modified by the interconnection weights (W_{ji}) and summed (S_j) to form a net input to the hidden processing element (unit). This sum includes the input from the processing element (unit) called the bias unit, which has an output of one at all times and is otherwise treated as any other processing element.

$$\text{Sum, } S_j = \sum_{i}^{N} X_i * W_{ji} + bW_{jo} \qquad (18.1)$$

where S_j is the summed weighted signal for the jth processor in the current layer; W_{jo} is the weight associated with the bias interconnections; and b is the output of the bias node ($b = 1$).

The summed (net) input, S_j, is then passed through an activation function, $f(S_j)$, which can be as simple as a threshold function, or as complex as a nonlinear continuous function such as the Sigmoid or Hyperbolic Tangent function (Lippmann 1987; Rumelhart and McClelland 1989). In any case, the activation function response is transmitted along the output path to become the inputs to other processing elements, or sent to external sources for integration. Typical nonlinear functions used in the nodes are:

(a) The logistic function (sigmoid):

$$f(S_j) = \frac{1}{1 + \exp(-S_j)} \tag{18.2}$$

$f(S_j)$ ranging from 0 to 1.

(b) The hyperbolic tangent function

$$f(S_j) = \mathrm{Tanh}(S_j)$$
$$= \exp(S_j) - \exp(-S_j))/(\exp(S_j) + \exp(-S_j)) \tag{18.3}$$

$f(S_j)$ ranging from –1 to 1.

(c) The sine function

$$f(S_j) = \mathrm{sine}(S_j) \tag{18.4}$$

$f(S_j)$ ranging from –1 to 1.

Network training

There are two distinct phases in the operation of a neural network: training (learning) and recall (application). Training involves the process of iteratively adjusting the interconnection weights in such a way that the prediction errors on the training set are minimized. The equations for adjusting the interconnection weights are:

$$\Delta W_{ji}(n + 1) = \eta * \delta_j * Y_i + \alpha * \Delta W_{ji}(n) \tag{18.5}$$

where:

$\Delta W_{ji}(n + 1)$	is the current change in weight W_{ji} at step $n + 1$.
$\Delta W_{ji}(n)$	is the previous change in weight W_{ji} at step n, initially set to zero.
Y_i	is the output for the ith processor in the lower layer connected to the weight in question.
η	is the learning rate coefficient (set from 0 to 1).
a	is the momentum coefficient (set from 0 to 1).
n	is the presentation number.
δ_j	is the error to be propagated back for the jth output layer processor.

New weight value:

$$W_{ji}(n + 1) = W_{ji}(n) + W_{ji}(n + 1) \tag{18.6}$$

where $W_{ji}(n + 1)$ is the new weight W_{ji} value at step $n + 1$ (after adjustment), and $W_{ji}(n)$ is the previous weight W_{ji} at step n (before adjustment).

The backpropagation algorithm is applied to each pattern set, input and target, for all pattern sets in the training set. Since the learning process is interative, the entire training set will have to be presented to the network over and over again, until the global error reaches a minimum acceptable value. This, however, is greatly dependent upon the variance within the training sets. During the training phase, if the interconnection weights are always updated immediately after the presentation of each pattern set, the learning rule that is being implemented is called the *generalized delta rule*. However, if the required weight updates are not applied to the network immediately after the presentation of each pattern set but are accumulated until all the pattern sets have been presented and if in addition these accumulated weight changes are normalized (e.g. by dividing the accumulated weight changes by the square root of the number of pattern sets in the training data set), the learning rule that is implemented is referred to as the *normalized cumulative delta rule*. For small training data sets, and reasonable learning parameters, the training of a backpropagation network simulated on a personal computer can generally be accomplished in less than an hour. A detailed discussion of the training is provided in Okafor *et al.* (1991), Adetona (1994), and any standard text.

Training parameter selection

The important questions are: how many hidden layer processors (nodes), n_j, should be assigned to a given network? What values should be picked for the learning rate, η, and momentum rate, α? The selection of these training parameters is more art than science and is reported to be dependent on application (Eberhart and Dobbins 1990). It was stated in an earlier section that published literature recommend contrasting values for these parameters. For example, Rumelhart and McClelland (1989) frequently use values of 0.5 for η and 0.9 for α as a place to start. Eberhart and Dobbins (1990), however, found that for their application these values throw the network into oscillation or saturate their nodes. They reported that they often get good results with values of $\eta = 0.15$ and $\alpha = 0.075$ for the applications to electroencephalogram (EEG) spike detection. Freeman and Skapura (1991) recommend η to be a small number in the range of 0.05 to 0.25 to ensure that the network will settle to a solution, whereas Dayhof (1990) recommends η between 0.25 and 0.75. It was suggested that very large values can lead to instability in the network and unsatisfactory learning, while too small a value can lead to excessively slow learning. For the hidden layer nodes, Eberhart and Dobbins suggest that a reasonable number to start with can be obtained by taking the square root of the number of input plus output nodes and adding a few.

18.3 EXAMPLE

18.3.1 Experiments

To evaluate the individual effects of training parameters, about one hundred networks were trained and tested using actual machining data from a machining multi-sensor data characterization experiment. The multi-sensor data consists of: RMS of acoustic emission (AE), RMS of spindle quill vibration in the x-axis direction (Ax), cutting force vector in the horizontal plane (Fv), cutting time (T), surface roughness of the machined part (Ra), and bore tolerance (B-Tol), obtained during circular end-milling of 4140 steel.

The diagram of the machined workpiece is shown in Fig. 18.3b. The machining experiments were performed on a Bridgeport Series I CNC vertical milling machine. Details of the experiment are given in Okafor *et al.* (1993) and Ertekin (1993). In addition, a method of improving the quality of the training data set is introduced and employed.

18.3.2 Training the networks

For some applications (like machining), the input/output data contain a lot of random noise in addition to the actual signal. It is, therefore, logical that if a method can be found that helps reduce the noise levels in the training data, networks trained on this 'clean' data in addition to the noisy data should be able to learn the relationship between the inputs and the outputs relatively easily. Such networks could also be expected to perform better when tested with actual machining data containing random noise.

To extract the signal portion of the sensor signals from the raw data (actual sensor signals), least mean square error regression curves were fitted to the scatter plot of the raw data from each of the sensors against time. The equations of the resulting curves were used to generate the expected value of each of these signals at discrete intervals of time (work-piece number). These expected values are hereafter referred to as the trend data and they represent the signal portion of the raw data. For best results, these generated trend data should be mixed with actual raw data and the resulting data set used to train the networks. Ideally, at least two independent experiments should be performed to acquire data. The data from one of the experiments can then be used to train the networks while the data from the other experiment can be used to test the trained networks. However, due to the high cost of acquiring a second set of sensor data, our analysis is based on just one set of machining data.

For our purposes, nine data points out of a total of 17 trend data points corresponding to the odd numbered workpieces, i.e. workpieces 1, 3, 5, . . ., 17, were used to train the networks. None of the actual raw data signals were presented to the networks during the training phase. The input data were presented to the networks as a four-dimensional input vector, namely: RMS of acoustic emission (AE), RMS of spindle vibration (Ax), RMS of cutting force vector (Fv) and machining time (T). The predicted output pattern was a one-dimensional vector which was either surface roughness (Ra) or bore tolerance (B-Tol). The artificial neural network paradigm used was the Error Backpropagation Paradigm. The software used for modeling the networks was Neuralworks Professional II Plus. At least three replicates of each type of network were obtained because of the variations in their performances. The following analysis is, therefore, based on average values of the trained networks' performances during a specified number of replicates. Therefore whenever the performance evaluation criteria are shown in upper case, average values are being referred to.

Unlike in previous published research efforts in this field, cutting time was added as an additional input to the networks. This was done because the correlation between cutting time and the measured quality characteristics of the finished parts was found to be greater than the correlation between any of the sensor signals and these quality characteristics. In addition, this would be compatible with the current practice in industry where cutting time is used as an indirect measure of tool wear, and tools are sometimes replaced after being in use for a specified period of time even if their performance is still satisfactory. In practical environments, information on cutting time can be provided to suitably trained networks from electronic timing devices (or counters), and could therefore function alongside actual sensors. As expected, networks that were trained and tested with cutting time as an additional input had significantly lower prediction errors than networks that did not have cutting time as an input.

The input/output vector pairs were normalized from the real world values to the network ranges and were presented to the networks in a random sequence of vector pairs. The normalization has the effect of performing a mapping of the minimum (or maximum) value of the real world data set (in each vector field) to the minimum (or maximum) value of the network's range respectively and vice versa. The minimum and maximum values of each vector field was obtained from a combination of both the training data set (trend data) and the raw data set (testing data).

18.3.3 Results and discussions

Effect of the number of hidden layer nodes

Fully interconnected backpropagation networks with Hyperbolic Tangent Transfer Functions in all their processing elements were trained with learning rates fixed at 0.6 and momentum rates fixed at 0.3. The number of processing elements in their hidden layers were varied between two and 16 hidden layer nodes. Five replicates were obtained for each of these 14 network architectures. The networks were trained with 9 sets of trend data but tested with a separate set of 17 raw sensor signal data. The average performance during each of the 5 replicates in terms of maximum error, mean error and RMS error, was calculated and recorded. The results are plotted in Fig. 18.4.

From Fig. 18.4, it can be seen that as the number of nodes in the hidden layer increases, the MAX error generally decreases. Between 2 and 16 hidden nodes the range of MAX error was 17.88% and 21.21%. The lowest values of MAX error of 17.88% and 18.04% were obtained at 14 and 16 hidden layer nodes respectively. The highest MAX error of 82.23% (not shown in the figure) was obtained when there was only one node in the hidden layer (i.e. one quarter of the input layer nodes). For between 2 and 16 hidden nodes the range of RMS error values were between 8.38% and 9.62%. The lowest RMS error values of 8.38% and 8.39% were obtained at 8 and 14 hidden layer nodes respectively. The highest value of RMS error of 45.61% was obtained for one hidden layer node (not shown in the figure). The lowest MEAN error of 7.03% was obtained for one hidden layer node (not shown in the figure). The lowest MEAN error of 7.03% was obtained when the number of hidden nodes was 7. The highest MEAN error of 40.5% was obtained when there was only one node in the hidden layer (not shown in the figure). The range of the MEAN errors for hidden nodes between 2 and 16 was from 7.03% to 8.4%. All the values of MEAN error were lower than their corresponding RMS error values.

In Fig. 18.5, it can be seen that for fixed learning parameters, the number of training CYCLES reduces initially with increases in the number of hidden nodes from 1 to 4, then it starts to show a significant rise with further increase in hidden layer nodes. The minimum number of 3483 CYCLES was obtained for 4 hidden nodes. The largest number of CYCLES of 100 779 (not shown in the figure) was obtained when there was only one node in the hidden layer. From this analysis, 7 hidden layer nodes seems to give the best compromise of minimum prediction errors and minimum training cycles, and is therefore chosen as the optimum number of hidden layer nodes.

Effect of the learning rate

Fully interconnected backpropagation networks with hyperbolic tangent transfer functions in all their processing elements were trained with the number of hidden layer nodes fixed at the optimum of seven. The momentum rates were also fixed at 0.3. Learning rates were increased in steps of 0.1 starting from 0.1 to 0.9. Three replicates were obtained for

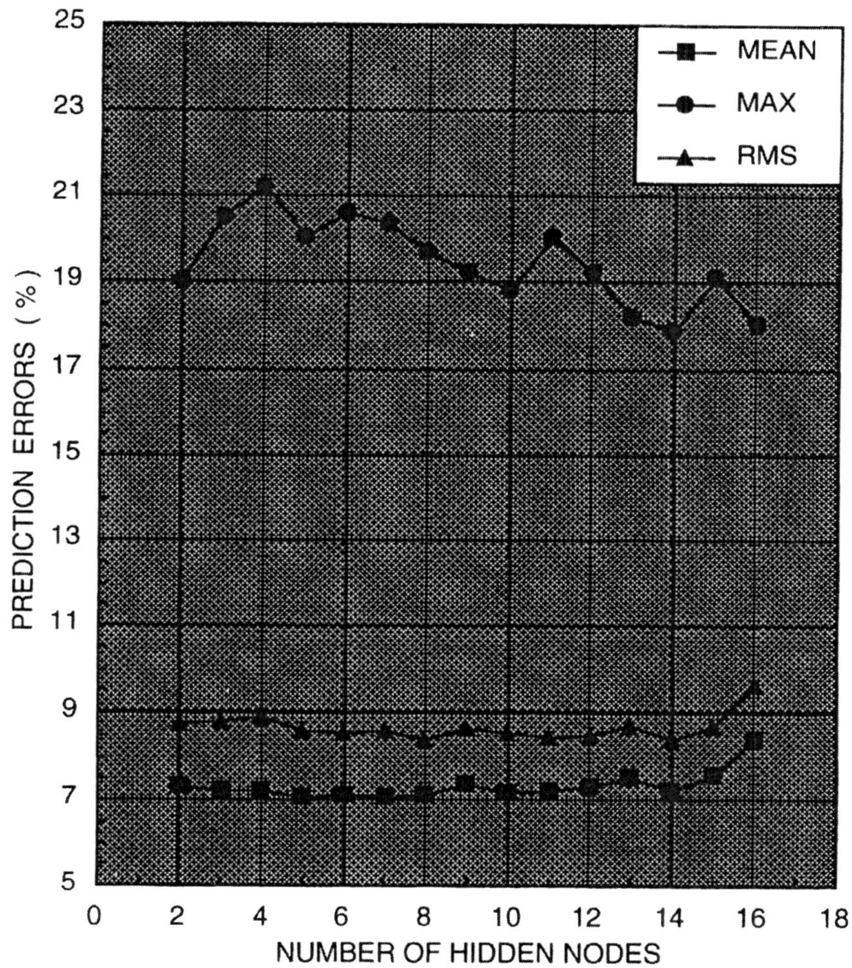

Fig. 18.4 Prediction error vs number of hidden nodes.

each of the networks. The networks were tested with 17 raw sensor signal data and the average performance during each of the three replicates in terms of the four parameters mentioned earlier were calculated and recorded. The results are plotted in Fig. 18.6.

From Fig. 18.6, as the learning rate is increased from 0.1 to 0.4 the MAX error increased slightly from a minimum of 18.29% at 0.1 to 20.78% at 0.4. However, for learning rates between 0.4 and 0.7 it varies between 20.29% and 20.85%, almost constant. When the learning rate is increased to 0.8 and 0.9, a sharp increasing trend was observed in the MAX error which gave a peak value of 49.79% at a learning rate of 0.9. The RMS

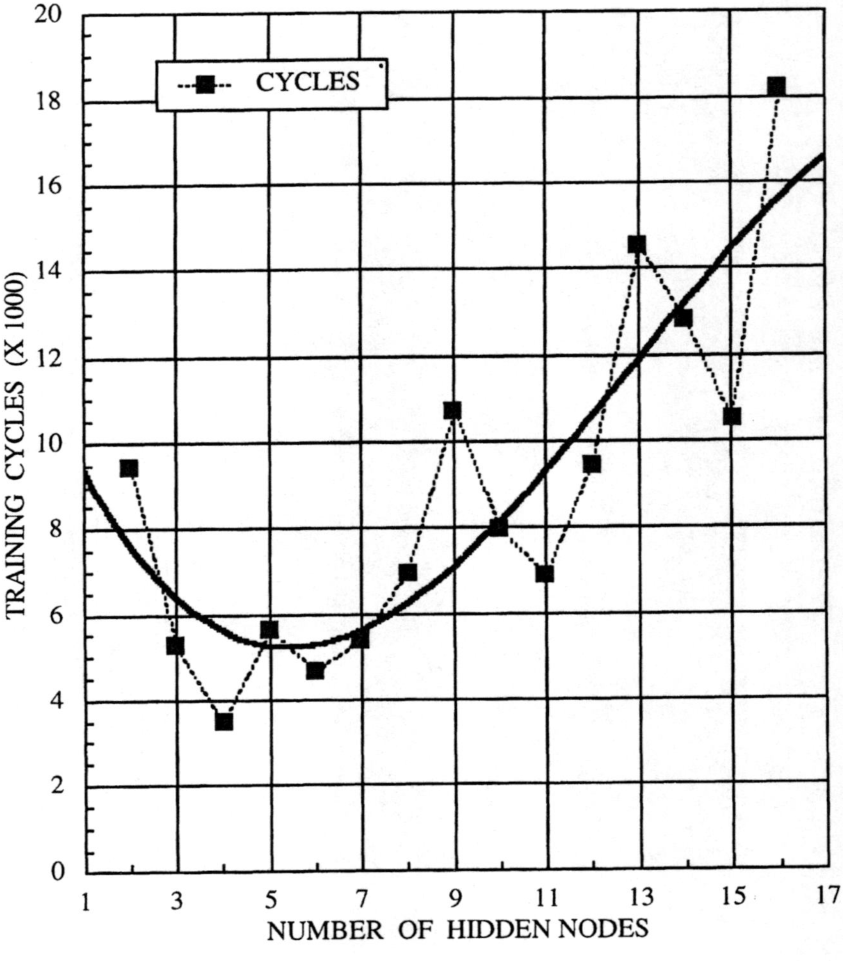

Fig. 18.5 Training cycles vs number of hidden nodes.

error remained fairly constant between learning rates of 0.1 and 0.7. The lowest RMS error value of 8.51% occurred at a learning rate of 0.3 and a sharp increase in the RMS error was observed at learning rates beyond 0.7. The MEAN error behaved in essentially the same way as the RMS error. It remained fairly constant for learning rates between 0.1 and 0.7. It dropped from 7.92% at 0.1 learning rate to a minimum of 7.05% at a learning rate of 0.6. It was observed that just as was the case for the MAX and RMS errors, when the learning rate was increased to 0.8 (or greater), the MEAN error began to rise very rapidly. As shown in Fig. 18.7, the learning rate has a very strong effect on the number of training CYCLES

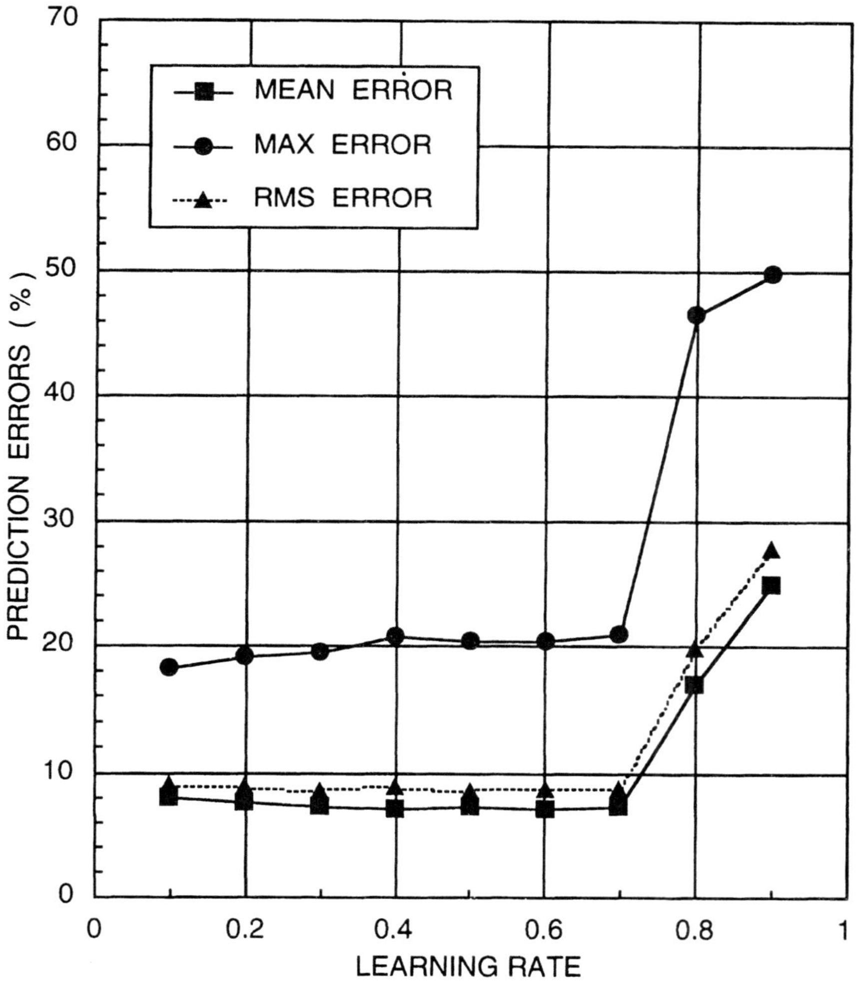

Fig. 18.6 Prediction error vs learning rate.

required for the networks to converge. At a learning rate of 0.1, the number of CYCLES required was 37 914 and dropped to 5109 CYCLES when the learning rate was increased from 0.1, in steps of 0.1, to 0.5. From 0.5 to 0.7 a slight increase in the number of CYCLES was observed after which there was a rapid rise in the number of CYCLES required for convergence beyond 0.7. Optimum learning rate, in terms of lowest prediction errors and lowest number of training cycles, seems to be between 0.3 and 0.7.

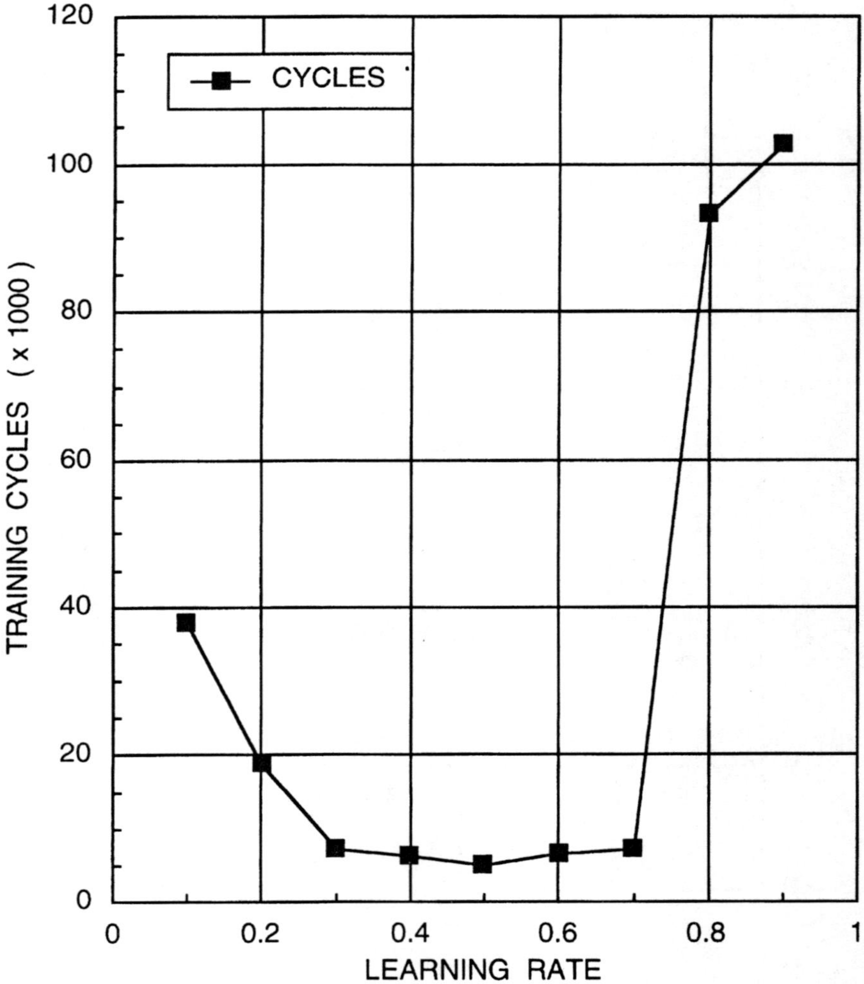

Fig. 18.7 Training cycles vs learning rate.

Effect of the momentum rate

To study the effect of momentum rate on performance, a number of backpropagation networks were trained with learning rates fixed at 0.6 while the momentum rate was increased from 0.1 to 0.6, in steps of 0.1. The number of nodes in the hidden layer of these networks was fixed at seven which had been found to be the optimum. The hyperbolic tangent transfer function was used in all the processing elements of these networks. A relatively high value of learning rate was chosen for all these networks in order to make the network performances more sensitive to variations in the momentum rate. The results obtained are plotted in Fig. 18.8.

Looking at Fig. 18.8, it can be seen that using a momentum rate of 0.5 gave the lowest MAX error of 18.91% while a momentum rate of 0.6 caused a sharp increase in the MAX error to 50.33%. The RMS error remained fairly constant for momentum rates between 0 and 0.5. The minimum RMS error of 8.2% was obtained at a momentum rate of 0.5. When the momentum rate was increased to 0.6, there was a considerable increase in the RMS error from 8.2% at 0.5 to 19.50% at 0.6. The MEAN

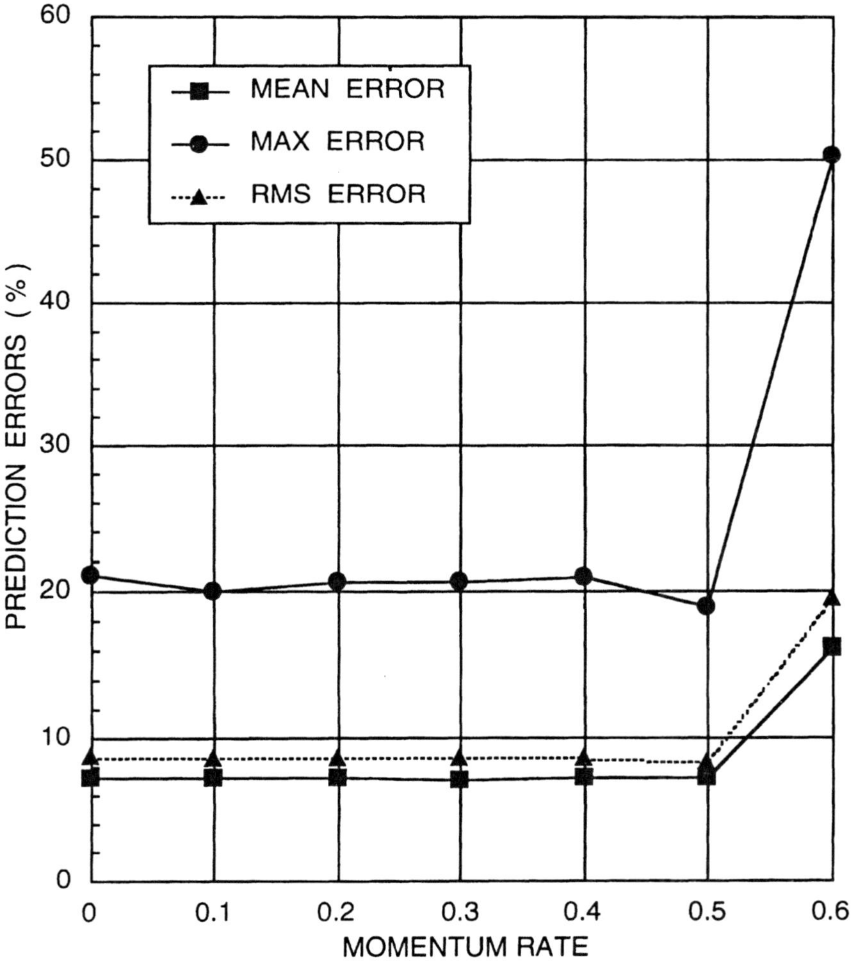

Fig. 18.8 Prediction errors vs momentum rate.

error behaved essentially like the RMS error except that minimum MEAN error of 7.13% occurred at a low momentum rate of 0.3.

Momentum rates between 0.1 and 0.4 led to a slight reduction in the number of CYCLES compared to networks that were trained without a momentum term (Fig. 18.9). It was observed that momentum rates of 0.5 (or greater) had a negative effect on the number of CYCLES since they caused an increase in the number of CYCLES. When the momentum rate reached 0.6 there was an astronomical increase in the number of training cycles for those networks that were able to converge within the preset

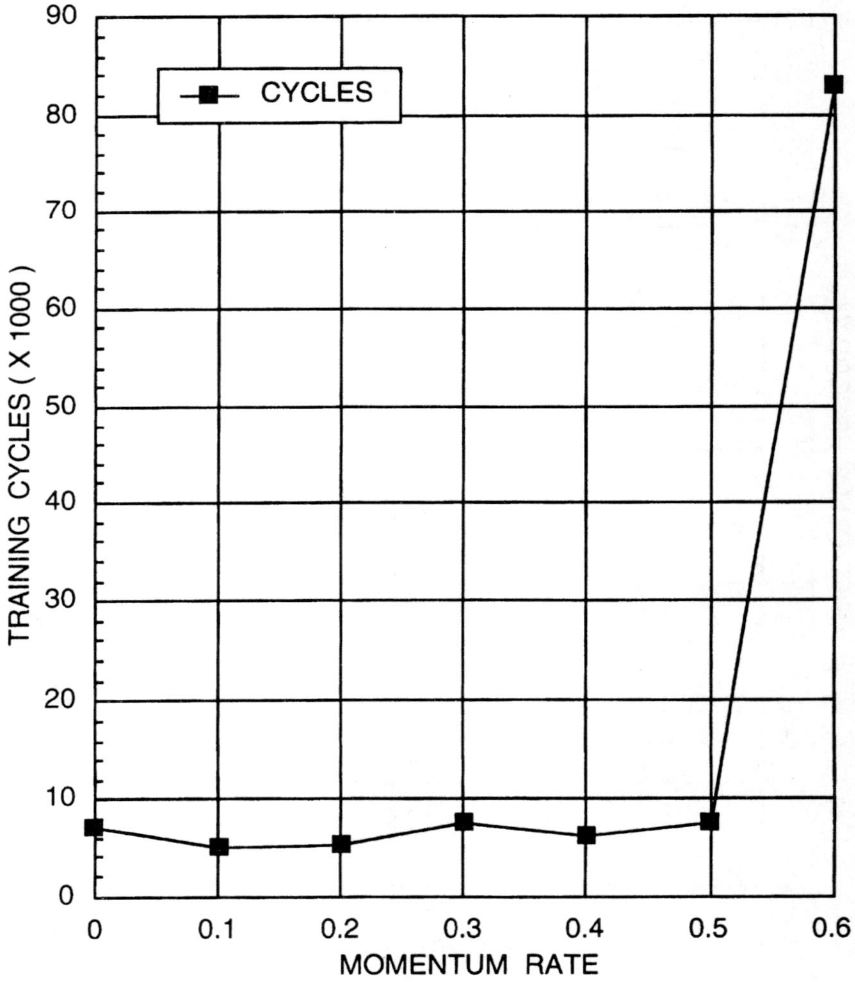

Fig. 18.9 Training cycles vs momentum rate.

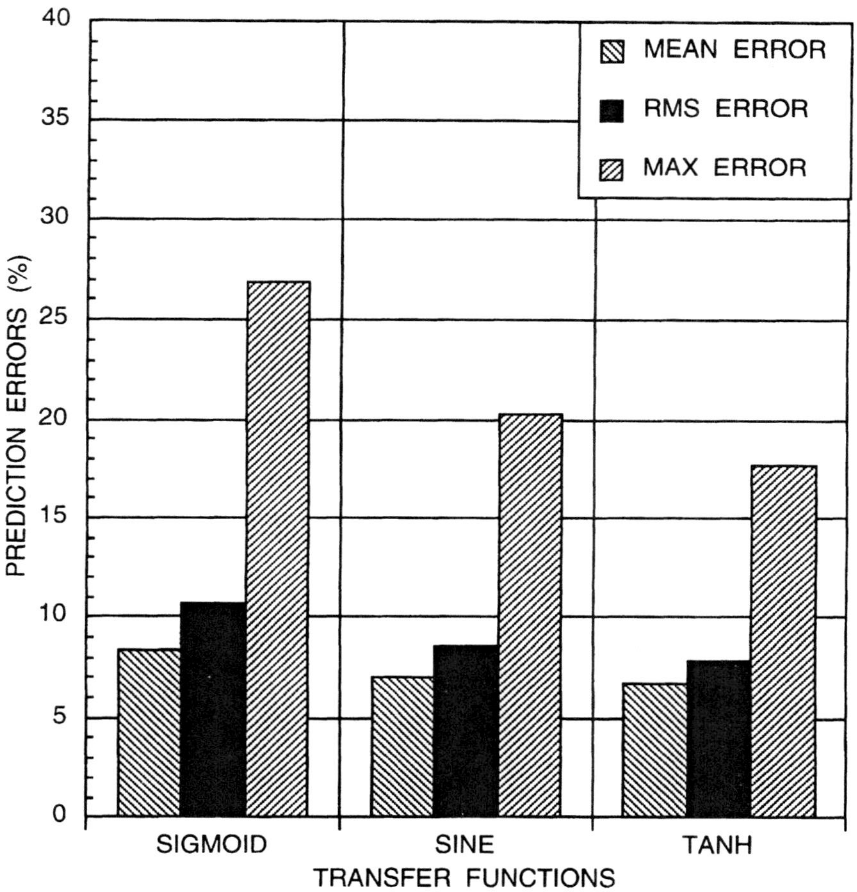

Fig. 18.10 Prediction error vs transfer function.

limit of 150 000 cycles. The minimum number of CYCLES obtained was 5079 at a momentum rate of 0.1.

Effect of transfer functions

To evaluate the effect of the transfer functions in the processing elements of the backpropagation networks, several networks with seven hidden layer nodes, learning rates of 0.6 and momentum rates of 0.3 were trained. Three replicates were obtained using either the hyperbolic tangent, sine, or sigmoid transfer functions in all of each network's processing elements. The results obtained are shown in Fig. 18.10.

In Fig. 18.10, it can be seen that for MEAN, MAX and RMS errors, the

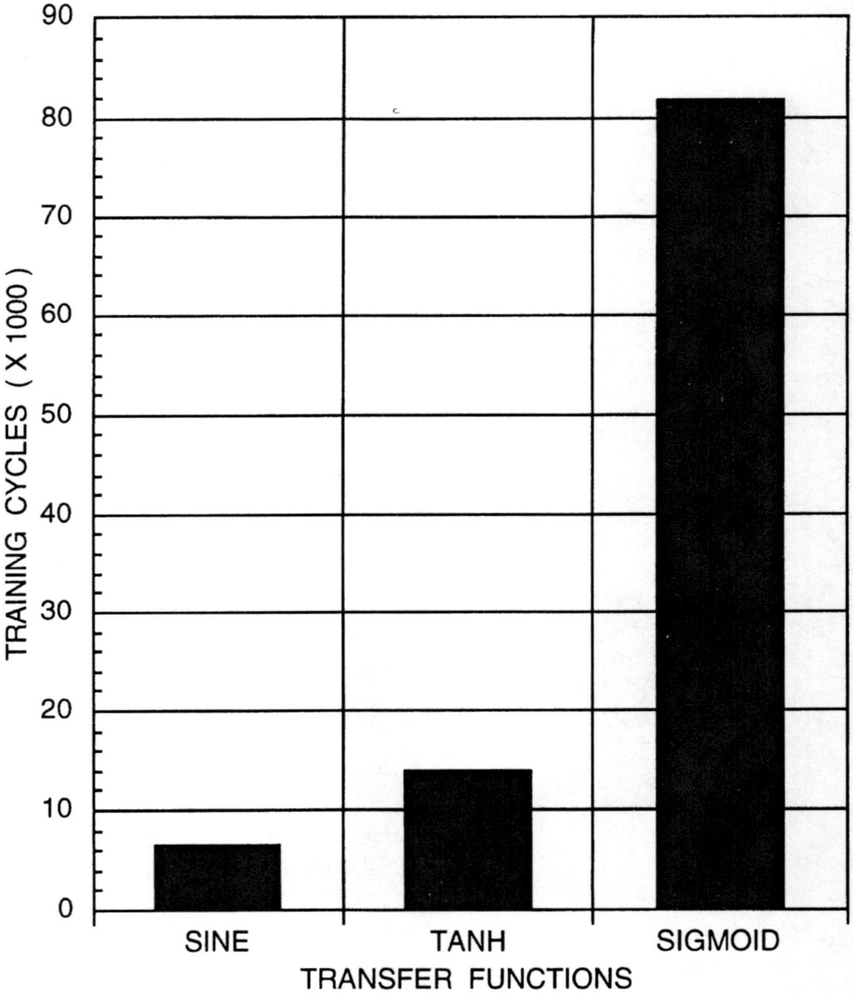

Fig. 18.11 Training cycles vs transfer function.

lowest prediction errors were obtained with the hyperbolic tangent transfer function (Tanh) while the highest were from the sigmoid. The MEAN, MAX and RMS errors obtained for the Tanh function were 6.73%, 17.73%, and 7.88% respectively. The values of the MEAN, MAX and RMS errors obtained for the sine transfer function were 4.75%, 14.44% and 9.52% respectively, higher than those for the Tanh, while the corresponding values for the sigmoid were 25.11%, 51.49% and 35.15% respectively, higher than those for the Tanh transfer function. Fig. 18.11

shows that the number of CYCLES was lowest for the Sine transfer function and highest, by a wide margin, for the sigmoid function. The number of training CYCLES required for the sine function was 6489. The number of training CYCLES required with the hyperbolic tangent were 115.158% greater than those for the sine while the corresponding number of CYCLES for the sigmoid was 1159.5% higher than those for the sine function. The hyperbolic tangent (Tanh) was able to outperform the popularly used sigmoid function because the output of the transfer function is used as a multiplier in the weight update equations. Therefore since the sigmoid has a range of 0 to 1, this leads to a smaller multiplier when the summation is small and a larger multiplier when the summation is large. This can lead to bias for learning desired outputs approaching 1. The Tanh function on the other hand has a range of –1 to 1 and therefore gives an equal weight to high and low (normalized) values.

Effect of learning rule

Six fully interconnected backpropagation networks were trained with both the learning and momentum rates fixed at 0.3, which had earlier been found to give a satisfactory performance. The transfer functions in the processing elements of the networks was the hyperbolic tangent which had also been found to be very good. Three of these six networks were trained using the generalized delta rule while the other three were trained using the normalized cumulative delta rule. An examination of the average performance of these networks led to the following observations.

From Fig. 18.12, it can be seen that the networks trained with the normalized cumulative delta rule had MEAN, MAX and RMS errors that are 59.97%, 78.13% and 68.27%, respectively, higher than those trained with the generalized delta rule. In Fig. 18.13, the networks trained with the normalized cumulative delta rule required 1440.45% more training CYCLES than those trained with the generalized delta rule.

Prediction of quality characteristics using networks trained with optimum and non-optimum parameters

The predictions of a network trained with the generalized delta rule, the hyperbolic tangent transfer function, and optimum parameters of seven hidden layer nodes, learning rate (η) of 0.3 and a momentum rate (α) of 0.3 are shown in Fig. 18.14. The predictions of a non-optimum network trained with a non-optimum learning rate of 0.9, but with identical learning parameters, are also shown in the same figure. The superior performance of the optimum network is obvious. Figure 18.5 shows the bore tolerance predictions from an optimum network trained with the same

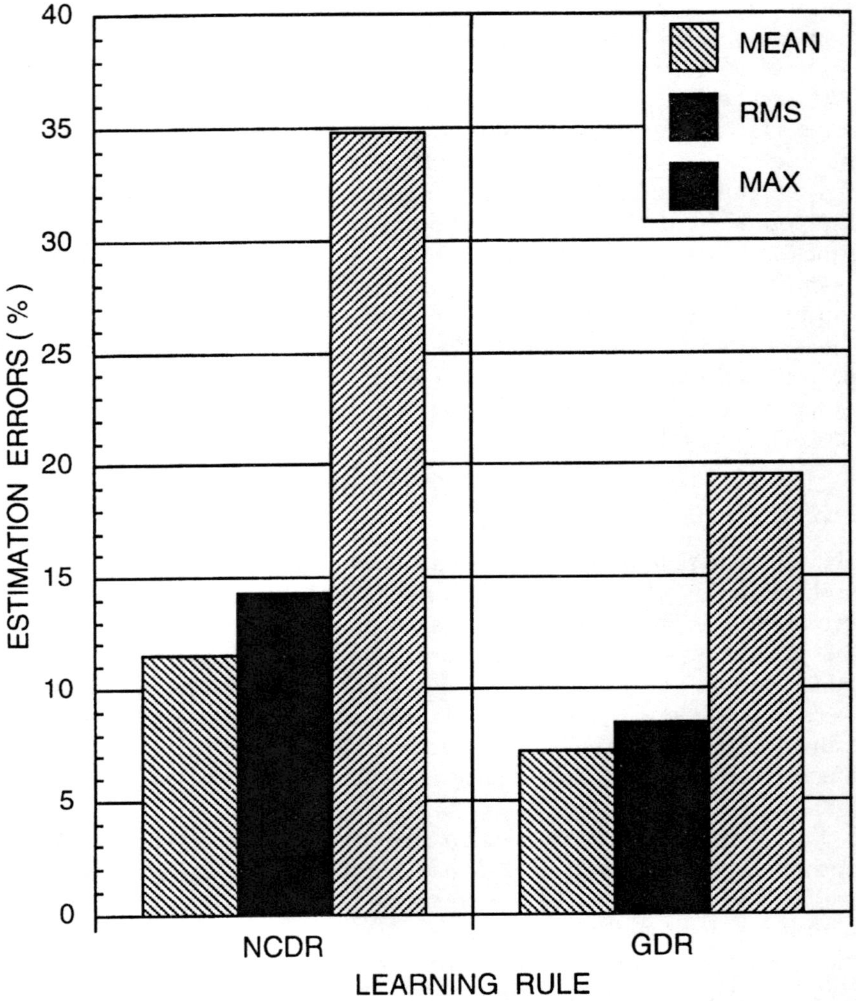

Fig. 18.12 Prediction error vs learning rule.

parameters at the optimum network in Fig. 18.14. Its excellent performance is also apparent from the figure.

18.4 CONCLUSIONS

From our results the following conclusions can be made:

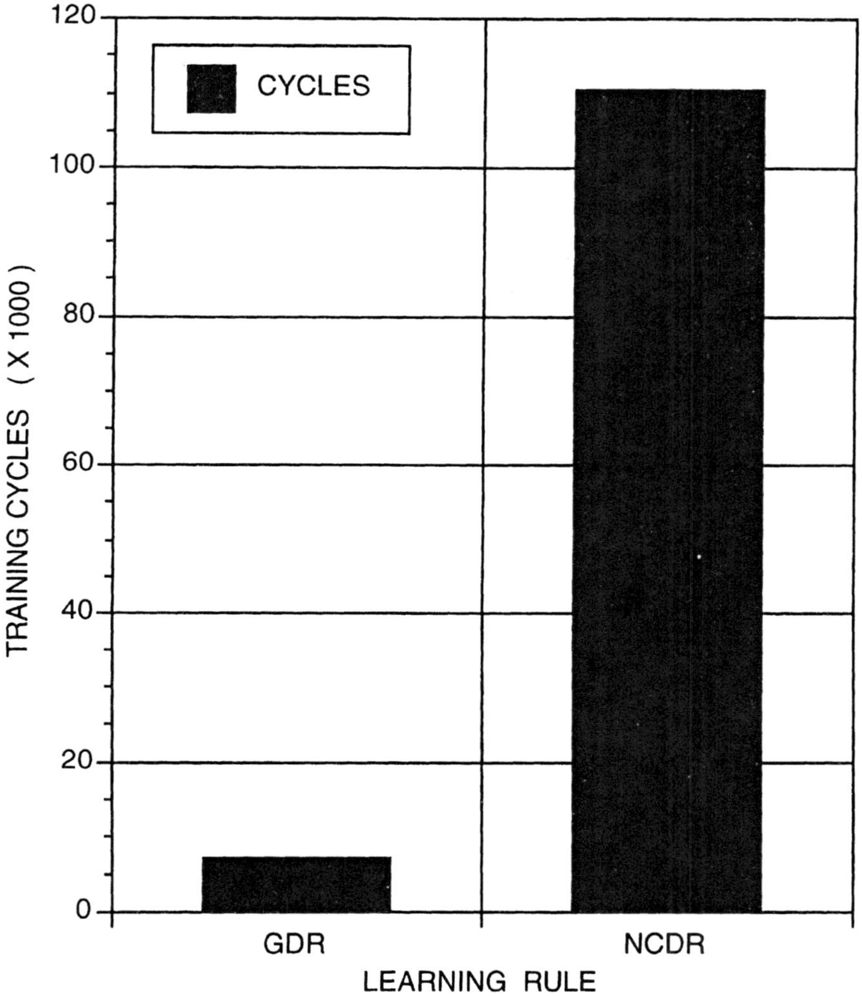

Fig. 18.13 Training cycles vs learning rule.

1. Learning rates below 0.7 give acceptable prediction errors while learning rates must be between 0.3 and 0.7 to minimize the number of training cycles and obtain low prediction errors. Therefore, learning rates that will give an overall optimum performance are any value between 0.3 and 0.7. This range is in very close agreement with the range recommended by Dayhof (1990).
2. To minimize the estimation errors, momentum rates between 0.1 and 0.5 are good. However, the momentum rate should not exceed 0.4 if the number of training cycles is also to be minimized.

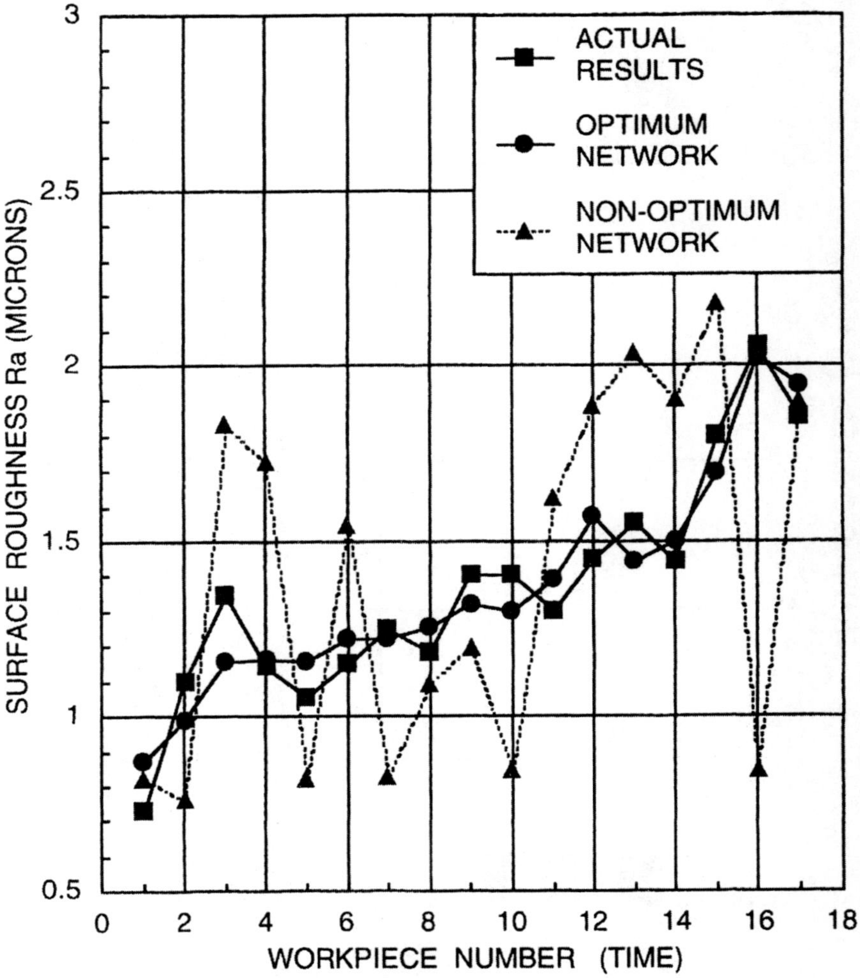

Fig. 18.14 Surface roughness vs workpiece number.

3. The optimum number of hidden layer nodes is 7 or 8. Networks with between 2 and 12 hidden layer nodes, other than 7 and 8, also performed fairly well but result in higher training cycles.
4. Networks trained with the hyperbolic tangent (Tanh) transfer function in all their processing elements give the least prediction errors, while those employing the sigmoid and sine give the highest and moderate prediction errors respectively.
5. Networks that employ the sine function require the lowest number

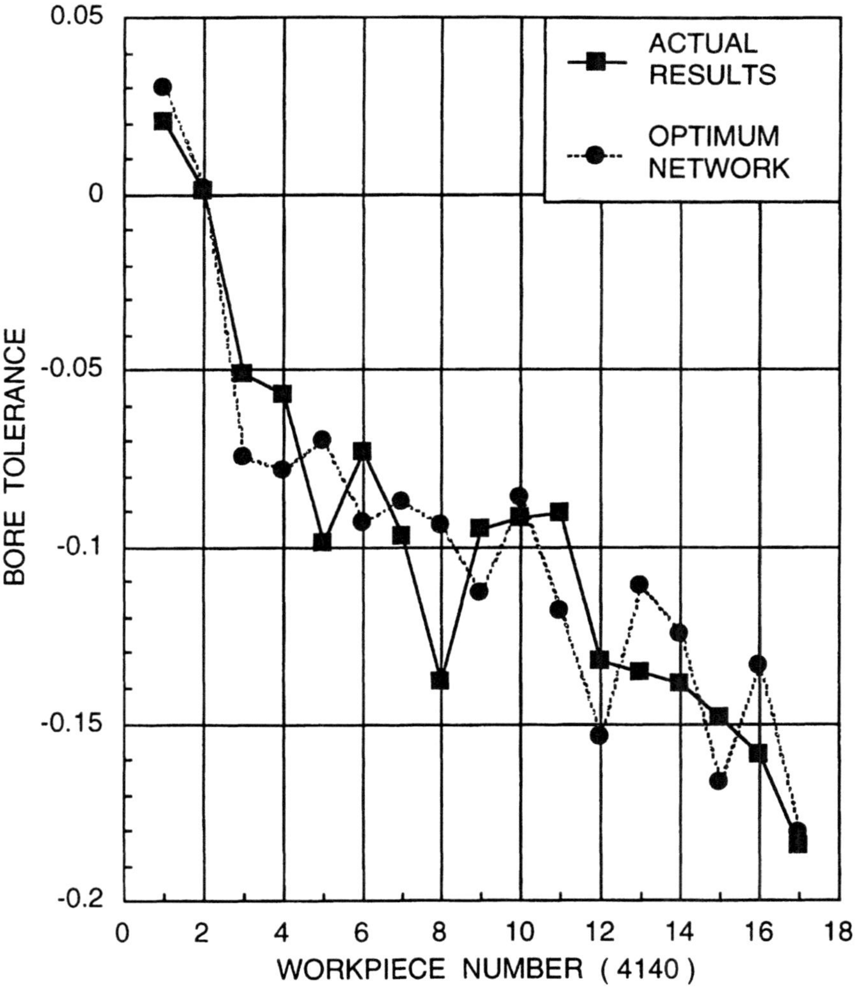

Fig. 18.15 Results from a network trained with an optimum architecture, learning rate and momentum rate.

of training cycles followed by the hyperbolic tangent, while those that employ the sigmoid require the highest number of training cycles.

6. Networks trained with cutting time as an additional input perform better than those trained without cutting time.
7. Networks trained with optimum training parameters give the best overall performance.

ACKNOWLEDGEMENTS

The financial support of this work by the National Science Foundation, under Grant No. DDM-7111712, is gratefully acknowledged. This research is also partially supported by Cincinnati Milacron, Inc. and the Manufacturing Research and Training Center at the University of Missouri-Rolla. The authors thank former graduate student, Yalcin Ertekin, who performed the machining experiments and multi-sensor data characterization.

REFERENCES

Adetona, O. (1994) *Multi-Sensor Integration Using Neural Networks for Predicting Quality Characteristics of End-milled Parts: Effect of Training Parameters and Learning Rules*, MS Thesis, University of Missouri-Rolla, April.

Altintas, Y., Yellowley, I. and Tlusty, J. (1985) The detection of tool breakage in milling. *Sensors and Controls for Manufacturing*, WAM of the ASME, Florida, November 17–22, 101–10.

Chryssolouris, G. and Domroese, M. (1988) Sensor integration for tool wear estimation in machining. *Sensors and Controls for Manufacturing*, WAM of the ASME, Chicago, Ilinois, November 27–December 2, 115–28.

Colwell, L.U. (1975) Cutting temperature versus tool wear. *Annals of the CIRP*, **24**(1), 73–6.

Dayhof, T. (1990) *Neural Network Architectures: An Introduction*, Van Nostrand Reinhold.

Dornfeld, D.A. (1991) Monitoring of the machining process by means of acoustic emission sensors. *Acoustic Emission: Current Practice and Future Directions*, ASTM STP 1077.

Eberhart, R.C. and Dobbins, R.W. (1990) *Neural Network PC Tools: A Practical Guide*, Academic Press, Inc.

Elanayar, S., Young, V.T., Shin, C. and Kumara, S. (1991) Machining condition monitoring for automation using neural networks. *Monitoring and Control for Manufacturing Processes*, ASME Winter Annual Meeting, Dallas, Texas, 85–100.

Ertekin, Y.M. (1993) *Extended Multi-Sensor Data Characterization and Prediction of Surface Roughness and Bore Tolerance in Circular End Milling*, MS Thesis, University of Missouri-Rolla, May.

Filippi, A.De. and Ippolito, R. (1969) Adaptive control in turning: cutting forces and tool wear relationship for P10, P20, P30 carbides. *Annals of the CIRP*, **17**, 337–85.

Freeman, J.A. and Skapura, D.M. (1991) *Neural Networks: Algorithms, Applications, and Programming Techniques*, Addison-Wesley Publishing Co.

Kim, S. and Klamecki, B.E. (1990) Milling cutter wear monitoring using spindle shaft vibration. *Monitoring and Control for Manufacturing Processes*, WAM of the ASME, Dallas, Texas, November 25–30, **44**, 57–74.

Lan, M.-S. and Dornfeld, D.A. (1982) Experimental studies of tool wear via acoustic emission analysis. *Proceedings of the 10th NAMRC*, SME, 305–311.

Lan, M.-S. and Yngve, N. (1986) In-process detection of tool breakage in milling. *Journal of Engineering for Industry*, **108**, 191.

Lee, L.C., Lee, K.S. and Gran, C.S. (1989) On the correlation between dynamic cutting force and tool wear. *International Journal of Machine Tools Manufacturing*, **29**(3), 295–303.

Lippmann, R.P. (1987) An introduction to computing with neural nets. *IEEE ASSP Magazine*, April, 4–22.

Liu, M. and Liang, Y.S. (1990) Monitoring of peripheral milling using acoustic emission. *Transactions of NAMRI*, SME, 120–7.

Liu, T.I. and Ko, E.J. (1991) On-line artificial neural networks. ASME WAM, Dallas, Texas, 1091–110.

Masory, O. (1991) Monitoring machining processes using multi-sensor readings fused by artificial neural networks. *Journal of Materials Processing Technology*, **28**, 231–40.

Mehta, N.K., Pandy, P.C. and Chakravarti, G. (1983) An investigation of tool wear and the vibration spectrum in milling. *Wear*, **91**, 219–34.

Micheletti, G.F., Koenig, W. and Victor, H.R. (1976) In-process tool wear sensors for cutting operation. *Annals of CIRP*, **25**, 483–96.

Moriwaki, T. (1980) Detection of cutting tool fracture by acoustic emission measurements. *Annals of CIRP*, **29**(1).

Okafor, A.C. (1993) Artificial neural networks: Theory and applications in manufacturing. In *Lean Manufacturing in the Automotive Industries, Proceedings of 26th International Symposium on Automotive Technology and Automation*, Aachen, Germany, 383–90.

Okafor, A.C. and Adetona, O. (1994) Multi-sensor integration using neural networks for predicting quality characteristics of end-milled parts. Part II: Interaction effects of training parameters. *Proceedings of the Second International Conference on Intelligent Materials*, Craig A. Rogers and Gordon G. Wallace (eds), June 5–8, Colonial Williamsburg, VA, 946–57.

Okafor, A.C. and Adetona, O. (1995) Predicting quality characteristics of end-milled parts based on multi-sensor integration using neural networks: Individual effects of learning parameters and rules. *Journal of Intelligent Manufacturing* **6**, 389–400.

Okafor, A.C. and Adetona, O. (1996) Predicting quality characteristics of end-milled parts based on multi-sensor integration using neural network: 6061-T6 aluminum. *Proceedings of the Third Africa-USA Conference on Manufacturing Technology*, Accra, Ghana, August 12–15, 145–152.

Okafor, A.C. and Chyou, Y.J. (1988) A comparative investigation for one-line end-mill wear and breakage detection for unmanned machining. *Sensors and Controls for Manufacturing*, WAM of the ASME, Chicago, Illinois, November 27–December 2, 105–14.

Okafor, A.C., Ertekin, Y.M., and Arora, S. (1993) Intelligent machining monitoring and diagnostic system for quality assurance based on multi-sensor integration. *Proceedings of the 1993 NSF Design and Manufacturing Systems Conference*, **2**, 1631–8.

Okafor, A.C., Marcus, M.S. and Tipirneni, R. (1991) Multiple sensor integration via neural networks for estimating surface roughness and bore tolerance in circular end milling: Part 1 – Time domain. *Condition Monitoring and Diagnostic Technology Journal*, **2**(2), 49–57.

Rangwala, S. and Dornfeld, D.A. (1987) Integration of sensors via neural networks for detection of tool wear state. *Intelligent and Integrated Analysis and Synthesis*, ASME Winter Annual Meeting, Boston, MA, December 12–13, 109–20.

Rumelhart, D.E. and McClelland, J.L. (1989) *Parallel Distributed Processing*, 1, MIT Press.

Tlusty, J. and Andrews, G.C. (1983) A critical review on sensors for unmanned manufacturing. *Annals of CIRP*, **32**(2), 563–72.
Weck, M. (1986) Development and application of a flexible, modular monitoring and diagnostic systems. *Computers in Industry*, **7**, 45–51.
Wright, P.K. and Bourne, D.A. (1988) *Manufacturing Intelligence*, Addison-Wesley Publishing Company, Inc.
Yee, K.W. and Evans, L. (1985) Drill-up an alternative for on-line determination of end-mill wear. *Proceedings of the 13th NAMR Conference*, May 19–22, 304–9.

Index

Lightning Source UK Ltd.
Milton Keynes UK
06 June 2010

155134UK00009B/6/A

9 780412 629709